Hans G. Natke · James T. P. Yao

Structural Safety Evaluation
Based on System Identification Approaches

Hans G. Natke · James T. P. Yao

Structural Safety Evaluation Based on System Identification Approaches

Proceedings of the Workshop at Lambrecht/Pfalz

Friedr. Vieweg & Sohn   Braunschweig / Wiesbaden

This volume contents the proceedings of the Workshop on Structural Safety Evaluation Based on System Identification Approaches, held in Lambrecht/Pfalz June 29th to July 1st 1987, Organized by H. G. Natke, Curt-Risch-Institut, TU Hannover.

1988
All rights reserved
© Friedr. Vieweg & Sohn Verlagsgesellschaft mbH, Braunschweig 1988

Printed and bound by Lengericher Handelsdruckerei, Lengerich

ISBN 978-3-528-06313-9   ISBN 978-3-663-05657-7 (eBook)
DOI 10.1007/978-3-663-05657-7

Contents

Preface

H.G.Natke, J.T.P.Yao

Preface

The idea of this workshop originated during a collaborative
research project by the Editors, who share a common interest
in safety evaluation based on theoretical and experimental
structural analysis. We believe that "Safety Evaluation
Based on System Identification Approaches" is an important
topic in engineering applications and needs a detailed
and critical consideration. Civil engineering applications
were emphasized at this workshop. Nevertheless, the inter-
disciplinary nature of this subject matter requires communication among experts with various backgrounds and experience.

It was not felt necessary to give an introduction to
the topics as covered herein. For a detailed description
of these topics, the interested reader can refer to several
textbooks and surveys as listed at the end of this preface.
The role of system identification is shown in Fig. 1
and an extended identification methodology for safety
evaluation is illustrated in Fig. 2 (from the paper by
H. G. Natke and J. T. P. Yao in the Proceedings).

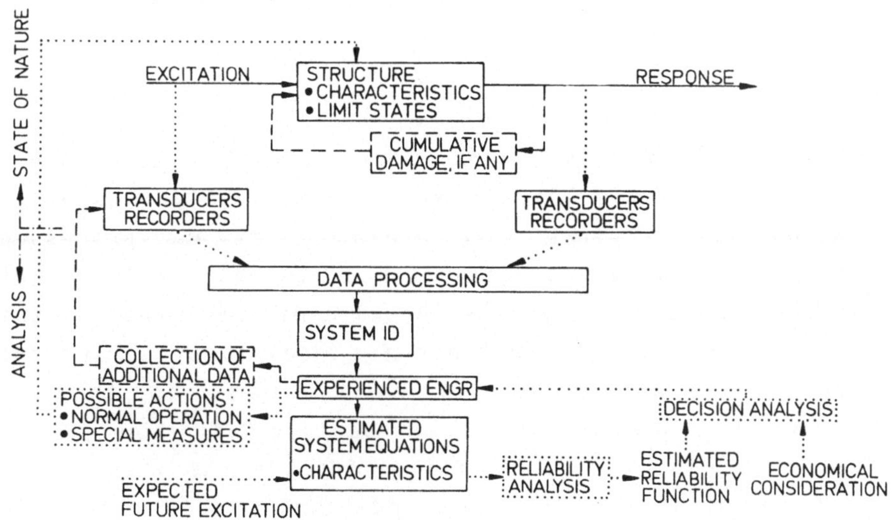

FIG. 1 SYSTEM IDENTIFICATION METHODOLOGY

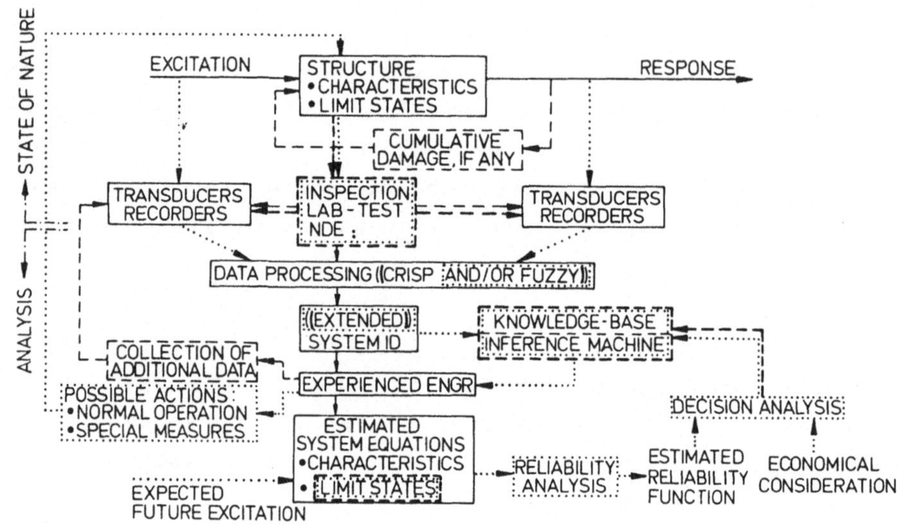

FIG.2 EXTENDED SYSTEM IDENTIFICATION METHODOLOGY
FOR COMPLEX AND HIGHLY NONLINEAR STRUCTURES

In Fig. 1, the current system identification methodology
and its application are shown schematically. Usually,
excitation and response data are measured and processed
for use in system identification studies. Results of
these studies are used by experienced engineers, who use the
resulting mathematical equations for additional structural
analyses including reliability calculations. In the extended system identification methodology as shown schematically in Fig. 2, we propose to include the identification of limit states with the additional input data such
as inspection results, laboratory test data, and results
of nondestructive evaluations. Such data may be either
crisp or fuzzy, which may require new techniques for data

processing. In addition we propose the development of expert systems including data bases and inference machines as consultative aid to the experienced engineers who are decision makers. An additional emphasis is placed on the need for an improved model with known confidence.

Fig. 3 contains a list of the topics treated. Initially we should know the definition and description of damage. Looking at the usual mathematical models in their discrete (with respect to local coordinates) form such as finite elements or multi-body models, any changes in their parameters (physical or modal) may serve the purpose for the detection of structural modifications, which are caused by damage. Some of the required parameters cannot be measured directly. How certain are we in making such statements? What are the most suitable quantities? What are the optimum testing (including measurement) conditions? These questions should be addressed in detail before the tools for the detection and localization of structure modifications by identification methods are discussed in general.

The discussion of system identification approaches is included in Item No. 2 in Fig. 3. Section 2.1 may be conceived as direct modal identification. Updating mathematical models (2.3) using measured or identified quantities is an important topic in order to obtain models with known confidence. Here the a priori knowledge coming from system analysis can be included. Instrumentation and data processing are important parts of identification, since the quality of the results depends on them.

The interaction between the results of system identification (estimated modifications) and damage evaluation is discussed in Item No. 3. Here probabilistic methods and related mathematics such as fuzzy sets are presented.

**WORKSHOP
ON STRUCTURAL SAFETY EVALUATION
BASED ON SYSTEM IDENTIFICATION APPROACHES**

1. DAMAGE DESCRIPTIONS AND BASIC REQUIREMENTS

    1.1 STRUCTURAL CONSIDERATIONS
    1.2 PARAMETERS (SENSITIVITY)
    1.3 DAMAGE INDICES
    1.4 TEST REQUIREMENTS

2. SYSTEM IDENTIFICATION

    2.1 IDENTIFICATION OF LINEAR SYSTEMS
    2.2 IDENTIFICATION OF NONLINEAR SYSTEMS
    2.3 UPDATING OF ANALYTICAL MODELS
    2.4 INSTRUMENTATION, DATA PROCESSING

3. INTERACTION BETWEEN SYSTEM IDENTIFICATION
   AND DAMAGE EVALUATION

    3.1. STOCHASTICS
    3.2 FUZZY SETS
    3.3 PATTERN RECOGNITION, EXPERT SYSTEMS

4. CONCEPTS

    4.1 LINEAR MODELS
    4.2 NONLINEAR MODELS

Fig. 3   List of the topics

Under Item No. 4 in Fig. 3, concepts of damage evaluation are discussed. Ideally both linear and nonlinear models should be considered. However, only linear models are presented herein.

In the above-mentioned categories the papers are classified in an arbitrary manner. Several papers cover more than one item. The reader will find several topics covered by many authors, and other topics without coverage. This may be interpreted as an indication of the state of the art.

The aim of the workshop was to bring together several researchers in different fields of experience in order to

(a) document the state of the art,
(b) discuss the topics from different viewpoints, and
(c) obtain recommendations regarding future work.

We wish to express our sincere gratitude to

    Stiftung Volkswagenwerk, Hannover, FRG,
    National Science Foundation, Washington, D.C., USA,
    Rheinland-Pfalz, Staatskanzlei, Mainz, FRG,
    Deutsche Lufthansa AG, Ingenieurdirektion, Hamburg, FRG,
    Dornier GmbH, Friedrichshafen, FRG,

for their encouragement and generous support, without which the workshop would not have been possible.

In addition the organizers want to thank all participants for their active interest and valuable contributions.

References

1. J. T. P. Yao: Safety and Reliability of Existing
                      Structures; Pitman Publ. Inc.,
                      Boston, London, Melbourne, 1985

2. H. G. Natke (Ed.): Identification of Vibrating Structures; CISM Courses and Lecture No. 272, Springer, Wien, New York, 1982

3. H. G. Natke: Einführung in Theorie und Praxis der Zeitreihen- und Modalanalyse; Vieweg, Braunschweig, Wiesbaden, 1983

4. H. G. Natke (Ed.): Application of System Identification in Engineering; CISM Courses and Lecture No. 296, Springer, Wien, New York, will appear

5. F. Kozin, H. G. Natke: System Identification Techniques; Structural Safety 3 (1986) 269 - 316.

Demage Descriptions and
Basic Requirements

Cracked Cross Section Measurement in Rotating Machinery
B. O. Dirr, D. Hartmann, B. K. Schmalhorst

Cracked Cross Section Measurement in Rotating Machinery

B.O. Dirr , D. Hartmann , B.K. Schmalhorst
Institut für Mechanik, Universität Hannover

<u>0. Nomenclature</u>

a = actual depth of the crack
a_o = initial (reference) depth of a crack or notch
d = shaft diameter
f = frequency
g = acceleration of gravity
h = grid mesh width
i,j,k = indeces
n = relaxation number
r = amplitude
t = time
v = vertical shaft deflection
w = horizontal shaft deflection
x,y,z = cartesian coordinates
$2y$ = test point distance
$\hat{A}$ = horizontal amplitude
$\hat{B}$ = vertical amplitude
F = loading force
M = bending moment
Q = shear-force
V = potential difference (crack depth = a)
V_o = initial potential difference (notch depth = a_o)
$V_{i,j,k}$ = potential of a grid node
α = angle of rotation
ξ = dimensionless coordinate
σ_o = initial stress
φ = slope of neutral axis
Ω = angular speed
 = relaxation coefficient (1.0 .. 2.0)

1. Introduction

The assessment of the operating condition of machines and complicated structures is often based on the analysis of the vibration behaviour of the structure. Refering to the trouble free system, differences between the structural answers and the known reference signature are used as an indicator for a defect in the system. In rotor dynamics information from vibration monitoring can usually be used but must be analysed with respect to the type of trouble. The case of a cracked rotor has been reported by Mayes and Davies [1], Grabowski [2], Mahrenholtz [3] and other authors. Mahrenholtz [3] has given a comprehensive survey in the field of rotor dynamics and especially on the calculation of cracked rotors. It is the common aim of all the investigations to establish a significant and - if possible - unique relation of the crack to the vibration behaviour so that one can conclude the crack configuration from measurements and vice versa. In case of rotors in operation it is known from experience where cracks appear, for instance at the root of the blade discs in turbo machinery or at the notches of generator rotors in electric power plants etc. The published investigations often deal with modelling a cracked rotor for calculation purposes in order to find out the rate of change of the transverse crack growth. Mayes and Davies [4] showed this for real rotors in a power plant, Grabowski and Pöppel [5] used a laboratory experiment to measure the crack growth. Fig. 1 shows, for example, measuring results and calculations of the Grabowski rotor, that are well known from literature [5].

The good agreement of these results at the beginning had to be altered after breakage of the rotor, because the real crack topology has not been in sufficient correspondence with the modelling of the structure. On the other hand, it is quite difficult to get a better modelling of the cracked structure since it is impossible to have correct information on the crack form and the vibration shape of the rotor during operation. In this context the presented paper describes a three-dimensional crack model which is able to characterize

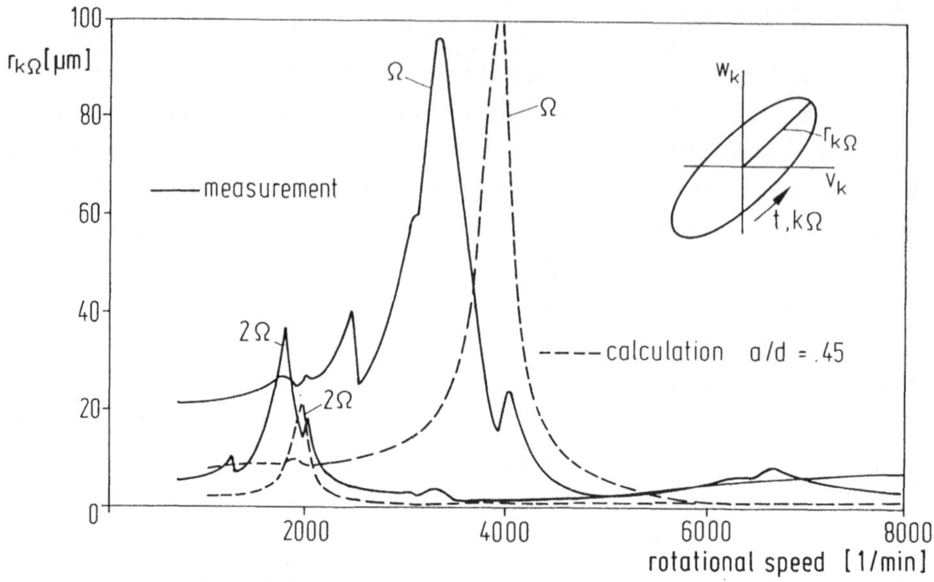

Fig.1. Amplitudes vs. rotational speed for
Grabowski rotor [5]

transient crack growth from a global point of view based on
measurement results for the single edge crack. Beach marks are
produced during test operation so that, after breakage of the
shaft, an evaluation of crack depth and the size of the
cracked area is possible. This can be used for modelling the
cracked part of the shaft by means of 20-node finite elements.
The beach marks are also used as a reference for the Direct
Current (DC)-potential method which is a non destructive
method. Together with a numerical solution of Laplace's
differential equation it is possible to describe in detail the
shape of the growing crack, before the structure is totally
separated. Naturally these results can be introduced to the
FE-procedure. In this way, calculations can be based on
measurement results without breaking the rotor.

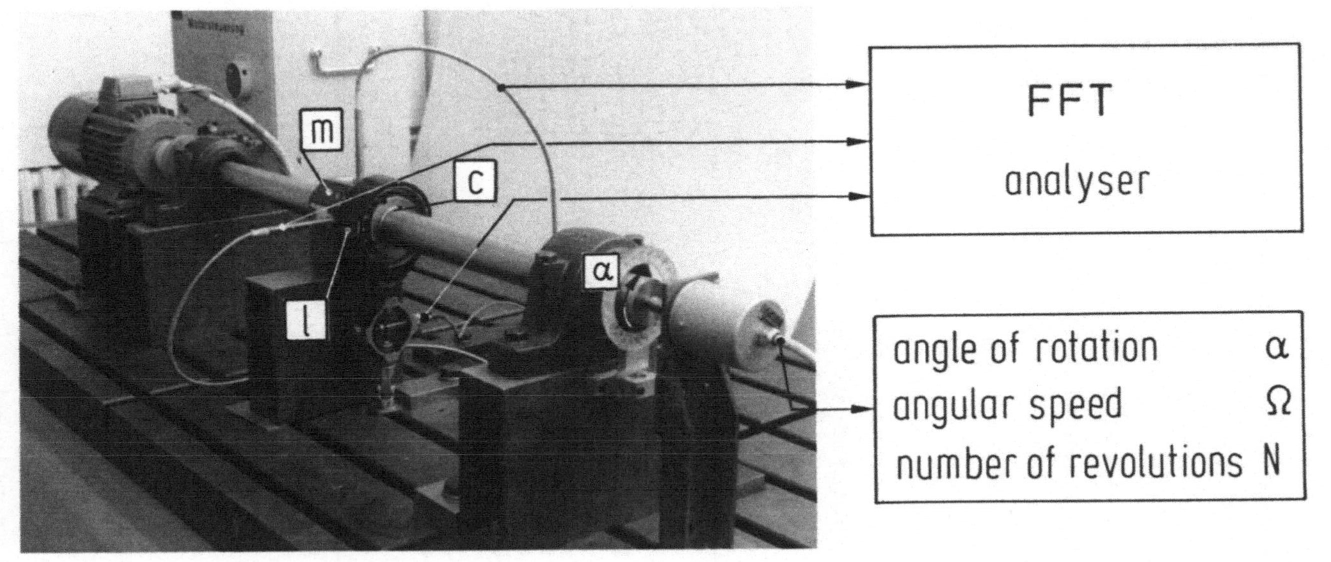

l = bending force position

c = crack position

m = measurement plane

shaft ⌀40x900 , 26 Ni Cr Mo V 14 5

Fig. 2. Testing equipment

2. Testing of cracked rotors

2.1 Testing device

The rotary specimen (ϕ 40mm x 900mm) is loaded by three-point bending with an initial stressing force acting in plane (ℓ), see Fig. 2. The rotor material is 26 Ni Cr Mo V 14 5 (1.6957) which is frequently used for real rotors. The vibration measurement is separately made using eddy-current pick-ups for both, the vertical and horizontal plane (m). The pick-up signals are analysed together with the bending force signal by an FFT-analyser. A cut, with a depth a_0 and a 150 μm radius at its straight front, is made, using an electro erosion technique, to initiate the crack at position (c). Operating the rotor with constant rotational speed leads to a fatigue crack after a number of revolutions, depending on the loading level.

2.2 Execution of experiments and their results

The vibration response indicates the beginning of the crack growth by increasing amplitudes, see Fig. 3. It can be seen from these curves, that a different sensitivity exists for different harmonics. The amplitude growth for the 2nd harmonic can, as in this case, change more rapidly than the 1st.

The rotor is stopped after a number of revolutions (denoted by N) suitable for compliance measurement and subsequent registration of vibration frequencies and vibration shapes of the shaft in a 'free-free' suspension condition. In this experiment the soft spring suspensions are positioned in the nodes for each of the vibration shapes, which are shown in Fig. 4 up to the third order.

Now, the process is started again, but now without loading. The number of revolutions is chosen so that a beach mark, which characterizes the cracked area of the cross section, is set. The experiment is continued with the same loading condition as before the interruption. When a convenient number

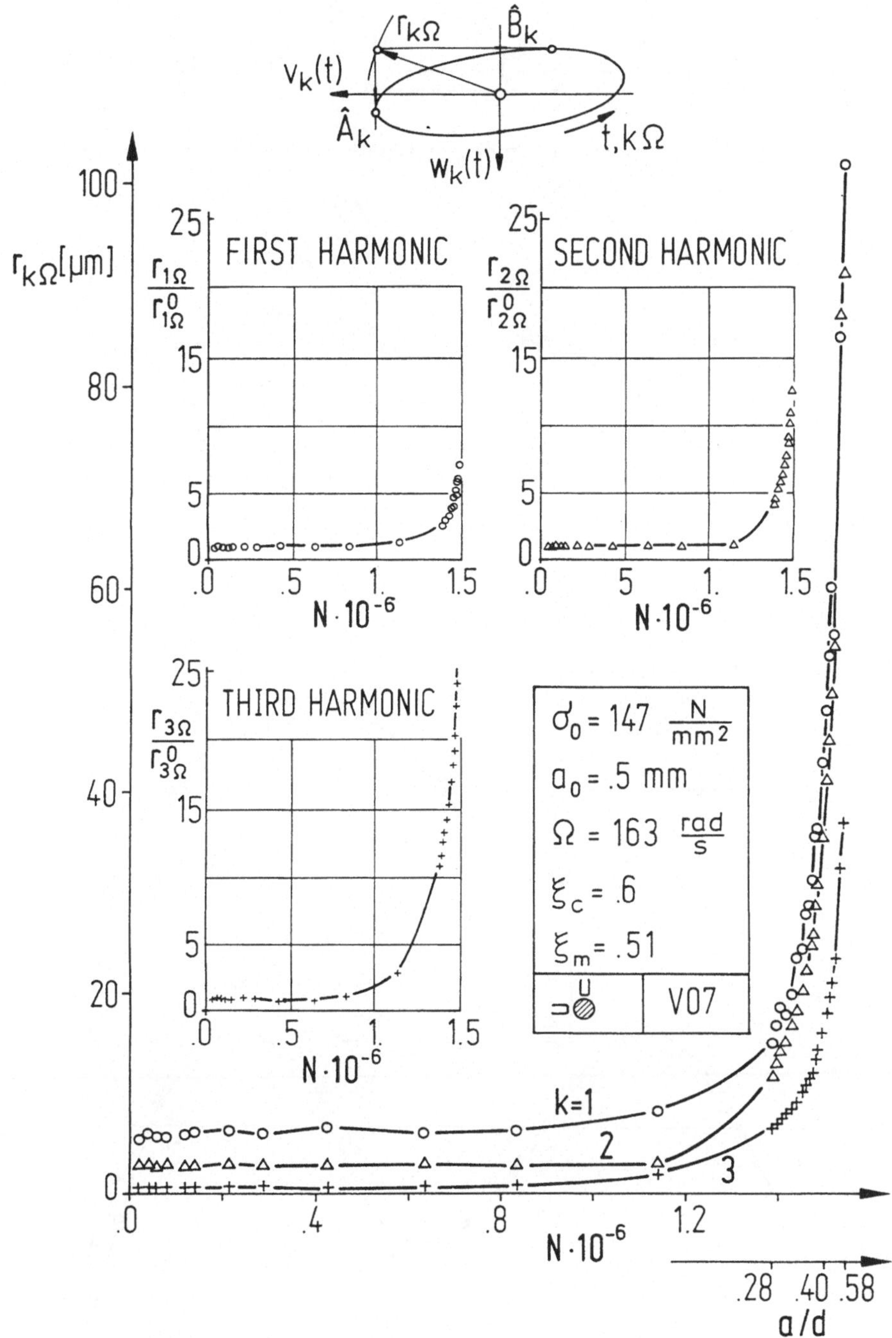

Fig. 3. Measured vibration response (V07)

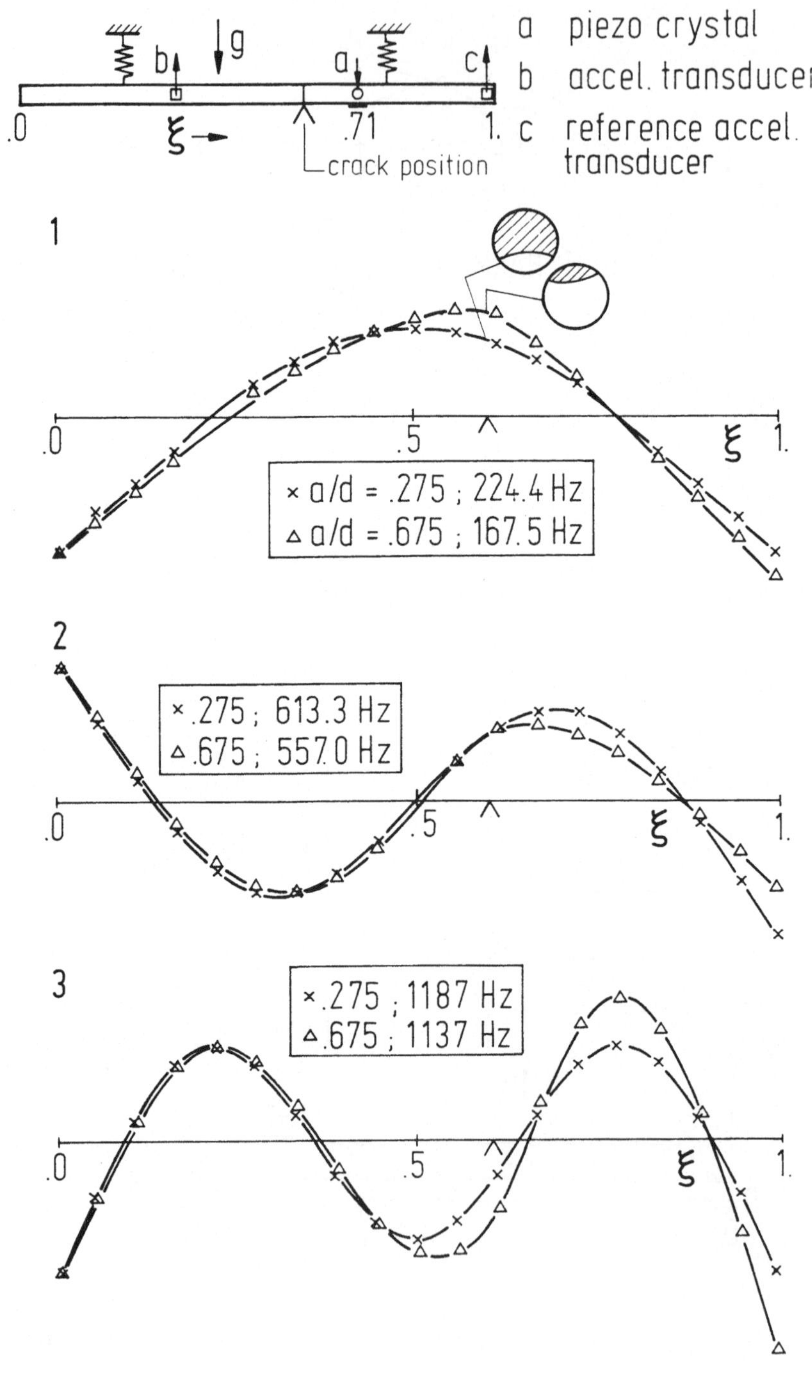

Fig. 4. Measured vibration shapes (V07, horizontal plane)

of revolutions is reached a new stop is made, etc.

After shaft breakage beach marks are visible on the cracked surface. Fig. 5 shows a photograph of the cracked surface and beach marks drawn from this.

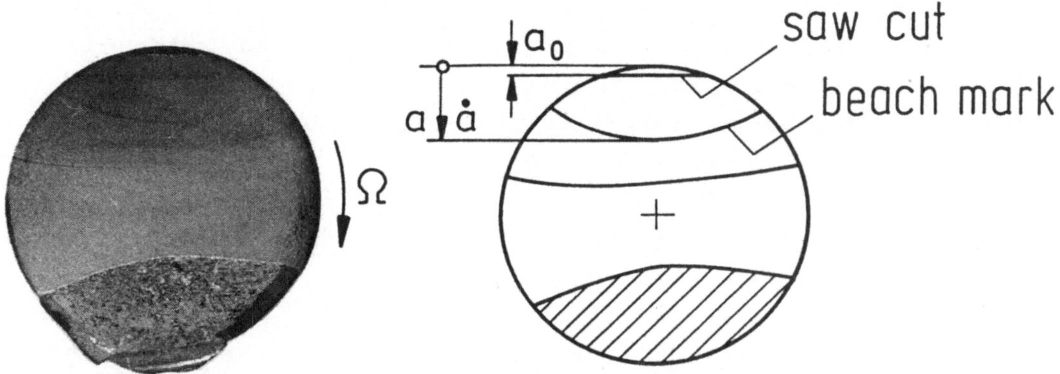

Fig. 5. Beach marks from a surface of fracture (V05)

3. The DC - potential method

If a material conducts an electric current, an electric flow-field exists inside the material. The shape of this field depends only on the conductivity of the material. Normally, a crack can be interpreted as a part of the boundary line. At the surface of the structure an electric potential difference can be measured. This difference can be used to establish the geometry of the crack.

The Laplace differential equation $\Delta V = 0$ describes the correlation between the electric flow-field and it's boundary line. As an elliptic boundary value problem, with a fixed potential at one part of the boundary (Dirichlet-problem) and a fixed normal derivation at the other part (Neumann-problem), it can be solved when the geometry of the boundary line is known. When the crack is modelled as a part of the boundary line, it's size can be calculated from the measured potential differences taken at the surface of the structure.

The following scheme defines the steps involved:

1) A direct current flows through the whole structure.
2) The potential drop is measured at a sufficient number of test points on the surface of the structure.
3) A boundary line (including the crack) is estimated.
4) The corresponding boundary value problem is solved.
5) The solution of step 4) is compared with the measurement of step 2).
6) The boundary line (at the crack) is corrected by a suitable strategy.

Steps 4) to 6) are repeated until the difference between two iterations is sufficiently small.

The simplest example for the application of the DC potential method is a long strip with a single notch at one side (CT - compact specimen; see Fig. 6). This is practically relevant in the testing of materials (refer BACHMANN 1983 [6]) to measure the crack growth velocity and determine the stress intensity factors in tensional and/or bending specimens.

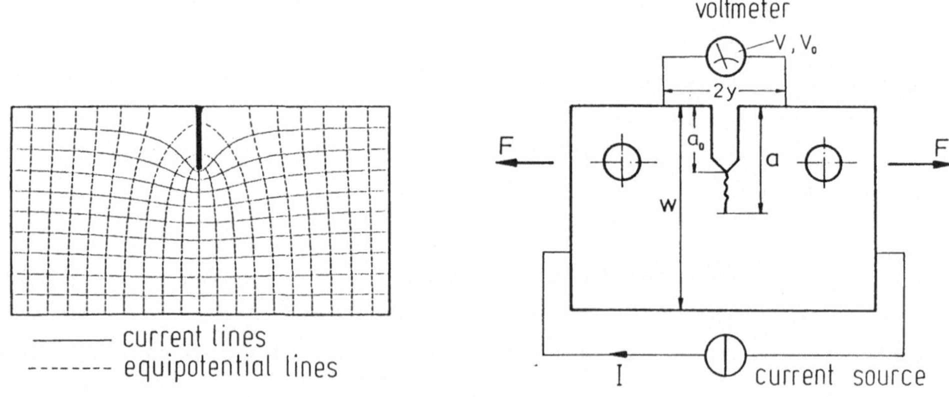

Fig. 6. Flow field and geometry of a CT-compact specimen

3.1 Measuring technique of the DC - potential method

The field of the potential drop must be known completely before the geometry of the crack can be calculated exactly. But it can be shown that the knowledge of only a small part of the field, at the surface of the structure near the crack, is sufficient to determine a satisfactory approximation.

The basis for all measurements and calculations is a sufficiently large constant direct current flowing through the whole structure. The significance of the measurement is directly proportional to the current density in the specimen. Densities of 50 mA/mm² are suitable. The transducer for the potential drop should have a resolution of less than 1 μV. Thermoelectric voltage and contact resistance must be avoided. The following equipment (see Fig. 7) was used for the investigations. It consists of

- a constant current source (100 A / 1 kW),
- a potential transducer with spring loaded electrodes
    (made from the same material as the shaft) and
- a preamplifier with a resolution better than 1 μV.

The post processing is done with a 5½ digit system voltmeter (DVM), a real-time computer and finally at the computing center of the university (RRZN).

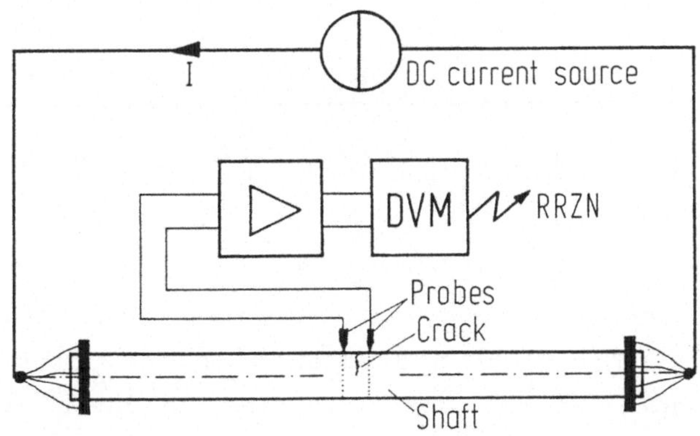

Fig. 7 Testing equipment for the DC-potential method

3.2 The finite difference method (FDM)

To solve a field problem, described by a partial differential
equation, it is convenient to use the following scheme:

1) Convert the differential equation into a difference
   equation (normally using a Taylor series).
2) Construct an adapted grid in the relevant area.
3) Formulate the difference equation at each grid node.
4) Formulate special equations at the boundary nodes.
5) Solve the generated set of linear equations.

These steps are now explained in detail.

The Laplace differential equation $\Delta V=0$, using three
dimensional cartesian coordinates, can be written using a
second order Taylor series expansion in a non equidistant
cartesian grid (refer to COLLATZ 1966 [7] and Fig. 8).

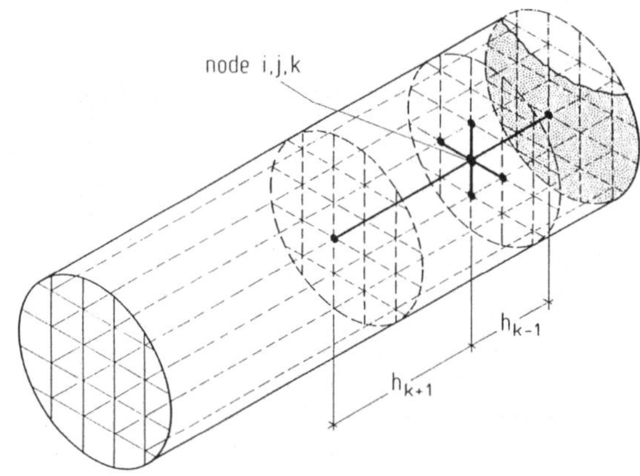

Fig. 8. Problem adapted grid for cylindrical specimen

To choose a problem adapted grid, it is helpful to know a
little about the solution. Fig. 6 shows the solution of the
plane problem. The highest gradients of the potential will be
near the crack. There, the grid has to be closely spaced.
Further from the crack, the gradients will be small and nearly
constant. Wide meshes should be used to reduce the number of
equations to be solved. So, it is convenient to use a grid

developed from a geometric series in the symmetry axis of the shaft and a partly constant mesh width in the other directions. Those widths should increase with the distance from the crack (see Fig. 8). A cartesian grid is more convenient than a cylindric because of it's equidistant discretisation along the radius.

The difference equation has to be formulated at each inner grid node. This equation determines the potential at the node as a weighted average of the potential of it's neighbours. Those neighbours, on the other hand, are the centres of other nodes.

At the boundary line, one must distinguish between the following two cases:

a) Dirichlet problem. The potential at this part of the boundary line has a fixed value. The potential of the grid nodes in this boundary line must be set to that fixed value. If the boundary line does not include the grid nodes, a linear interpolation must be made to determine the potential of the nearest nodes.

b) Neumann problem. The equipotential line runs perpendicular to the boundary line. The potential of imaginary nodes beyond the boundary line must be set to the same value or a linear interpolation of their nearest inner opposite nodes.

Using this information, one can construct a set of linear equations which describe the potential field. If the number of equations increases, it is convenient to use iterative methods to solve the system. In 1976 and 1981 respectively, MARSALL [8] and YOUNG and HAGEMANN [9] published work on successive overrelaxation methods. Their advantages are that they require less computer memory than standard algorithms (Gauss / Cholesky...) and that an approximation of the result can be obtained very early in the calculation. The equation

$$V_{i,j,k}^{(n+1)} = (1-\Omega)\, V_{i,j,k}^{(n)} + \frac{\Omega}{2\left[1 + 1 + \dfrac{1}{h_{k-1}\, h_{k+1}}\right]} \,\star$$

$$\star \left[V_{i-1,j,k}^{(n+1)} + V_{i+1,j,k}^{(n)} + V_{i,j-1,k}^{(n+1)} + V_{i,j+1,k}^{(n)} \right.$$

$$\left. + \frac{2\, V_{i,j,k-1}^{(n+1)}}{h_{k-1}\left(h_{k-1} + h_{k+1}\right)} + \frac{2\, V_{i,j,k+1}^{(n)}}{h_{k+1}\left(h_{k-1} + h_{k+1}\right)} \right]$$

defines the expression for a successive overrelaxation (SOR), where Ω is the relaxation coefficient. YOUNG and HAGEMANN [9] published an adaptive method to determine the best relaxation coefficient during the calculation.

3.3 Numerical check of a crack growth experiment

A shaft with a small notch (perpendicular to the symmetry axis of the shaft) is loaded as described above (Chapter 2.1). After a few million revolutions a steady crack growth is established. Then, the experiment is interrupted several times to produce beach marks and to examine the shaft using the DC-potential method.

Using the measured potential curves, a crack size is estimated and the corresponding potential field is calculated by the FDM. The solution is compared with the measured potential curve and the estimation is corrected. Since a suitable identification strategy is still being developed, the update is done by hand. After a few iteration steps a satisfying result is obtained. Fig. 9 shows the results compared to the measured potential curve (dashed line). The beach marks and the best numeric estimations of the crack size are shown in Fig. 10 . The good agreement confirms the performance of the method.

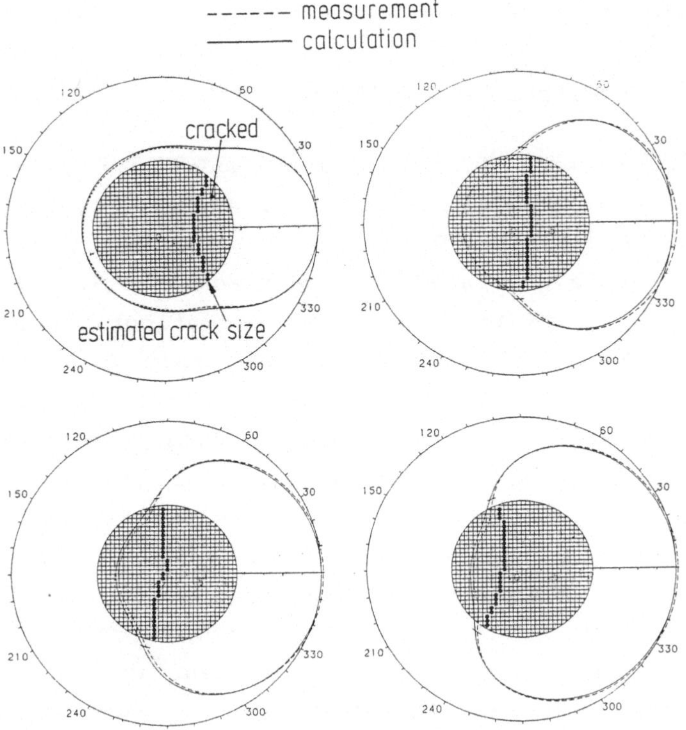

Fig.9. Measured and computed potential curves

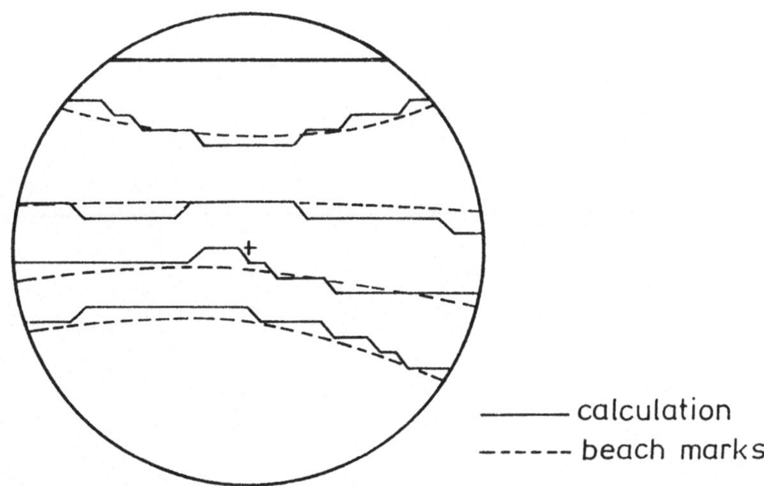

Fig.10. Comparison between experimental and
calculated crack sizes

4. Modelling of a crack

4.1 3D-FE model for the single edge crack

A three-dimensional finite element model was established on the basis of 20-node elements implemented in the computer program ADINA (1984) [10]. Fig. 11 shows the finite element network, that is generated for the cracked region of the shaft in accordance with the experimental configuration. Modelling of the crack itself is achieved by applying a double node technique. This 'global' modelling only enables a vibration analysis, whereas fracture mechanics questions need a proper 'local' modelling of the crack tip line. Since the cracked cross section can only be loaded by pressure forces triangular contact elements are added to the contact surfaces. This nonlinear model allows rotor 'breathing' to be calculated (see for example Fig. 12, where for $\alpha = n \cdot \pi/2$, n=1,3,5... the compressed part of the cracked surface derived from calculation is shown). The computer run for this is quite expensive, because an iteration procedure has to be performed to find out a kinematically correct shape of the cracked surface.

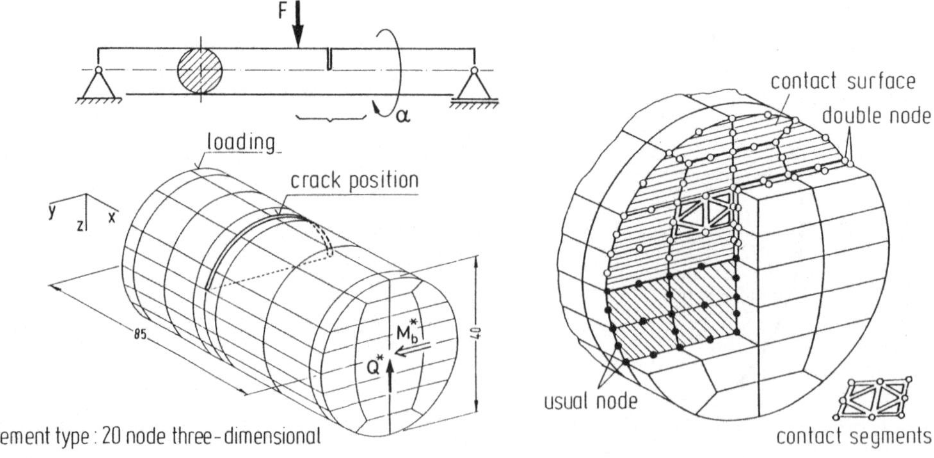

Fig. 11. 3D-FEM model of cracked region

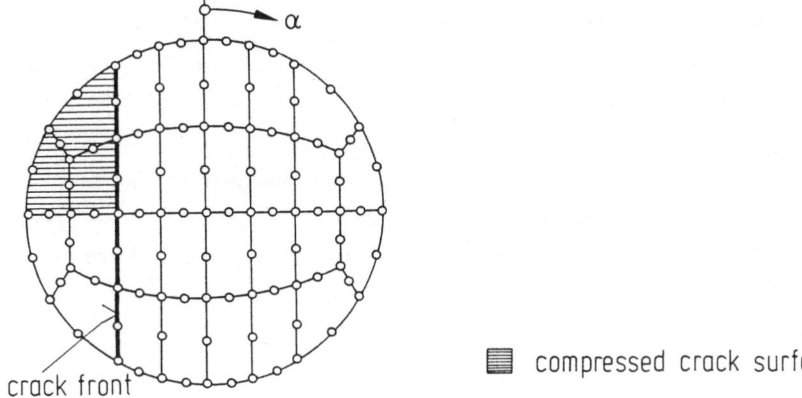

Fig.12. Compressed crack surface from FE-calculation ($\alpha = 3\pi/2$)

4.2 Calculation of eigenfrequencies

The total stiffness of the 3D-model shall be transformed, for convenience, to a conventional analysis for the calculation of the rotor's eigenfrequencies. Fig. 13 shows the one-dimensional model for the analysis of bending vibration in the principal plane. The mass distribution is assumed unaffected by the crack and therefore the mass matrix can be derived using the Timoshenko beam theory. The stiffness matrices of the non-cracked rotor sections are also deduced from the beam theory. The total stiffness of the cracked region, however, is calculated from the 3D-model strictly using loading and displacement quantities at its boundary cross sections. Thus, a two-node finite element is established. Assuming equilibrium of forces and moments and rigid body motions for the non-supported element leads to the general expression for the cracked region of the shaft (see Fig. 13):

$$
\begin{bmatrix}
A & B-Al & -A & -B \\
C-Al & Al^2-Bl-Cl+D & -C+Al & -D+Bl \\
-A & -B+Al & A & B \\
-C & -D+Cl & C & D
\end{bmatrix}
\begin{bmatrix}
w_1 \\ \varphi_1 \\ w_2 \\ \varphi_2
\end{bmatrix}
=
\begin{bmatrix}
Q_{z1} \\ M_{by1} \\ Q_{z2} \\ M_{by2}
\end{bmatrix}
$$

The matrix in the above equation represents the generalized stiffness of a finite beam element. The coefficients A,B,C and D are determined by using two independent sets of loadings and generalized displacements derived from the 3D-FE model. The symmetrization of the element stiffness matrix is done by averaging B and C. Fig. 14 shows measured and calculated bending frequencies of the VO1 rotor in free-free condition. The results are in good agreement for this simple slender shaft.

5. Conclusion

Using beach marks, the crack growth of a single edge cracked slender rotor of uniform cross section is investigated in a laboratory experiment. The DC-potential method is used as a tool which is able - accompanied by a numerical solution of Laplace's differential equation - to evaluate the depth and form of the crack without destroying the rotor. In this way one can calculate the actual bending stiffness of the cracked region at every instant during the experiment by applying a 3D-finite element modelling to the problem. Then a condensation leads to a numerical treatment of the rotor, based on measurements of a single edge crack. The calculated frequencies agree very well with the measurement results. No doubt the method can principally be applied to more complicated rotor structures.

6. Acknowledgement

The reported work was sponsored by Deutsche Forschungsgemein-schaft under contract number SFB 211, TP C2. The authors wish to express their thanks for this grant.

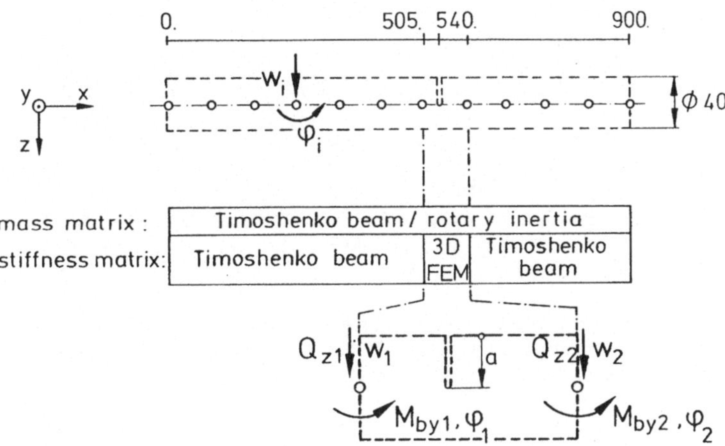

Fig. 13. Mass and stiffnes matrices derived from beam theory resp. 3D-FEM

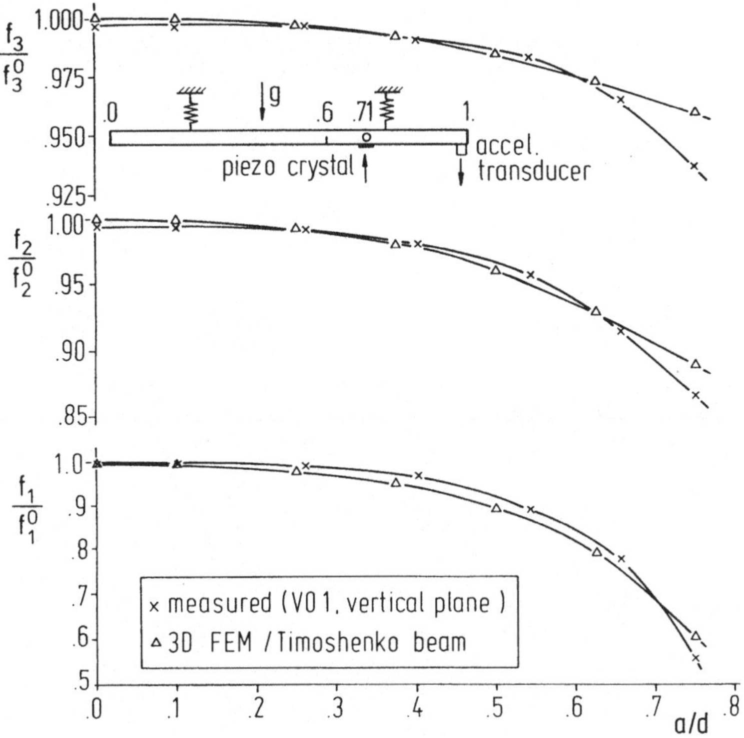

Fig. 14. Measured and calculated vibration frequencies

<u>7. References</u>

[1] Mayes, I.W., Davies, W.G.R., 1976: The Vibrational Beha-
    viour of a Rotating Shaft System Containing a Transverse
    Crack. Conf. on Vibrations in Rotating Machinery, Univer-
    sity of Cambridge, pp. 53-64.

[2] Grabowski, B., 1980: The Vibrational Behaviour of a
    Rotating Shaft Containing a Transverse Crack. Course on
    Dynamics of Rotors, International Centre for Mechanical
    Sciences, Udine, Italy.

[3] Mahrenholtz, O., 1983: The Dynamical Behaviour of Cracked
    Rotors. Proc. Sixth IFTOMM Congress, Theory of Machines
    and Mechanisms. New Delhi, pp. 35-40.

[4] Mayes, I.W., Davies, W.G.R., 1980: A Method of Calcula-
    ting the Vibrational Behaviour of Coupled Rotating Shafts
    Containing a Transverse Crack. Second International Conf.
    Vibrations in Rotating Machinery, University of
    Cambridge.

[5] Grabowski, B., Pöppel, R., 1982: Das Schwingungsverhalten
    eines Rotors mit Querriß - experimentelle und theoreti-
    sche Ergebnisse. VDI-Bericht 456, Düsseldorf, 167-175.

[6] Bachmann, V., 1983: Messung des Beginns der stabilen Riß-
    verlängerung an Kompaktproben mit Hilfe der Gleichstrom-
    potentialmethode. Int. Bericht, Inst. f. Werkstoff-For-
    schung, Köln.

[7] Collatz, L., 1966: The Numerical Treatment of Differen-
    tial Equations. 3. Aufl., Springer-Verlag, Berlin.

[8] Marsall, D., 1976: Die numerische Lösung partieller
    Differentialgleichungen. BI-Verlag, Mannheim.

[9] Young, D.M., Hagemann, L.A., 1981: Applied Iterative
    Methods. Academic Press, New York.

[10] ADINA, 1984: A Finite Element Program for Automatic
     Dynamic Incremental Nonlinear Analysis, ADINA Engineering
     Report AE 84-1, Watertown, Mass.

System Identification

Utilization of Experimental Investigations in the Process of Tower Structures Dynamic Identification

R.Ciesielski and J.Kawecki

1. Introduction

Identification is understood as determination, on the basis
of a known imput and output, of appartenance to a determined
class of systems with respect to which the given system is
equivalent (acc. to Zadeh [1]).
Equivalency of systems is defined by the so called identification criterion, i.e. function of errors defined on the output (exit) signal of process (y) and model (y_M):

$$\varepsilon = \varepsilon\left(y,\ y_M\right) \tag{1}$$

In consequence of identification a model arises; it contains
three kinds of information:
- on the structure (presented in form of mathematical iden-
  tities, block diagrams, graphs, connection matrices),
- on parameter values (i.e. independent variables),
- on values of dependent variables (i.e. coordinates of state given in discrete time moments or in time function).

In majority of cases a certain amount of information on the
model structure can be collected basing upon general knowledge of the process and at least partial determination of
its functioning. In such cases the information that has to
be aquired is limited to calculations of numerical values of
parameters or numerical values of state variables. The task
of identification is, then, limited to process parameter
estimation or estimation of state variables.
Eykhoff [2] presented the scheme of identification process
realization as transition from the process to its model,
showing the relation between the knowledge a priori (structural) and knowledge a posteriori (measureable). This scheme
with the authors further complements of little significance
for further considerations, was presented in fig.1.

PROCESS
MEASUREMENTS
MEASUREMENT NOISE
MEASUREMENT RESULTS
PHYSICAL LAWS
MODELLING ERRORS
PARTIAL DIFFERENTIAL EQUATIONS (NONLINEAR)
ELABORATION OF DATA
LINEARIZATIONS
LINEARIZATION ERRORS
ELABORATION OF RESULTS
PARTIAL DIFFERENTIAL EQUATIONS (LINEAR)
REDUCTION
NORMAL DIFFERENTIAL EQUATION
STRUCTURE
STATE ESTIMATORS
PARAMETER ESTIMATORS
ESTIMATION ERRORS
STATES
ESTIMATION ERRORS
PARAMETERS
MODEL

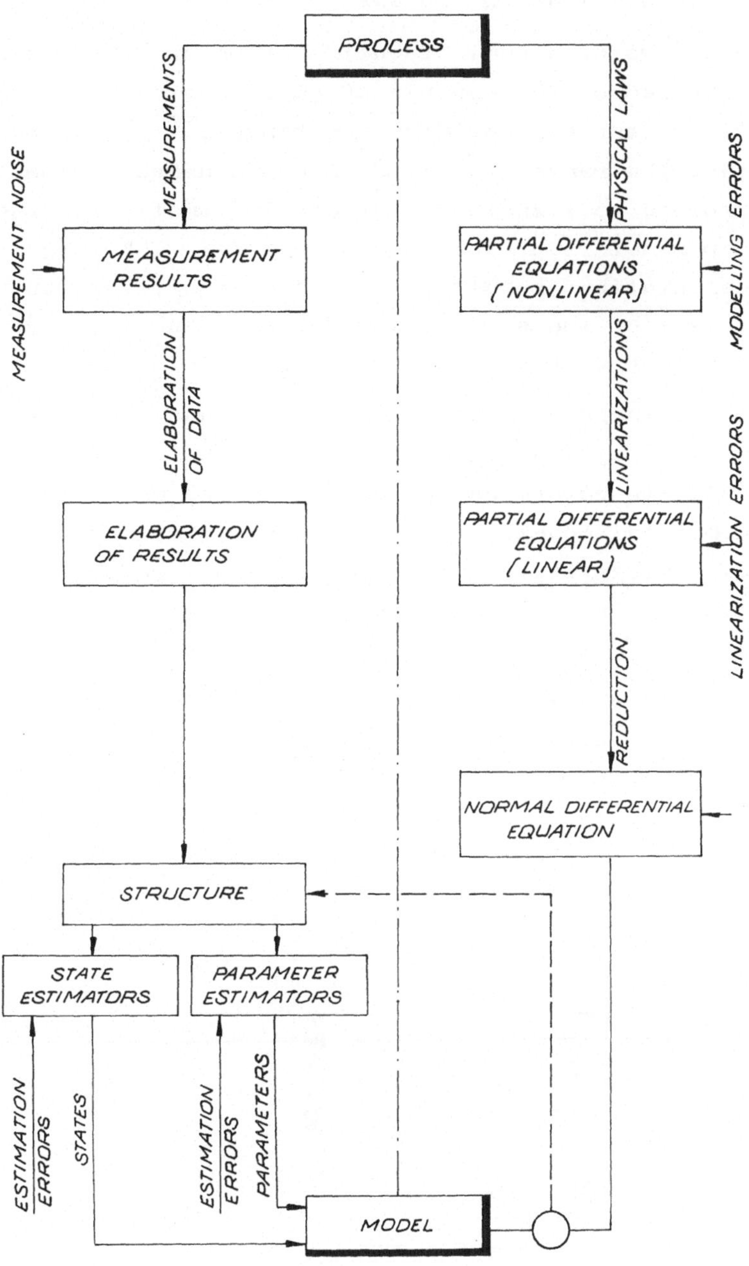

FIG. 1.

2. Utilization of specific characters of tower structures in the identification process

In [3] problems of dynamic identification of tower buildings were considered on the example of guyed masts. A conventional definition of three identification ranges: α , β , and γ was introduced. The range   permits in consequence of identification process realization to obtain information on the structure and parameters of the structure model. The range β includes calculations resulting from the identification process of parameters of the structure model adopting its structure on the basis of a priori information.

Specific features of tower structures cause the fact that adoption of a physical model of the object is practically equivalent with determination of parameters and structure of its mathematical model. The above observation was used in defining the identification range γ which leads to chosing the best model from the analysed in the process of identification.

Realization of the identification process in the range γ was presented schematically after [3] in fig.2.

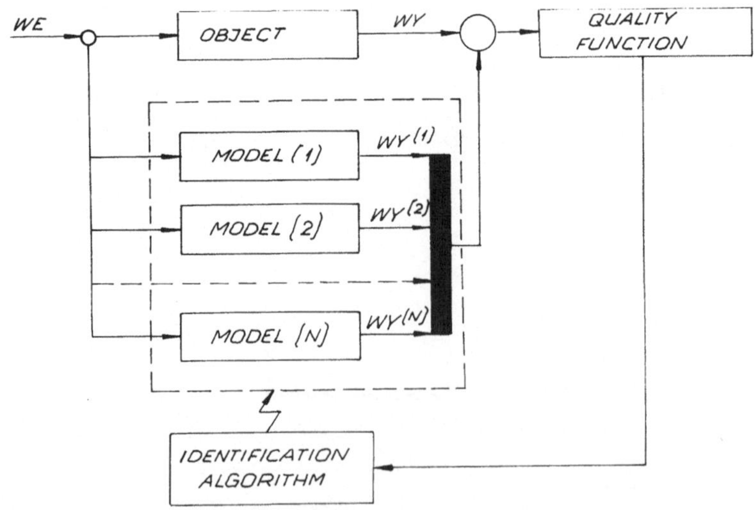

FIG. 2.

Following this scheme in [3] a mathematical model of tall guyed masts was carried out.

The realization scheme of the identification process in the range β was presented in fig.1. (A priori information permitted to define the model structure. This in combination with experimental investigation results was subsequently used for estimation of model parameters). The presented identification scheme was applied respecting tower structures in [4] and tall chimneys in [5].

In the presented schemes (figs 1 and 2) two parallelly used ways of the identification process are noticeable. One of them concerns operations on the structure model, and the other - investigations on the object. In the further part of the paper problems connected with realization of the latter way of obtaining information on the model object, indispensable in the identification process, were considered.

3. Succesive realization phases of the experiment on tower structures

The most significant phase of identification process realization concerning the experiment on the object was presented in fig.1 in a concised form, defining it as "measurements". Here, on the other hand, in fig.3 successive elements of this phase were described.

Planning the experiment the following was established:

a/ character of the experiment (active, passive),

b/ technics of the experiment (choice of apparatus, way of results recording and analysis),

c/ conditions of the experiment (accuracy, duration),

The measuring system (for measurement and recording of input - WE and output - WY) is a sequence of successive measurement results including always: a measurement gauge, amplifier and recorder (or analyser). In elaboration of the measurement way first of all information on the transmitted frequency band and amplitude range of measured values should be considered. Information on the range of low and very low frequencies characteristic of tower structures, obtained on the basis of measurements, is absolutely necessary.

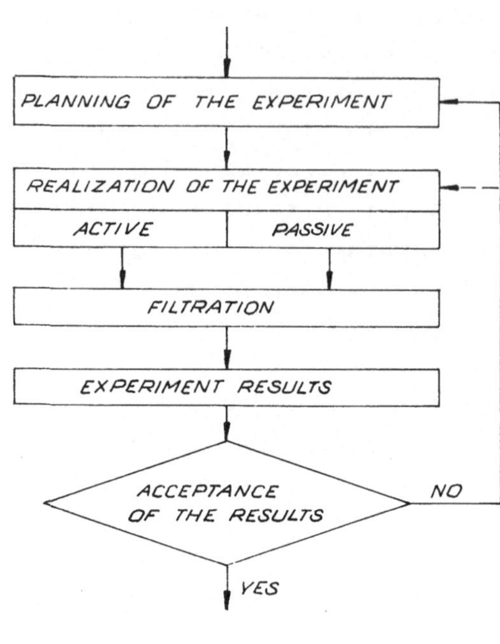

FIG.3.

Adoption of the form in which investigation results will be
presented (and utilized) is a crucial problem which should
be considered when planning the experiment. In present investigations, carried out by the authors of the present elaboration, dynamic characteristics of the structure were obtained
in consequence of experiments.
These were expressed as:
- sets of successive frequencies and forms of free vibrations,
  coefficients of damping (logarytmic decrement of damping,
  coefficients of critical damping),
- amplitude - frequency characteristics (transmittance),
- responses to standard forcings (in time).

4. Applied forcing (WE) of structure vibrations

Investigations were carried out both within the framework of
the active and passive experiment.
During active experiments a series and parallel methods were
applied.

In the series method many difficulties were encountered,
especially when attempting to apply it to tall structures.
Significant difficulties were:
- a long measurement time,
- changing conditions in which the structure remains during
  investigations (occurrence of considerable measurement
  noises),
- necessary application of a vibrator of great mass with
  regard to excitation of very low frequency vibrations,
- a long stabilization time of forced vibrations (large
  object).
Information on application of the series method in investigations of medium heigh masts (comp. Dietze [5]), high structures etc. is known. The authors applied this method in investigations on laboratory models of masts (comp. [3])and
towers.
In the parallel method standard forcings such as: impuls
jump, impact etc. are applied. In investigations tower structures the authors usually used a jump forcing. This was rea-
lized by sudden release of a rope prestressed by a previously
measured force. The rope was fixed to the structure in a
strictly determined place. Release of this rope (an additional rope) was obtained by a brittle break of a joining element or release of a special mechanism. Best results were
obtained by application of a special mechanism making the
force in rope drop suddenly. This way of forcing was applied
by the authors in investigations of guyed masts (comp. [3]),
tall structures, chimneys etc.
In order to obtain information on free vibration frequencies
of tower structures periodical, controlled, horizontal move-
ments of people (1 or 2 people), placed on the top of the
structure, were applied as initiation of vibrations during
active experiments. Good results were obtained in examinations of steel chimneys, buildings and aerials fixed on guyed
masts.
In passive experiments week wind gusts or paraseismic loads
(explosions in quarries, vibrations from traffic etc) were
the usual forcing. Some investigation results were descri-

bed in [7]. The basic difficulties are: a long recording
period and lack of comprehensive information on vibration
forcing (a large number of measuring points for recording
forcings is absolutely necessary).

5. Recording of structure response (WY)

In the phase of dynamic investigation planning some crucial
problems connected with the structure dynamic response should
be solved. The solutions are an answer to the questions:
- what parameter values should be recorded,
- what measurement apparatus should be used,
- where to appoint the measuring points.
It may be generally stated that in investigations of tall
structures: a/ accessory apparatus should transmit possibly
lowest frequencies b/ measurement gauges should be placed
in characteristic modes in points of marked incontinuity,
on the way of forcings etc.
In measurements concerning free vibrations, measurement of
displacements or deflections could be advised. Recording of
displacements can be disturbed by a strong influence of high
frequency noises.

6. Examples of utilization of experimental investigation
   results in the identification process.

6.1. Investigations on guyed masts (comp. [3 , 9])

Dynamic examinations of guyed masts were carried out within
the framework of realization of the identification process
of high masts. Identification within the range γ was applied.
A few mast models were subjected to primary considerations.
In the final elaboration stage of a priori information three
physical mast models were considered. In those models the
shaft was constituted by a continuous beam on linear-elastic
supports (comp. fig.4). In model "A" the (representative)
transmitted stiffnesses of the supports were adopted to be
independent on vibration frequency. Whereas, in model "B"
stiffnesses of supports were adopted to be dependent on vibration frequency (ω) . In both models the mass of the shaft

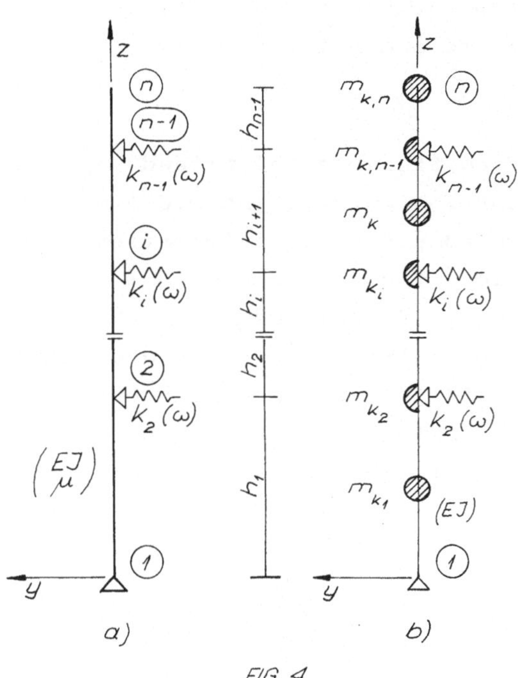

FIG. 4

rods is distributed along their length. In model "C" the mass
of the shaft is concentrated in joints; stiffnesses of supports being as in model "B". All three considered models are
linear.

It was adopted in the report that the system presented in
fig. 4b which is described by equation of motion:

$$[M]\,\{\ddot{y}\} + [C]\{\dot{y}\} + [K]\{y\} = \{p\} \qquad (2)$$

can be a physical model of the mast.

In the given mathematical model (2) the matrices $[M]$, $[C]$ and
$[K]$ are matrices of inertia, damping and stiffness respectively, $\{y\}$ is the vector of the model response to the forcing
described by the vector $\{p\}$.

The matrix of stiffness can be described as a sum of matrices
of stiffness of the rods of the shaft $[K_p]$ and stiffness $[K_s]$
of elastic supports formed by guys i.e.

$$[K] = [K_p] + [K_s] \qquad (3)$$

In a similar way, the matrix of damping $[C]$ can be presented as a linear combination of the matrices of inertia and stiffness according to the relation:

$$[c] = \bar{\alpha}\,[K] + \bar{\beta}\,[M] \tag{4}$$

$\bar{\alpha}$ and $\bar{\beta}$ are coefficients determined experimentally. Calculations carried out for the model described with equation (2) require complemented information concerning determination of matrices (comp. formula (3)) and coefficients $\bar{\alpha}$ and $\bar{\beta}$ occurring in formula (4).

The element k_1 of matrices $[K_s]$ corresponding to the elastic support "1" consisting of three guys can be determined using the formulae given in paper [3] which were presented here in fig.5 together with the necessary description of the magni-

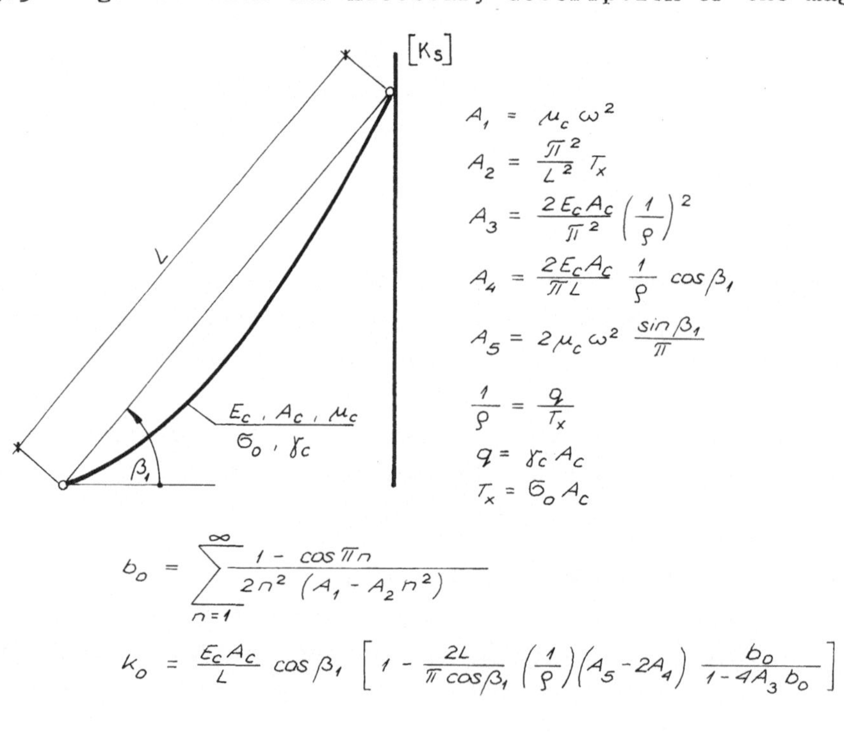

$$A_1 = \mu_c\,\omega^2$$

$$A_2 = \frac{\pi^2}{L^2}\,T_x$$

$$A_3 = \frac{2 E_c A_c}{\pi^2}\left(\frac{1}{\rho}\right)^2$$

$$A_4 = \frac{2 E_c A_c}{\pi L}\,\frac{1}{\rho}\,\cos\beta_1$$

$$A_5 = 2\mu_c\,\omega^2\,\frac{\sin\beta_1}{\pi}$$

$$\frac{1}{\rho} = \frac{q}{T_x}$$

$$q = \gamma_c A_c$$

$$T_x = \sigma_o A_c$$

$$b_o = \sum_{n=1}^{\infty} \frac{1 - \cos\pi n}{2n^2\,(A_1 - A_2 n^2)}$$

$$k_o = \frac{E_c A_c}{L}\,\cos\beta_1\left[1 - \frac{2L}{\pi\cos\beta_1}\left(\frac{1}{\rho}\right)(A_5 - 2A_4)\,\frac{b_o}{1 - 4A_3 b_o}\right]$$

$$k_1 = 1{,}5\,k_o\,\cos\beta_1$$

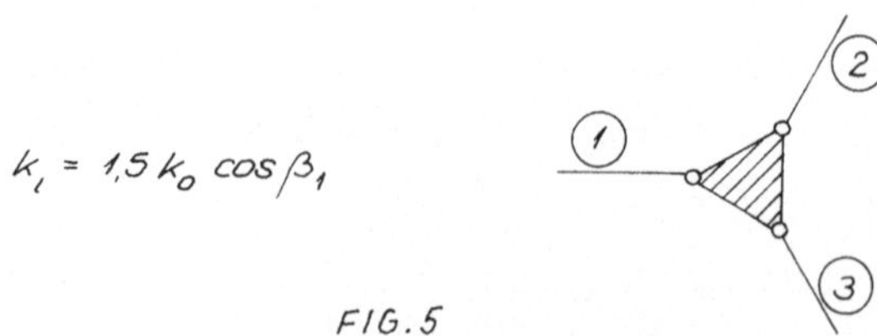

FIG.5

tudes occurring in the formulae.
It results from the listed formulae that stiffness k_1 is a
function of vibration frequency of the model.
In a similar way fig.6 presents information and formulae on
the basis of which the value of coefficients $\bar{\alpha}$ and $\bar{\beta}$ can
be determined. Information on damping properties of the structure are given as values of the logarytmic decrement of damping Δ . On the basis of information on values Δ corresponding to two different free vibration frequencies of the sy-
stem f_a and f_b (comp. fig.6) the values $\bar{\alpha}$ and $\bar{\beta}$ can be determined.

$$[C] = \bar{\alpha}\,[K] + \bar{\beta}\,[M]$$

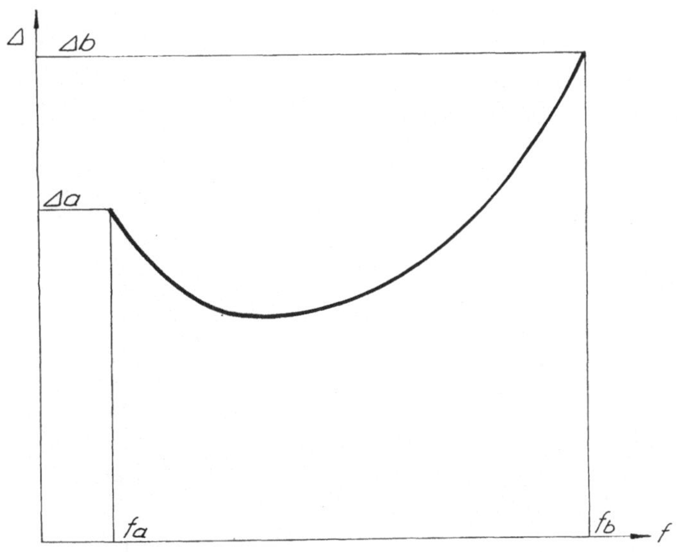

$$\bar{\alpha} = \frac{f_a\,\Delta_a - f_b\,\Delta_b}{2\pi^2\,(f_a^2 - f_b^2)}$$

$$\bar{\beta} = 2f_a\,\Delta_a - (2\pi f_a)^2\,\bar{\alpha}$$

FIG 6

Dynamic investigations were carried out on four masts with
guys:
I-mast at Chorągwica 262,5 m high, with four guy levels,

II-mast at Łagów, 288,75 m high, with four guy levels,
III-mast at Trzeciewiec of hight and guys as in mast II,
IV-mast at Konstantynów, 644,3 m high (whole 646 m), with
   six guy levels.
In experimental investigations first of all the active experiment was applied i.e. controlled vibration excitation (forcing). This forcing was executed by rupture of an additional
guy mounted to the shaft in its lowest span as shown in fig.7
(mast at Chorągwica). Investigation results obtained during
the passive experiment were treated as complementary results.
Wind gusts were then the vibration excitations. Measurement
of the applied forcing value was taken by means of a dynamo-
metr (active experiment)and anemometers (passive experiment).
Measurement of mast response to the applied forcings was taken by use of various sets of apparatus, recording: displacements, strains or accelerations in chosen measurement
points (acc. fig.8).

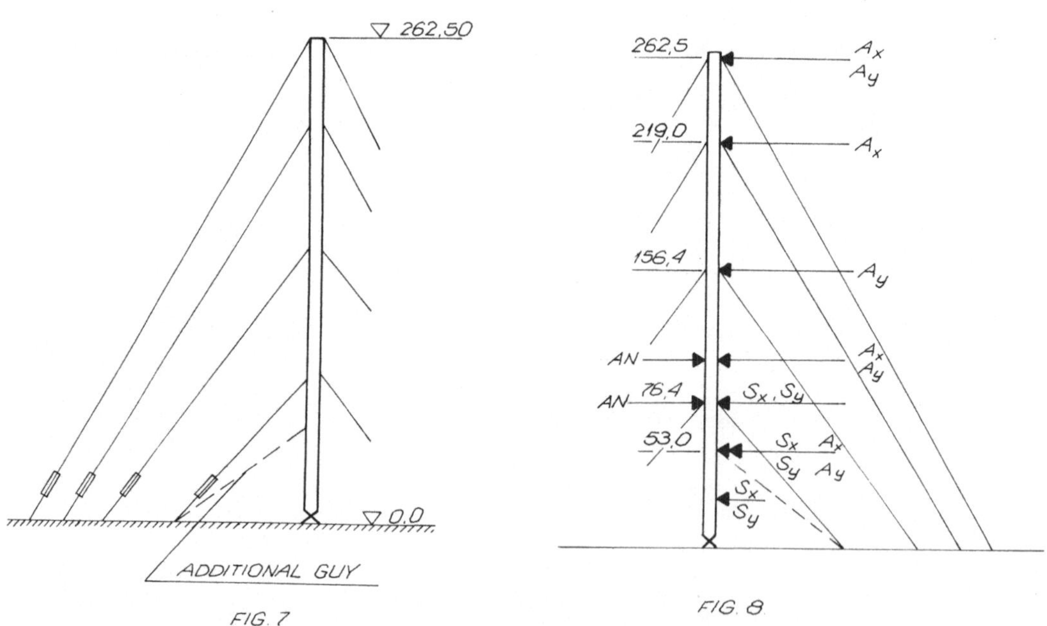

Measurement of displacements was performed by using of a seismic
apparatus set (gauge S5S, mouthpiece Szk-2, oscyllograph

POD-12M).Acceleration measurement was carried out by use of accelerometers (piezoelectric gauge 8306, electric feeder 2805, normalizing load amplifier 2626, four track tape-recorder 7003) of Brüel-Kjaer make.
Measurement of strains (deformations) was performed by use of a tensometric apparatus set (resistance tensometer RB, tensometric amplifier UM131 or KWS-6A, four track tape-recorder 7003 or recorder RP 2AB). Measurement points, where responses to forcings were measured, were appointed on the mast, first of all in close vicinity of guy mounting points, and additionally in the lowest span of the shaft for displacement measurement. Apparatuses used permitted measurements of very low frequency vibrations. Measurement results were elaborated, in the first place, in an apparatus set permitting analogue analysis (tape recorder 7003, impuls reproducer 5626, heterogenous analyser 2010, level recorder 2307)of Brüel-Kjaer make. Analysis was performed in order to obtain successive values of free vibration frequencies of masts. Similar information was obtained independently in result of digital experiments carried out on the analysed models A and B.

Similarly information on vibration forms corresponding to determined free vibration frequencies were obtained from investigations on objects and calculations made for the models.
All this information was used in examining the quality function value for each of the considered models.
In the described identification process it was adopted that the quality function with respect to vibration frequency will be written as follows:

$$I_m^{(f)} = \left| \frac{fm - fm^{(N)}}{fm} \right| \leqslant 0,10 \tag{5}$$

where: fm - is a successive m-th free vibration of the
             mast in natural scale,
        $f_m^{(N)}$ - is a successive m-th free vibration frequency
             determined for the model N $\left(N \in A, B \right)$.
Requirements written in the above formula can be out of practical reasons restricted to some lowest free vibration frequencies (e.g. m = 1,2, ... 8). With reference to vibration form the following criteria were adopted as complementary to

the foregoing:

- horizontal displacements of the joints of the mast shaft, corresponding to the m-th vibration form determined for the object (Ym) and model $(Ym^{(N)})$ should bear the same signs (joints whose displacements are very small are here an exception).

- the greatest relative difference of horizontal displacements of the k-th mast shaft node for the m-th frequency should satisfy the unequality:

$$T_m^{(N)} = \max_{k,m} \left| \frac{Ymk - Ymk^{(N)}}{Ymk} \right| \leqslant 0,25 \qquad (6)$$

Calculation and measurement results expressed as successive free vibration frequencies were presented in form of example in Table 1.

Table 1

Mast Nr	Model or measurement	Free vibration frequencies f_m [Hz]				
		1	2	3	4	5
I	"A"	0,39	0,63	0,86	1,10	
	"B"	0,22	0,28	0,31	0,34	0,39
	"C"	0,22	0,28	0,29	0,31	0,36
	measurement	0,20	0,28	0,31	0,34	0,39
II	"A"	0,38	0,62	0,84		
	"B"	0,20	0,26	0,28	0,31	0,38
	"C"	0,20	0,26	0,29	0,32	0,40
	measurement	0,21	0,26	0,30	0,33	0,39

In consequence of realization of the identification process (acc.fig.2) it was found that the so far applied model "A" cannot be considered as a proper representation of a guyed mast. Model "B" and "C" take into consideration parameters influencing significantly the description of dynamic parameters of guyed masts.

6.2. Investigations of objects in laboratory scale

Dynamic investigations of objects in laboratory scale were carried out within the framework of realization the identi-

fication process. Identification in the range β was applied.
The scheme of the investigations object was presented
in fig.9. A pipe shaft section 1 m high was the basic element
of the object. These elements could be joined with one another obtaining a shaft 1 to 5 m high. The bottom support of the shaft was a hinged bearing. In each joing node of the pipe element three guys were mounted, their other end being anchored by adjustment into the hall floor. Investigations were carried out as an active experiment. Forcing was obtained by repture of an additional guy fixed to the object on various levels, as shown with a broken line in fig.9. Measurement points were established in joints

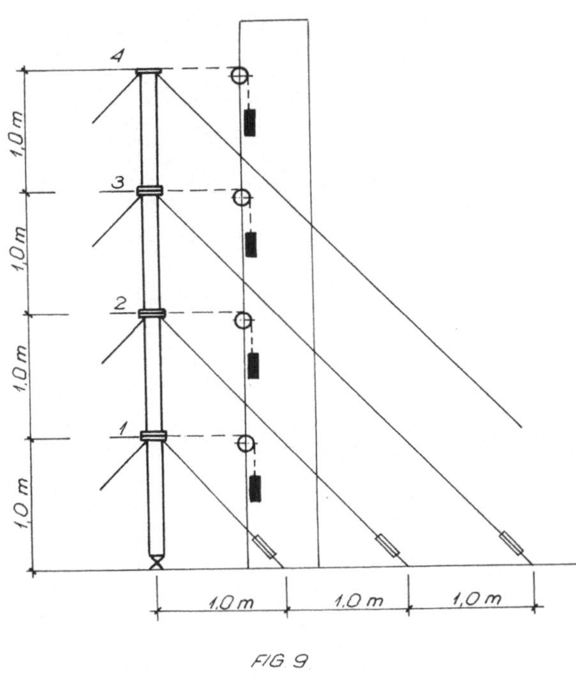

FIG 9

placing there displacement gauges (CT 12-1 or OT 44-5).
These gauges cooperated with measuring instruments (RF 01T).
The signal was recorded on a recording tape (four track tape-
-recorder). Recorded changes of horizontal displacements of
measurement points were subjected to analysis using the
apparatus set presented in 6.1 in result of analysis ampli-
tudinal - frequency characteristics of the object were obtained. Chosen results of analysis for a 3 m high object model
were presented in fig.10. The results presented in such a
way served for determination of parameters of the object model whose structure was adopted on the basis of a priori
information in form (2)

$$[\text{M}]\{\ddot{y}\} + [\text{C}]\{\dot{y}\} + [\text{K}]\{y\} = \{p\}$$

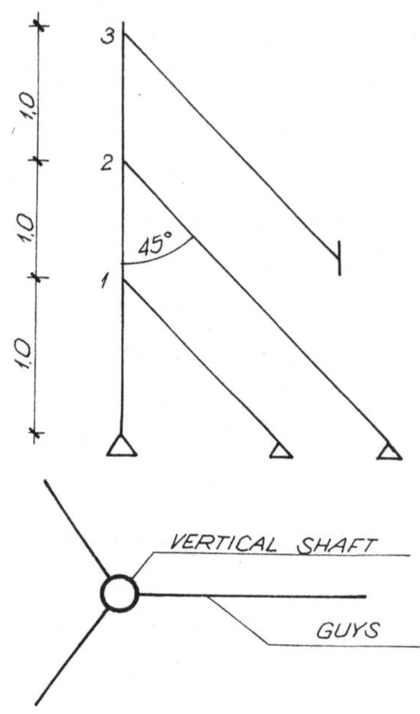

a) SCHEME OF MODEL

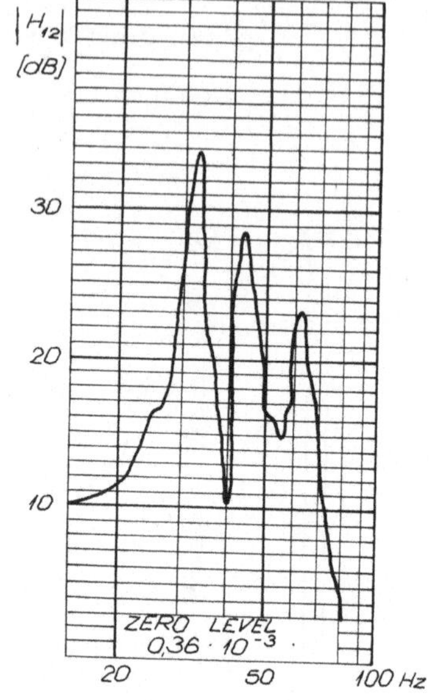

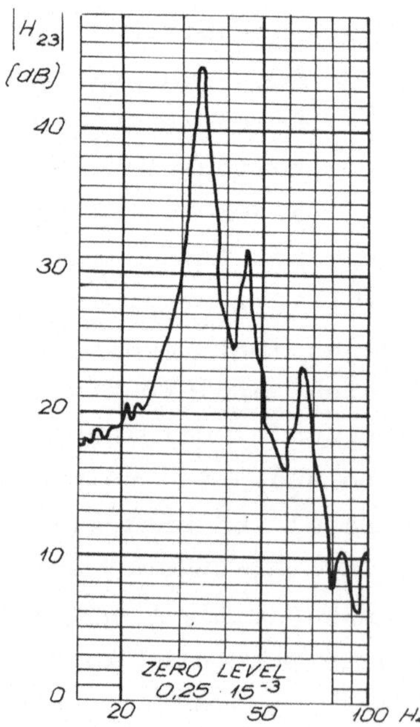

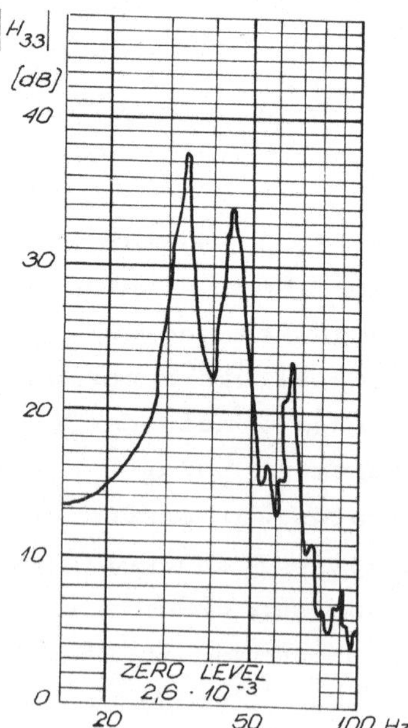

b) SPECTRAL ANALYSIS OF RESULTS

FIG 10

The adopted magnitudes dimension of mass $[M]$, damping $[C]$, and stiffness $[K]$ matrices equaled (3×3). Description of the metrices and their detailed discussion was given in $[3]$. Additionally it was adopted that matrix of damping can be described with the formula (4)

$$[C] = \bar{\alpha}\,[K] + \bar{\beta}\,[M]$$

Using the investigation results presented in form of charactereistics $|H_{ij}|$ the elements of matrix $[K]$, $[M]$, and coefficients $\bar{\alpha}$ and $\bar{\beta}$ were obtained. In this way the identification problem for the object considered was solved. Applying a similar procedure a solution of the identification problem for laboratory model of a chimney made of plexiglass, 3 m high, was presented in $[5]$.
In consequence of realization of the identification procedure for model "C" (comp. chapter 6.1) following parameters of the model were obtained (fig. 11) :

$$[M] = \begin{bmatrix} 3,33 & 0 & 0 \\ 0 & 3,20 & 0 \\ 0 & 0 & 1,49 \end{bmatrix}, \ \text{kg}$$

$$[K] = \begin{bmatrix} 340 & -77 & 24 \\ -77 & 232 & -41 \\ 24 & -41 & 91 \end{bmatrix}, \ \text{kNm}^{-1}$$

$$\bar{\alpha} = 0,00024$$

$$\bar{\beta} = 2,8 \div 3,2$$

FIG. 11.

7. Evaluation of reliability of the mast structure

The above given considerations concern one of the basic problems of dynamic diagnostics of a mast i.e. evaluation of dynamic properties. This is essential in all dealings both in analysis (dimensioning), examination of satisfying all additional conditions apart from strength (e.g. stiffness, stability and in cases of diagnosis of the state of the existing objects. Experimental dynamic identification carried out at various moments of the "life of the structure" can show differences in results which indicate to a change in the technical state of object. Quantitative results are the basis for evaluation of the kind of these changes.

Evaluation of the dynamic response of the structure can show some imperfectness of this structure in the first examinations already.

The results of dynamic identification can be utilized also at classical evaluation of structure realiability (comp. [10]). Assuming the necessity of mast work maintainance in the elastic range and treating attainance of yield point as the limit state for an undamaged mast the following formulae can be used for determination of the damage probability of the structure:

$$\sigma_{lim} = \sigma \left(\mathcal{H}_k, W, \alpha \right) \tag{7}$$

$$p \left(\sigma_{lim} \right) = p \left[\sigma \left(\mathcal{H}_k, W, \alpha \right) \right] \tag{8}$$

$$P_m = \int\limits_{\mathcal{H}_k} \int\limits_{W} \int\limits_{\alpha} p \left[\sigma \left(\mathcal{H}_k W, \alpha \right) \right] d\sigma \cdot p(w) \, dw \; p(\alpha) \, d\alpha \tag{9}$$

where: $\mathcal{H}_k$ — is the dynamic response of the structure obtained on the basis of dynamic identification

$p(\sigma_{lim})$— probability of reaching the limit state of stress

W   — defined normal wind load according to the assigned distribution

$p(W)$ — probability of occurrence of this load

α — angle (spatial) of wind action and other phenomena of wind action

$p(\alpha)$ — probability of their occurrence.

The considered case assumes lack of a concomitant action of other loads, apart from wind load, and does not particularize all the assistant elements serving at determination of the three, most important, mentioned factors.

Performing calculations according to the given formulae is troublesome since it requires a previous recognition or assum-

ption of load data which can proceed on the basis of statically elaborated meteorological data and only exceptionaly
per analogia to other cases.
As a whole, however , dynamic identification of the mast
structure permits determination, basing upon it, of an adequate dynamic model; it is, as it was said previously, a key
element of every case of diagnostic analysis and evaluation
of safety and reliability of the structure.

8. Summary

Recently the identification process is more and more often
used for obtaining information on physical and mathematical
models of building structures. Identification methods are
specially useful in determination of model of untypical structures or of new structural solutions, e.g. [8]. Realization of
the identification process is connected with experiments
investigations of objects in natural scale. The present paper is focussed on problems connected with carrying out experimental investigations. It was presented how they were solved carring out experimental investigations on engineering
structures. Examples given at the end are a practical visua-
lization of the previously described solutions.

References.

1. Zadeh L.A.; From circuit theory to system theory, Proc.
   IRE 1962, vol.50, pp. 856-865.

2. Eykhoff P.; System identification. Parameter and state
   estimation, John Wiley and Sons, London 1974.

3. Kawecki J.; Dynamic identification of engineery objects
   of the guyed masts type (in Polish), Politechnika Krakow-
   ska, Zeszyt Naukowy Nr 5 z. 56 , Kraków 1978 .

4. Kawecki J.; Estimation of parameters of the mathematical
   model of high buildings on the basis of dynamic measurement results (in Polish), Inżynieria i Budownictwo, No
   11-12, 1983, pp. 468-473.

5. Ciesielski R., Kawecki J.; Dynamical experimental identification of chimneys, IV International Symposium on Industrial Chimneys. The Hague, 1981.

6. Dietze F.; Vibration measurements on anchored masts. IASS - Symposium Book Bratislava, 1966.

7. Ciesielski R., Kawecki J., Maciąg E., Pieronek M.; Methods of determination of free vibrations of tower - type structures, Euromech Colloquium 112, Akademiai Kiado, Budapest 1980. pp. 81-104.

8. Ciesielski R., Kawecki J.; Experimental method for evaluation of stiffness of multistory building,5-th Symposium CIB, Madrid 1980. Symposium book. pp. 289-306

9. Kawecki J.; Dynamic experimental investigations of masts with guys (in Polish), Archiwum Inżynierii Lądowej, vol. XXXII, No 1/1986, pp. 167-180.

10. Stephens J.E., Yao T.P.; Damage assesment using response measurements. Journal of Str. Engineering, vol. 113,No 4/1987 p.p. 787-801.

On the Determination of the Number of Effective Modes from Vibration Test Data

Michael Link *)

1. Introduction

With phase seperation techniques modal data are determined from vibration tests using single or multipoint exciter configurations which are not appropriated to a certain mode. Typical for such situations is that some of the modes contained within a frequency range of interest may be very weakly excited. The identification of weakly excited modes always yields problems since the signal to noise ratio is low in such cases and since their response may highly be superimposed by the response of neighbored more strongly excited modes. Essential for many phase separation techniques is their ability to identify not only the strong modes but also as many weak modes as possible and to supply some indication on their accuracy, i.e. to identify a maximum number of modes effective in a given frequency range.

2. Theoretical Background

In recent times methods for the direct identification of the system matrices in the discrete equation of motion

$$(-\omega^2 \underline{M} + j\omega \underline{D} + \underline{K})\ \underline{u}\,(j\omega) = \underline{f}\,(j\omega) \tag{1}$$

$$\underline{M},\ \underline{D},\ \underline{K}\ \ = \text{system matrices}$$
$$\text{(mass, damping, stiffness),}$$
$$\underline{u}\,(j\omega)\ = \text{complex frequency response } (j = \sqrt{-1}),$$
$$\underline{f}\,(j\omega)\ = \text{complex excitation force,}$$

*) Professor for light weight structures,
   University of Kassel, FR Germany

have gained interest /5/, /6 / mainly because of two objectives:

(1) to obtain the basis for a direct comparison of the
    mass and stiffness matrices of the related analytical
    model (order n)

(2) to obtain not only the complex modal data of the damped
    structure but also the real modal data of the related
    undamped structure. This is necessary because the real
    modal data represent the basis for the indirect verifi-
    cation of the analytical model which generally is un-
    damped due to the lacking possibily to establish ana-
    lytical damping matrices for complex structures.

At first sight it looks straightforward to identify the
system matrices from writing eq. (1) down for i = 1,2...m
frequency points (spectral lines) to arrive (after premultiplying by $\underline{M}^{-1}$ to account for single point excitation and
after splitting into real and imaginary parts) at the
following identification equation

$$\underline{A}\,\hat{\underline{K}}^{*T} - \underline{\Omega}\left[\underline{B} \quad \vdots \quad -\underline{\Omega}\,\hat{\underline{P}}_{re}\right]\begin{bmatrix}\hat{\underline{D}}^{*T} \\ \hat{\underline{M}}^{-1}\end{bmatrix} = \underline{\Omega}^2\,\underline{A} \qquad (2a)$$

$$\underline{B}\,\hat{\underline{K}}^{*T} + \underline{\Omega}\left[\underline{A} \quad \vdots \quad \underline{\Omega}\,\hat{\underline{P}}_{im}\right]\begin{bmatrix}\hat{\underline{D}}^{*T} \\ \hat{\underline{M}}^{-1}\end{bmatrix} = \underline{\Omega}^2\,\underline{B} \qquad (2b)$$

where

$$\underline{A},\ \underline{B} = \begin{bmatrix}\ddot{u}(\omega_1)_1 & \cdots & \ddot{u}(\omega_1)_p \\ \vdots & & \vdots \\ \ddot{u}(\omega_m) & \cdots & \ddot{u}(\omega_m)_p\end{bmatrix} = \text{measured}$$

$$(m,p) \qquad\qquad\qquad\qquad\qquad \text{frequency}$$
$$\text{acceleration}$$
$$\text{response}$$

$$\hat{\underline{P}}re,im \atop (m,\ p_f) = \begin{bmatrix} p(\omega_1)_1 & \cdots & p(\omega_1)_{p_f} \\ \cdot & & \\ \cdot & & \\ p(\omega_m)_1 & \cdots & p(\omega_m)_{p_f} \end{bmatrix} = \begin{array}{l} \text{measured} \\ \text{forcing} \\ \text{functions} \end{array}$$

$$\underline{\omega} = \mathrm{diag}\ (\ \omega_1 \ldots \omega_m\ ) = \text{excitation frequencies}$$

m = no. of frequency points (spectral lines)

p = no. of measurement degrees of freedom (MDOF)

p_f = no. of excited MDOF's

Eqs. (2) present a set of overdetermined ($m > p$) equations which could be solved by standard least square procedures with respect to the parameter matrices

$$\overset{*}{\underline{K}}(p,p) = \underline{M}^{-1}\underline{K}, \quad \overset{*}{\underline{D}}(p,p) = \underline{M}^{-1}\underline{D} \quad \text{and} \quad \overset{*}{\underline{M}}^{-1}\ (p_f,p)\ .$$

The eigenvalue solution of the matrix pair $\overset{*}{\underline{K}}$, $\underline{I}_p$ yields the undamped (real) eigenvalues and eigenvectors whereas the eigenvalue solution of the pair

$$\begin{bmatrix} \underline{0} & -\underline{I}_p \\ & \\ \overset{*}{\underline{K}} & \overset{*}{\underline{D}} \end{bmatrix} \quad \text{and} \quad \underline{I}_{2p} \quad \text{yields the}$$

damped (complex) eigenvalues and eigenvectors.

When this procedure (first published in /1/) was applied to real measurement data severe limitations have been experienced due to the incompleteness properties of measured data. We denote a vibration test incomplete when the number r of excited modes is smaller than the number of measurement DOF's p ($r < p$). In this case the measurement matrices $\underline{A}$ and $\underline{B}$ theoretically must have the rank $r < p$. However, calculating the rank of measured $\underline{A}$, $\underline{B}$ no rank defect will be detected due to

- measurement noise (random and systematic),
- the influence of weakly excited modes,
- the inadequacy of the linearity assumption used in eq.(1).

The consequence is that a complete set of p modes would be calculated from $\underset{*}{\underline{K}}$ and $\underset{*}{\underline{D}}$ including p-r pure computational modes. Practical experience revealed that computational modes and structural modes could not always be seperated and influenced each other. Any successful direct matrix identification procedure therefore has to account for the incompleteness condition, i.e. the effect of computational modes has to be extracted from the identification equation, otherwise the method will fail for real world applications.

3. Procedures for rank estimation

In the following section two procedures for rank estimation will be described. The method used in /2/ and /3/ is based on a singular value decomposition (SVD) of the measurement matrices and subsequent reduction of the identification equation to r principal coordinates, e.g. the SVD of the measurement matrix B yields

$$\underline{B} = \underline{U}_B \; \underline{S}_B \; \underline{V}_B^T \tag{3a}$$

The diagonal matrix $\underline{S}_B$ contains the singular values

$$\underline{S}_B = \text{diag} (s_1 \; \ldots \; s_r, \; s_{r+1} \ldots \; s_p) \tag{3b}$$

$\underline{U}_B$ (m,p) and $\underline{V}_B$ (p,p) represent the modal matrices of $\underline{BB}^T$ and $\underline{B}^T\underline{B}$. The singular values beyond r are equal to zero only theoretically but not under real measurement conditions. Depending on the amount of measurement errors the magnitude of the singular values will drop more or less significantly beyond r. A typical curve obtained from measurement data is shown in fig. 1 for the ratios s_j/s_{j+1} of consecutive singular values, where the rank is indicated by a significant increase of that ratio (which tends to infinity for an ideal (measurement)

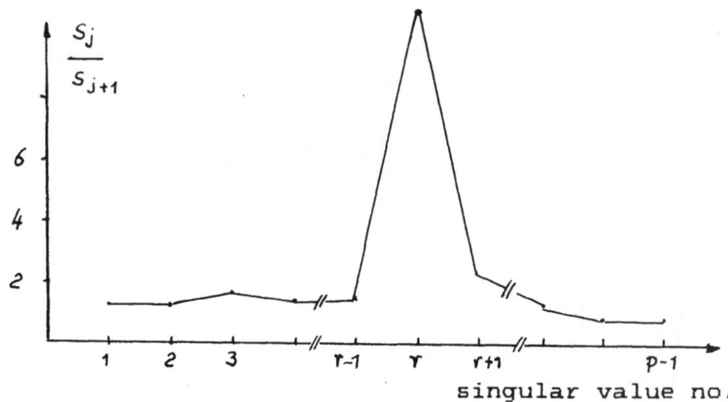

Fig. 1 : Ratios of consecutive singular values vs. singular value no.

Practical applications revealed that rank determination acc. to fig. 1 was not sufficient to obtain a rank leading to good identification results. Therefore a second procedure was introduced. With r estimated either from the first step or by the no. of frequency response peaks the singular value matrix is reduced to

$$\underline{\tilde{S}}_B = \text{diag} (s_1 \dots s_r) \tag{3c}$$

by artificially setting $s_i = 0$ for $i = r+1, \dots p$

Introducing $\underline{\tilde{S}}_B$ into eq. (3a) instead of $\underline{S}_B$ a modified "substitute" measuring matrix $\underline{\tilde{B}}$ ($\underline{\tilde{A}}$ accordingly) of rank $r < p$ is obtained:

$$\underline{\tilde{B}} = \underline{\tilde{U}}_B \ \underline{\tilde{S}}_B \ \underline{\tilde{V}}_B \tag{3d}$$

where $\underline{\tilde{U}}_B$ and $\underline{\tilde{V}}_B$ are submatrices of $\underline{U}_B$ and $\underline{V}_B$ belonging to the non-zero values of S. Now, the estimate of r is checked numerically by the rms-deviation of the substitute measurement matrices and the original matrices at each MDOF j:

$$\text{rms}_j = \sqrt{\frac{\sum\limits_i (\tilde{B}_{ij} - B_{ij})^2}{\sum\limits_i B_{ij}^2}} \tag{4}$$

$$(i = 1,2 \dots m , \quad j = 1,2 \dots p)$$

Practical experience showed that the identification was successful even when the rms_j-deviations were very large at some measurement DOF's depending on the magnitude of the signal. Therefore the rms-values of the complete matrix was also used

$$\text{rms} = \sqrt{\frac{\sum\limits_{ij} (\tilde{B}_{ij} - B_{ij})^2}{\sum\limits_{ij} B_{ij}^2}} \tag{5}$$

$$(i = 1,2 \dots m , \quad j = 1,2 \dots p)$$

where the influence of large errors on small signals is not so pronounced.
The rms-values are equal to zero in two limiting cases

(a)   in the case of ideal measurement data without noise and systematic errors when the singular values beyond r are equal to zero

$$S_i \ = \ 0 \qquad (\ i \ = \ r+1, \ \dots \ p \ )$$

(b)   in the case of non-ideal measurement data when in contrast to our intention no rank defect is introduced and all computational modes are contained in the system matrices.

For real measurement data the rms-values will be non-zero. If plotted versus differently estimated rank numbers r, fig. 2, the minimal number required to obtain a reasonable substitute measurement can be derived from the steepest change of slope of that curve acc. to fig. 2 .

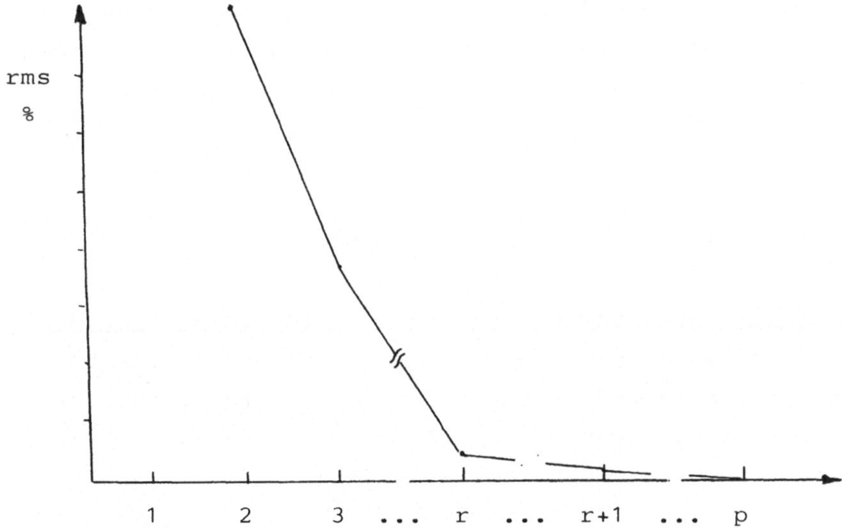

estimated rank of measurement matrix

Fig. 2 :   rms-deviation of substitute measurement vs.
estimated rank of measurement matrix

56

As shown in /3/ and /8/ the substitute measurement matrices can now be transformed to r principal coordinates by the transformation matrix $\tilde{\underline{V}}_B$ (p,r)

$$\hat{\underline{A}} = \tilde{\underline{A}}\ \tilde{\underline{V}}_B \qquad \text{and} \qquad \hat{\underline{B}} = \tilde{\underline{B}}\ \tilde{\underline{V}}_B \tag{6}$$

leading to a condensed identification equation (2) with respect to the unknown parameter matrices

$$\hat{\underline{K}}^* = \tilde{\underline{V}}_B^T\ \tilde{\underline{K}}^*\ \tilde{\underline{V}}_B \tag{7a}$$
$$(r,r)$$

$$\hat{\underline{D}}^* = \tilde{\underline{V}}_B^T\ \tilde{\underline{D}}^*\ \tilde{\underline{V}}_B \tag{7b}$$
$$(r,r)$$

$$\hat{\underline{M}}^{-1} = \hat{\underline{M}}^{-1}\ \tilde{\underline{V}}_B \tag{7b}$$
$$(p_f,r)$$

which allow the calculation of the damped and undamped eigenvalues, mode shapes, the modal damping parameters and also the modal masses. The mode shape matrix in the measured physical coordinates is calculated from that in principal coordinates by

$$\underline{Y} = \tilde{\underline{V}}_B\ \hat{\underline{Y}} \tag{8}$$
$$(p,r) \qquad (p,r) \qquad (r,r)$$

The final check whether the rank estimate was reasonable or not is obtained by using the identified modal data to recalculate the measured frequency response curves. The agreement beetween these curves indicates, at least globally, the accuracy of the identified modal data.

4. Applications

In this section two applications are presented:

*   a small scale 5-DOF test structure (fig. 3)

*   the solar array of a satellite (fig. 7) which was tested
    on a shaker table at ESA's test facilities during the
    qualification test phase of the project.

4.1 5-DOF test structure

The 5-DOF test structure shown in fig. 3 was designed and
fabricated within /9/ in order to study the ability of the
identification procedure to extract weakly excited
modes caused by increased damping, non-appropriated exciter
configuration and by non-linearity effects of viscous
damping layers. The test structure consists of 5 cantilever
steel beams with damping layers and aluminum face sheets
bonded on both sides. The beams are coupled by steel arch
springs. Five discrete masses have been concentrated at the
tips of the beams in order to have a discrete 5-DOF model
with the higher eigenfrequencies significantly seperated
from the first 5 eigenfrequencies. A single point harmonic
excitation was applied at DOF 5. The digitized measured
frequency response curves (imaginary parts) are shown in
fig. 4a. From these curves 4 peaks are detected indicating
4 strongly excited modes. Obviously the very small response
peak in the vicinity of the third peak is due to the 4th
mode which is excited only very weakly. To render the identification even more difficult the measured frequency band
was truncated at 61.42 Hz to establish the incompleteness
condition that the MDOF no. p=5 is larger than the effective
mode no. r. The ratios of consecutive singular values of the
im-part measurement matrix $\underline{B}$ (m=118 frequency points, p=5
MDOF's) plotted in fig. 5 clearly indicate that at least r=3

effective modes are active in that frequency range. To prove this the rms-deviations at the substitute measurement matrix $\tilde{\underset{.}{B}}$ acc. to eq. (5) have been calculated for different estimates of r. The plot in fig. 6 also indicates r=3 effective modes. Comparing the substitute response curves for r=2 (rms=40%) in fig. 4b with the measured curves in fig. 4a also reveals large deviations whereas the deviations of the curves for r=3 (rms=3.9%) are small and not visible from the plot.

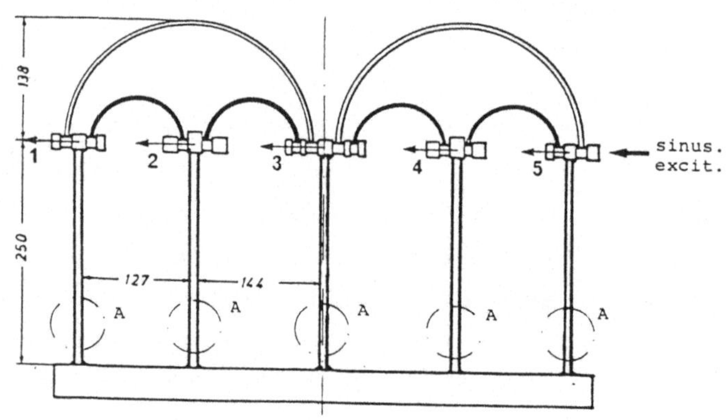

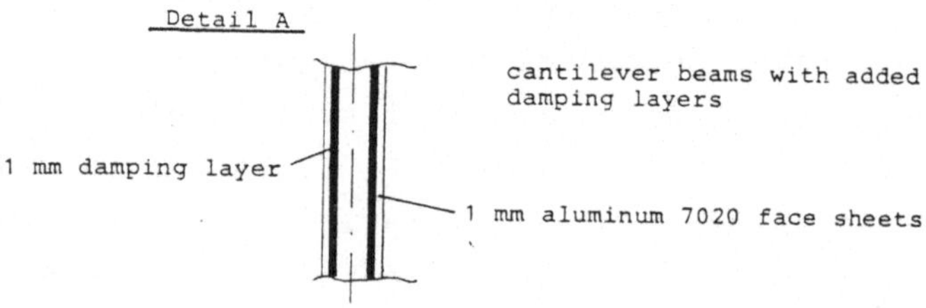

Fig. 3   5-DOF test structure with damping layers

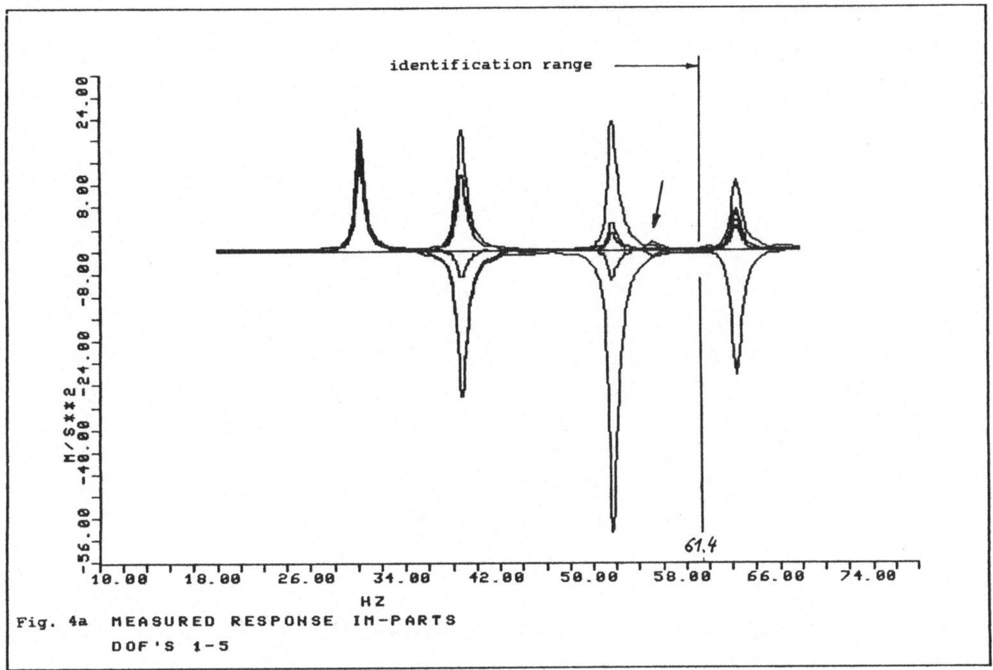

Fig. 4a MEASURED RESPONSE IM-PARTS
DOF'S 1-5

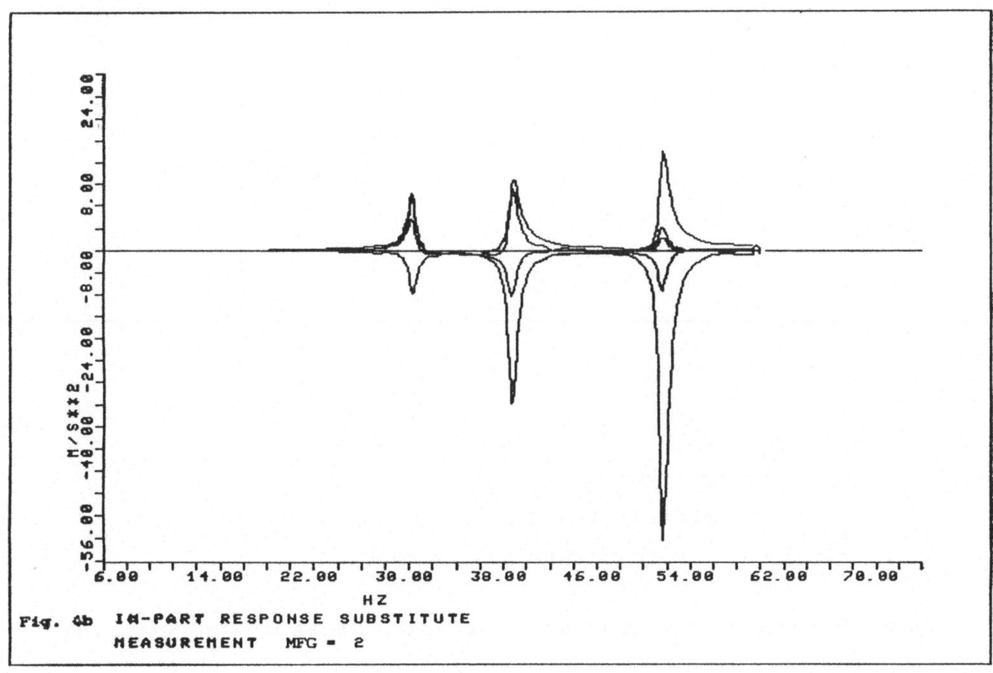

Fig. 4b IM-PART RESPONSE SUBSTITUTE
MEASUREMENT MFG = 2

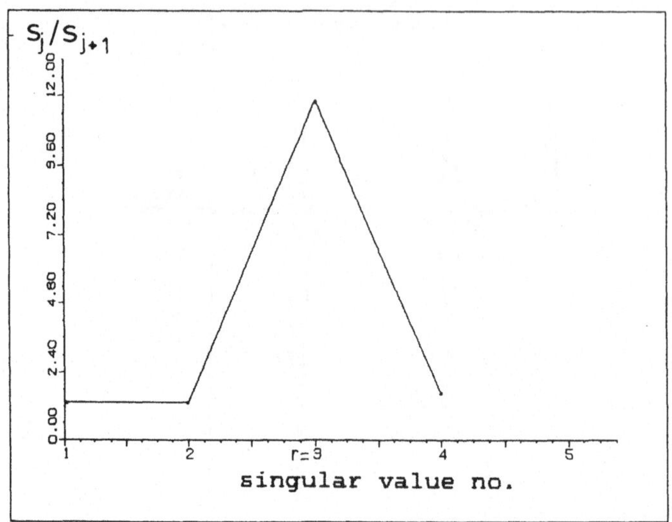

Fig. 5 : Ratios of consecutive singular values

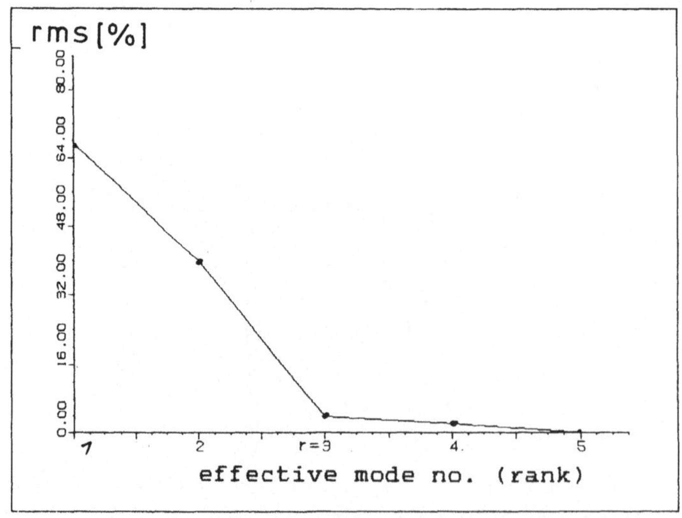

Fig. 6 : rms-deviation of substitute measurement matrix

In a next step the question was investigated whether increasing the matrix rank beyond r=3 would either allow the identification of the hidden mode or would only produce additional modes. The results presented in table 1 for the eigenfrequencies and modal masses indicate that

*   the hidden mode could not be identified,

*   increasing the rank estimate beyond r=3 yields pure computational modes,

*   the eigenfrequencies of the 3 effective modes remain stable with respect to r whereas changes of up tp 10 % were identified for the modal masses. The stability of the mode shape amplitudes not presented here was high for the large amplitudes and lower for the near zero amplitudes.

rank r	eigenfrequencies [Hz]			modal masses [kg]			no. of comp. modes
	f1	f2	f3	m1	m2	m3	
3	32.27	40.74	53.71	2.76	1.80	0.70	–
4	32.26	40.75	53.74	2.62	1.84	0.78	1
5	32.26	40.79	53.76	2.53	1.98	0.81	2

Tab. 1 : Identified eigenfrequencies and modal masses with respect to different rank estimates r

4.2   Satellite solar array

To extract the modal characteristics of a typical spacecraft
subsystem the identification procedure was applied using the
test data from a slow sinusoidal sweep test. This test was
part of a qualification test program during the development. Fig. 7a shows the 7-panel array in unfolded position.
The configuration tested was a folded 3-panel version fixed
together at 6 hold-down points. Fig. 7b shows the test set-
up at ESA/ESTEC's vibration laboratory. The excitation level
was 0.3 g in the range 5-150 Hz in direction perpendicular
to the panel plane. The acceleration response measured at 32
MDOF's is presented in fig. 8a for the range 50.0 - 84.6 Hz
(only im-parts). Due to unknown measurement error sources
and non-linearity effects the ratios of consecutive singular
values of the measurement matrices plotted in fig. 9 do not
indicate a unique effective mode number. Therefore the
identification procedure had to be run with different rank
estimates. From comparing the recalculated and the measured
response (figs. 8a and 8b) best agreement was obtained using
the rank estimate r=5. 4 eigenfrequencies and modes and one
computational mode have been extracted. Mode no. 4 is
presented in fig.10. Although several response peaks are
observed in the vicinity of the 3rd eigenfrequency in
fig.8a only one mode could be extracted in that range which
limits its accuracy. Identification runs with other rank
estimates led to a 3% scatter of the identified eigenfrequencies indicating the amount of measurement errors.
Assessing the results it may be concluded that it was
possible to extract the global modes responding to axial
excitation and to establish a best fit linear model for the
structure.

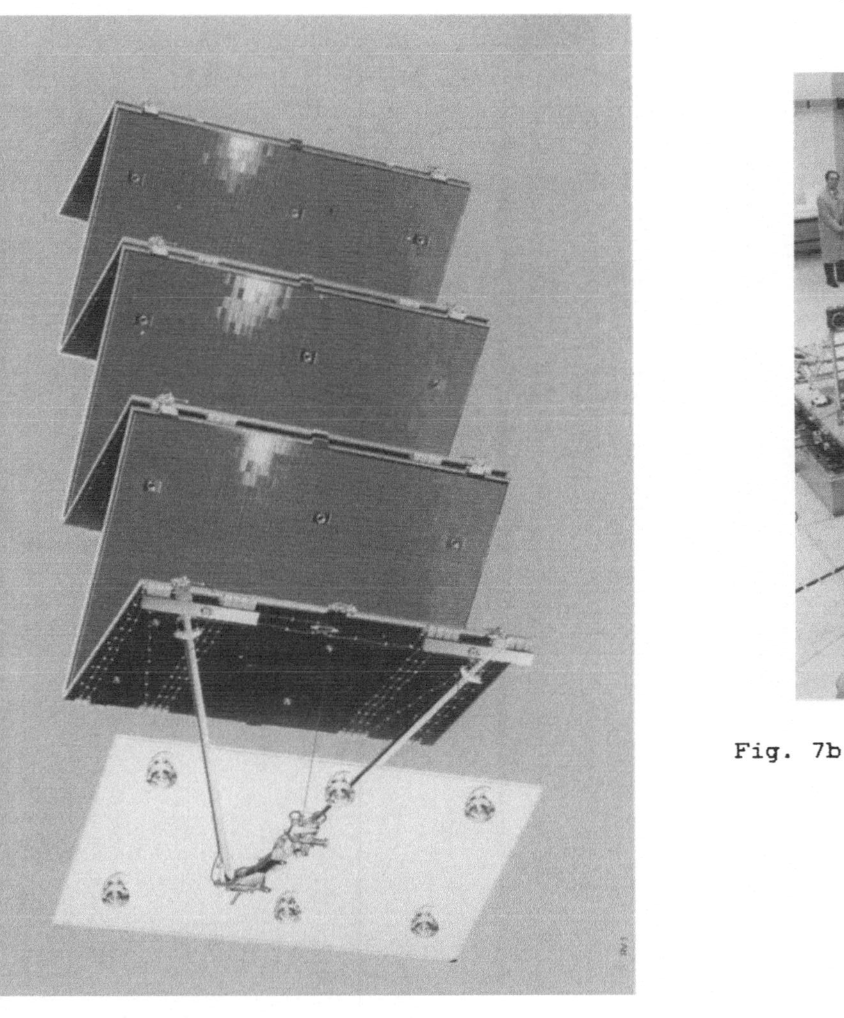

Fig. 7a : 7-Panel Solar Array (unfolded)

Fig. 7b : Test set-up at ESA/ESTEC's test laboratory
Fotographs: FOKKER B.V., Schiphol, Holland

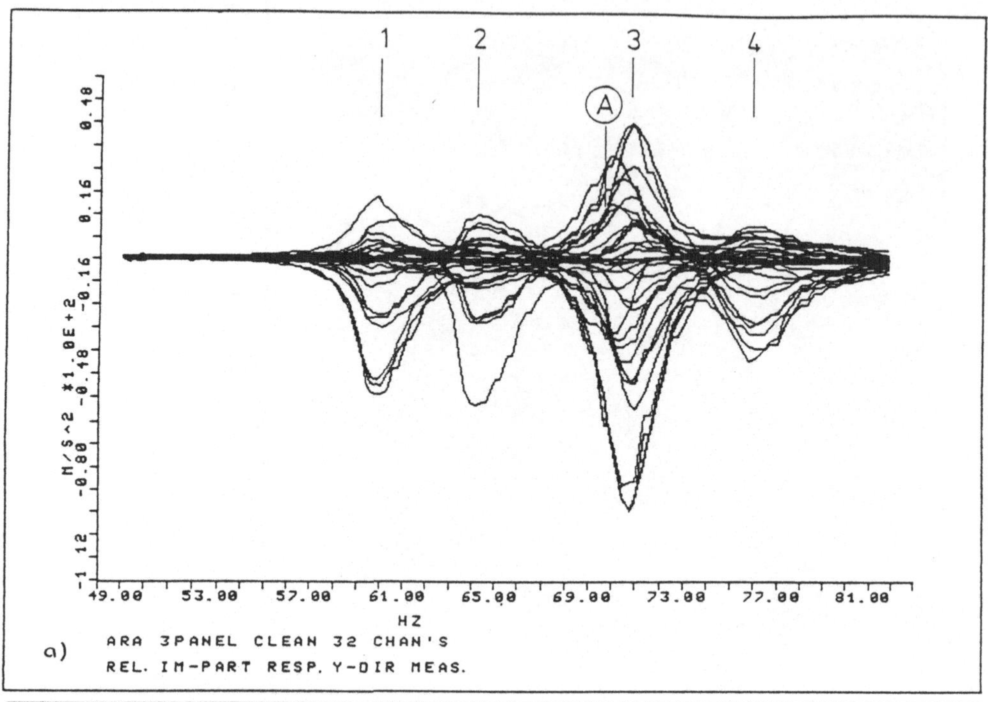

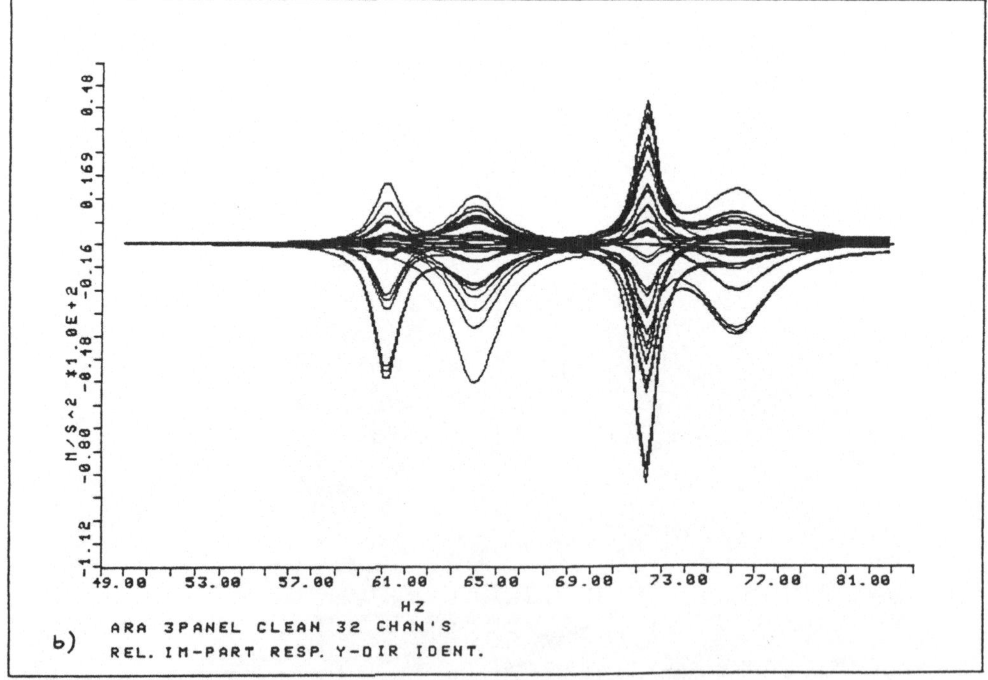

Fig. 8a,b Comparison of measured and recalculated response data
Im. part response   range 50...82 Hz

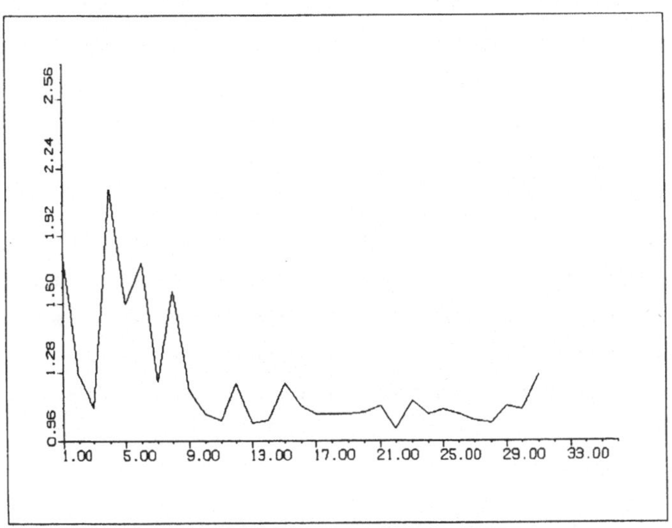

Fig. 9 : Ratios of consecutive singular values

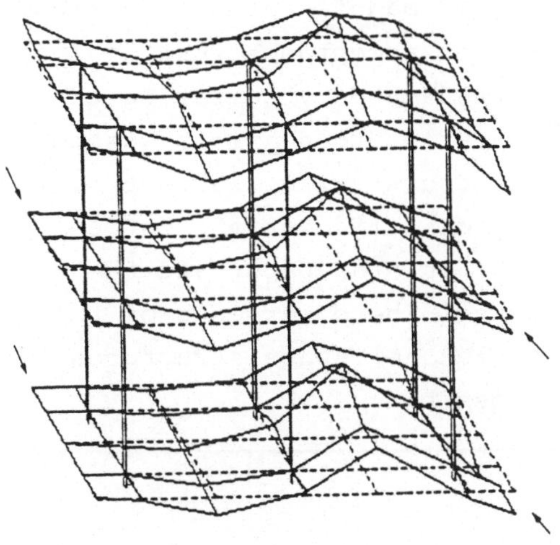

Fig. 10 : mode shape

5. Conclusions

The results, especially those of the solar array structure which was tested in real project environment, confirmed that the identified modal data depend on the different control parameters of the identification procedure.

The amount of scatter of the modal data depends on the type of excitation and of the amount of measurement noise and systematic deviations of the structure's behavior from the linearity assumption. The question discussed during the workshop whether system identification could be used for damage detection as a first step of safety evaluation cannot generally be answered.

One approach looks potentially applicable in industries where vibration tests are used as non-destructive acceptance tests for structures manufactured in series. System identifications as part of acceptance test evaluation could yield a means to detect damage by comparing the identified modal parameters. The question then arises whether the sensitivity of the modal parameters with respect to structural damage was high enough to differentiate between the inaccuracy effects discussed in the paper and the effects introduced by structural damage. The experience of the author up to now is that for complex aerospace structures this differentiation is not yet possible.

References

/1/   M. Link, A. Vollan : Identification of Structural
      System Parameters from Dynamic Response Data.
      Z. Flugwiss. Weltraumforsch. 2 (1978), pp. 165-174

/2/   M. Link : Theory of a Method for Identifying
      Incomplete System Matrices from Vibration Test Data
      Z. Flugwiss. Weltraumforsch. 9 (1985), H2

/3/   M. Link, M. Weiland, J. Moreno B. : Direct Physical
      Matrix Identification as Compared to Phase Resonance
      Testing. An Assessment Based on Practical Application.
      5th International Modal Analysis Conference (IMAC),
      London, England, 1987, pp. 804-811

/4/   J. Leuridan : Direct System Parameters Identification
      of Mechanical Structures with Application to Modal
      Analysis. Masters Thesis, Univ. of Cincinnati,
      Ohio, 1981

/5/   R. R. Craig, A. J. Kurdila, W. M. Kim : State-Space
      Formulation of Multi-Shaker Modal Analysis.
      5th International Modal Analysis Conference (IMAC),
      London, England, 1987, pp. 1069-1077

/6/   K. Sheye, M. Richardson : Mass, Stiffness, and
      Damping Matrix Estimates from Structural Measurement.
      5th International Modal Analysis Conference (IMAC),
      London, England, 1987, pp. 756-761

/7/   F. Lembreghts, R. Snoeys, J. Leuridan : Application
      and Evaluation of Multiple Input Modal Parameter
      Estimation. 5th International Modal Analysis Conference
      (IMAC), London, England, 1987, pp. 966-978

/8/ K. Badenhausen : Identifikation der Modellparameter
elastomechanischer Systeme aus Schwingungsversuchen.
Dissertation, Univ. Gesamthochschule Kassel, 1986

/9/ M. Link et al. : Updating of Dynamic Mathematical
Models, Test Phase. ESA/ESTEC Contr. 6262/85/nl/pb(sc),
Final Report Univ.GH Kassel,FGL 7/87, May 1987

Structural System Identification from Modal Information

Jean-Guy Beliveau*
Civil Engineering and Mechanical Engineering Department
University of Vermont
Burlington, Vermont, 05405

Abstract

Advances in modal testing and in partial eigenvalue and eigenvector
calculation of large matrix systems and in eigensensitivity make it
feasible to estimate actual physical parameters of finite element or
frame analysis models by comparing calculated modal information with
measured modal characteristics. The relevent mathematics required for
structural system identification is presented in this paper with regards
to linear structural systems.

* Adjunct Professor, Universite de Sherbrooke, Sherbrooke, Quebec

1. Introduction

The structural evaluation of existing structures is indeed a difficult yet necessary task for modern day structural engineers. The initial static testing of bridges before traffic was allowed to proceed has
traditionally been required right after construction. After many useful
years of service, the bridge and its major structural components have
surely suffered large loads and loads of a repeated nature causing
fatigue.

The economic question regarding the bridge's replacement or repair
is an important one. All too often very little quantitative information
is available to the strucutral engineer. A load test similar to the one
done right after construction would of course be feasible in evaluating
the strength of the structure but would necessitate some costs and some
structural risk.

Similarly, a building may have been affected by high winds or
large earthquakes. Is it structurally sound or does it meet current
building code requirements? If not, can it be retrofitted to do so? If
it is repaired, what will this do to the dynamic characteristics? Sim-

ilar questions arise for tunnels, dams, reservoirs, and all structures
that have been built and which now fall in the category of "our decaying
infrastructure".

Although there are qualitative inspection procedures and some quantitative techniques to evaluate local material properties, very little
has been developed to address the overall structural integrity. The
vibrations one observes while crossing a bridge, or in the upper stories
of a building on windy days suggest that dynamic response is measureable
and could be used in evaluating the structural stiffness, and ultimately
the stability and strength.

In this paper a methodology is discussed for incorporating these
types of measurements, which are relatively easy to acquire, into a
structural system identification and parameter estimation of physical
parameters. The actual time measurements must first be converted to
modal information such as resonant frequencies, damping ratios and mode
shapes of vibration. Current modal testing procedures are useful in
this context.

Once the modal information is available, system identification
procedures may then be used to determine physical parameters, such as
the equivalent moment of inertia of a cracked concrete girder, the
partial fixity of a connection, the soil-structure interaction of a
foundation etc. These are possible provided a good mathematical model
of the structure and/or structural component is available. Efficient
numerical techniques are needed to calculate the corresponding dynamic
characteristics of the structure and their sensitivity with respect to
the parameters being determined from the mathematical model.

This paper first gives a summary review of system identification as
it relates to modal information for damped linear structural systems.
Next, procedures are discussed for obtaining modal information from time
histories or frequency response functions. The topics of eigenvalue
analysis of the structural model and the sensitivity of these and the
eigenvectors with respect to physical parameters is then presented.

2.   Structural System Identification

For purposes of discussion, it can safely be stated that the modal
measurements will generally not be linear functions of the physical
parameters. Structural system identification in the broadest sense

implies that the model must be determined. It is assumed here that such
a model does exist, and that the assumptions made are reasonable. Once
parameter estimation is carried out, the structural engineer must decide
whether or not the model is adequate, or, if perhaps another model
should be considered, in cases where measurements are still very different from those calculated, for instance.

There are a number of ways to arrive at the normal equations, one
based on statistical considerations, another on optimization or minimization of squared errors, and one based on purely linear algebraic
reasoning. Only the latter is given here for nonlinear least squares
with a priori estimates of the pararmeters.

2.1 Sensitivity Matrix

Since the measured modal quantitites are nonlinear functions of the
parameters, we shall assume a first order Taylor series expansion for n
measurements which depend on m parameters.

$$d_i = d_i^k + \sum_{j=1}^{m} \frac{\partial di}{\partial pj} \, \delta j \tag{1}$$

where k represents the k-th iteration and δ is a correction. Similarly,
a first order Taylor series expansion of the parameters is exact if the
parameters are not functions of each other

$$p_j = p_j^k + \delta j \tag{2}$$

Rewriting these as matrix equations in δ and equating the data for d_i and
a priori estimates of the parameters for p_j these become

$$A\delta = d - d^k \tag{3}$$

$$\delta = p - p^k \tag{4}$$

in which the sensitivity matrix [A] has elements given by

$$A_{ij} = \frac{\partial d_i}{\partial p_j} \tag{5}$$

2.2 Normal Equations

There are more equations than parameters. Also, the data consists of quantitites of different nature such as frequencies and mode shapes some on the order of powers of ten and others on the order of one. In order to fit all the data it is necessary to appropriately scale or weigh each data. This is often done by diagonal matrices with the diagonal elements corresponding to the reciprocal of variance of the measurements and data. In order to yield a square symmetric matrix systems, the equations are then premultipled appropriately to yield the normal equations of nonlinear least squares or Bayesian estimation which are symmetric and square matrices of order equal to the number of unknown parameters.

$$[A'W_d A + W_p] \, \delta = A'W_d(d-d^k) + W_p \, (p - p^k) \tag{6}$$

in which ' represents the transpose and $[W_d]$ and $[W_p]$ are the weighing matrices for the data and the parameters respectively. If the calculated modal quantities were exactly equal to the measured values, the right hand side would be zero as would the correction to the parameters. At the first iteration the second term on the right is zero, but it is not for subsequent iterations. In order for the data to affect the a priori estimates of the parameters it must be appropriately weighed and it must be sensitive to the parameters, otherwise both first terms on each side of the equation are discounted and these result in the a priori estimates to the parameters.

3. Modal Information

The linear ordinary equations of structural dynamics are assumed to have constant coefficients and represent small motions about an equilibrium position. For a viscously damped structure, they may be expressed as:

$$[M]\{\ddot{x}\} + [C]\{\dot{x}\} + [K]\{x\} = \{f\} \tag{7}$$

in which x are degrees of freedom for both lateral and rotational motion, $\dot{x}$ and $\ddot{x}$ are the corresponding velocities and accelerations. M, C, and K are the mass, damping, and stiffness matrices respectively and f is a vector of non-conservative mechanical loads acting at the corresponding degrees of freedom.

3.1 First Order System

This second order system of equations may be represented as a first order system by a coordinate transformation involving both displacements and velocities.

$$\{\dot{y}\} = [Q]\{y\} + \{g\} \tag{8}$$

in which there are twice as many degrees of freedom, displacements and the corresponding velocities.

$$\{y\} = \left\{ \begin{array}{c} x \\ \hline \dot{x} \end{array} \right\} \qquad \{g\} = \left\{ \begin{array}{c} \{0\} \\ \hline [M]^{-1}\{f\} \end{array} \right\} \qquad [Q] = \left[\begin{array}{c|c} [0] & [1] \\ \hline -[M]^{-1}[K] & -[M]^{-1}[C] \end{array} \right] \tag{9-11}$$

The solution to this first order system with constant coefficients may be expressed as a function of the eigenvalues and eigenvectors of Q, and V respectively.

$$y = Ve^{\Lambda t} V^{-1} y_o + \int_o^t Ve^{\Lambda(t-\tau)} V^{-1} g d\tau \tag{12}$$

in which y_o is the vector of initial displacements and velocities and and are diagonal matrices.

$$V^{-1} QV = \Lambda \qquad y_o = \left\{ \begin{array}{c} x_o \\ v_o \end{array} \right\} \tag{13,14}$$

3.2 Step Relaxation

For initial displacements the response is given by the homogeneous portion

$$Y_o = \left\{ \begin{array}{c} x_o \\ o \end{array} \right\} \tag{15}$$

in which x_o are the initial displacements. Depending on this initial vector, {y} can be made to be any linear combination of the mode shapes.

3.3 Impact Testing

For impulsive type of loads at time t = 0 the particular portion of
the solution may be represented as follows

$$y = Ve^{\Lambda t} \int_{0}^{t} e^{-\Lambda \tau} \delta(\tau) d\tau \ V^{-1} \ g \tag{16}$$

which because of the nature of the Dirac delta function $\delta(\tau)$ and since
impact is equal to the change of momentum, gives

$$\int_{0}^{t} f(\tau) \ \delta(\tau) d\tau = M \ v_{o} \tag{17}$$

yielding the homogeneous solution with initial velocities, v_{o}

$$y_{o} = \left\{ \begin{matrix} o \\ v_{o} \end{matrix} \right\} \tag{18}$$

Again the response is a sum of complex exponentials as before, proportional to the modes. Combining initial displacements and velocities
there results

$$y = \sum_{i=1}^{2n} a_{i} e^{\lambda_{i} t} \ v_{i} \tag{19}$$

in which eigenvalues and eigenvectors generally occur in complex conju-
gate pairs and the constants for the linear combination a are given by

$$a = V^{-1} \ y^{o} \tag{20}$$

3.4 Frequency Response

The impulse response function [p] of the structural system is zero
for time less than zero and is given by the relationship (for time
greater or equal to zero)

$$p = Ve^{\Lambda t} \ V^{-1} \tag{21}$$

The frequency response function [P] is the Fourier transform of the
impulse response function {p} and for stable system may be represented

by the equation

$$P = V \ (iw - \Lambda)^{-1} \ V^{-1} \qquad\qquad (22)$$

Because of Fourier relationships in linear systems the Fourier transform
of the response Y is related to the Fourier transform of the loads, G,
by the frequency response function.

$$Y = PG \qquad\qquad (23)$$

Y is a linear combination of the eigenvectors, the combination being
different at each frequency. The complex division of the Fourier transform of one response and the load at another point gives the frequency
response function components.

3.5 <u>Second</u> <u>Order</u> <u>Frequency</u> <u>Response</u> <u>Function</u>

The second-order frequency response function H is often expressed
as an nxn matrix equal to the inverse of the impedance matrix $[H]=[Z]^{-1}$

$$Z = K -w^2 M + iwC \qquad\qquad (24)$$

This would correspond to the top right hand corner of the P matrix post
multiplied by the inverse of the mass matrix.   This is expressed as a
function of the eigenvalues and eigenvectors of the system.   The second
order system may be represented as a function of the original matrices.

$$[\lambda^2 [M] + \lambda[C] + [K]]\{u\} = \{0\} \qquad\qquad (25)$$

It may also be expressed as a function of the eigenvalues and eigenvectors {u} of the quadratic matrix system (1).   For damped systems the
eigenvalues Λ and eigenvectors V may be expressed as functions of λ
and u (2).

3.6 <u>Complex</u> <u>Exponential</u>

Modern modal testing techniques yield the frequency response function from impact tests, sweep tests, and random shaker loadings provided
both the acceleration and force in the exciter are measured. Measurement of acceleration and base acceleration can also yield this for base
excitation (3). Because of the Fourier relationship between the frequency response function and the impulse response function, the fre-

quency response function of a particular point loaded at another location can be converted to a sum of complex exponentials. This is also
true for step-relaxation tests having initial displacements or impulse
tests with one point impacted at a time. The accelerometers may be
located at different points to obtain a number of frequency response
functions as can the hammer or shaker.

In all these cases modal information in the form of damped natural
frequency, modal damping ratios, and complex mode shapes can be determined by fitting a sum of complex exponentials to the data.

4.   Eigenvalue

The mass and stiffness matrices obtained by the finite element
method or by frame analysis procedures are linear combinations of the
local element matrices, appropriately rotated and placed in the global
large degree of freedom matrices. The partial derivative of these
matrices with respect to specific physical parameters, such as the
moment of inertia, the area, Young's modulus, etc. is straight forward
since local elements matrices are added to form the global matrices. (4)

Thus, in order to estimate these parameters, the sensitivity of the
measured modal quantities in the mathematical model is based on the
sensitivity of the mass stiffness and damping matrices. The same parameters may be associated with one or many members, and as many parameters as required may be considered in the parameters estimation.

The eigenvectors of the elements corresponding to velocity of the
first order system are not independent and their correct sensitivity
would require extensive formulation. The quadratic eigenvalue problem
with real matrices has elements of the eigenvectors which are independent, however the left eigenvector, w, satisfies the transpose relation,

$$\left[\lambda^2 [M]' + \lambda[C]' + [K]'\right] \{w\} = \{0\} \tag{26}$$

4.1 Eigenvalue Sensitivity

Premultiplying the next to the last equation by $[w]^1$ and using the chain
rule results is a simple formula for the eigenvalue sensitivity which

depends only on the sensitivity of the mass, damping, and stiffness matrices and the corresponding eigenvalue and eigenvector.

$$\frac{\partial \lambda}{\partial p} = - \frac{\{w\}' \left[\lambda^2 \left[\frac{\partial M}{\partial p} \right] + \lambda \left[\frac{\partial C}{\partial p} \right] + \left[\frac{\partial K}{\partial p} \right] \right] \{u\}}{\{w\}' \left[2\lambda [M] + [C] \right] \{u\}} \tag{27}$$

For symmetric matrices $v = u = \phi$ the normal modes. If in addition there is no damping the eigenvalue is purely imaginary $\lambda = i\,\Omega$ and the sensitivity of the radial frequency Ω is given by

$$\frac{\partial \Omega}{\partial p} = \frac{1}{2\Omega} \phi' \left[\frac{\partial K}{\partial p} - \Omega^2 \frac{\partial M}{\partial p} \right] \phi / \phi' m \phi \tag{28}$$

ϕ are the normal modes and Ω is a resonant undamped natural frequency of vibration of classical vibration analysis.

$$\left[[K] - \Omega^2 [M] \right] \{\phi,\} = \{0\} \tag{29}$$

As can be seen, the eigenvalue sensitivity does not require knowledge of all the eigenvalues and eigenvectors of the matrix system. This is important for large structures, since partial eigenvalue routines may be used to determine only the frequencies in the frequency range of interest (5).

Thus, a parameter estimation strategy could conceivably be implemented to consider only resonant frequency information. It is, however, difficult to decide with which measured frequency an analytical one should be compared to in a best fit without knowledge of the mode shapes, of vibration, i.e. the eigenvectors.

4.2 Eigenvector Sensitivity

The matrix eigenvalue problem can be differentiated by the chain rule directly yielding the result

$$[\lambda^2 M + \lambda C + K] \frac{\partial u}{\partial p} = b \tag{30}$$

in which {b} includes terms involving the sensitivity of the mass, damping, stiffness matrices, and of the eigenvalue, determined in the previous section.

$$b = -\left[\lambda^2 \frac{\partial M}{\partial p} + \lambda \frac{\partial c}{\partial p} + \frac{\partial K}{\partial p} + (2\lambda M+C) \frac{\partial \lambda}{\partial p}\right] u \qquad (31)$$

The system is of course singular, and an additional constraint on the eigenvector must be imposed. The simplest normalization is to set the largest element equal to a constant [4]. The corresponding line and column of the matrix are then set equal to zero as well as the corresponding element on the right hand side, except for the diagonal element of the matrix. It should be of the same order as the other diagonal elements of the stiffness matrix in order to have a good condition number of the matrix.

This procedure maintains the properties of the matrices such as band structure, sparsity, and symmetry. The eigenvector sensitivity is then solved for each parameter, with only one inverse of the modified matrix at each eigenvalue. It is assumed that eigenvalues are not repeated here or that the rank of the subspace is equal to the multiplicity of the eigenvalue. A very important characteristic is that only the corresponding eigenvalue and eigenvector are required, and not all the eigenvalues and eigenvectors. For undamped structural systems with symmetric matrices, the sensitivity of the real modes appropriately normalized such that one element is a constant is given by the modified matrix equation

$$(\bar{K} - \Omega^2 M) \frac{\partial \phi}{\partial p} = \bar{b} \qquad (32)$$

in which K and M have been modified as before and

$$\bar{b} = -\left[\frac{\partial K}{\partial p} - \Omega^2 \frac{\partial M}{\partial p} - \frac{\partial \Omega^2}{\partial p} M\right] \phi \qquad (33)$$

except for the component corresponding to the element used for normalization, which is zero.

5. Discussion

5.1 Modal Testing

There are certain advantages to reducing experimental data to modal information. The data is countable, i.e. a number of mode shapes, resonant frequencies, and modal damping ratios. The actual number of data points in arriving at the information could have been enormous, but data acquisition and modal testing have reduced the time histories and/or frequency response functions to just a few modal characteristics. These are representative of the stiffness of the structure and of its dynamic response. They form an ideal basis for which to compare a model with measurements.

Secondly, it is not necessary to obtain modal vector information at all degrees of freedom of the finite element or frame analysis model. The location of the accelerometers and exciter is left to the engineer. Depending on what information is sought, whether for the whole structure or one major structural component, different strategies and even equipment may be required.

Recent advances in modal testing have generally been made for mechanical components and machines. Little has been dedicated to full scale civil engineering structures. Yet, ambient vibrations are easily measurable and from that frequencies, damping and mode shapes can be obtained (6). The equipment needed on the structure is easily transportable and can run on batteries. Even some spectrum analyzers are now battery operated. Thus, these developments simplify greatly the procedure of acquiring modal information, compared to sinusoidal testing with shakers, for instance (7).

Various levels of sophistication are easy to envision. For instance in an inspection context with the aim of determining which structures require further investigation, monitoring of the motion under ambient conditions may be adequate. For structures specifically targetted for structural evaluation impact testing or shaker testing may be required to ascertain its structural integrity. Information on damping as well as natural frequencies can be used as well as measured mode shapes.

5.2 Structural Model

With regard to the structural model, advances in eigenvalue analysis and in sensitivity calculations make it feasible to estimate physical parameters of the structure. It is true, however, that good parameter estimates will be obtained only if the measured data is sensitive to the parameters and if it is appropriately weighed. Good initial estimates and a good model of the structural behavior is of course required as well.

5.3 Damping

Another basic problem is damping. There are no generally accepted models of incorporating damping into finite element models. Thus, the sensitivity of modal characteristics with respect to damping parameters is not possible. One possibility is to convert complex modes to real modes and to treat modal damping ratios as "physical parameters" to be estimated within the structural system identification context (7). The use of damped complex modes has bewildered structural engineers for many years now. Perhaps, with the experimentally derived complex mode shapes, further development in damping and its modeling will follow, as well as the corresponding parameter estimation procedures.

5.4 Strength

Nevertheless, quantitative structural information is needed. The techniques should be nondestructive, if they are to be used on structures which are candidates for repair. The technique proposed herein essentially determines the stiffness and not the strength of the structure which is the ultimate goal. With proper modelling, however, a fairly good estimate of strength can be made based on the stiffness model obtained.

In spite of these remarkable achievements, much has to be done. The frequency response functions obtained by impact tests are fairly reliable provided sufficient energy can be transmitted to the structure without causing local damage. It could conceivably be used to evaluate the integrity of main members, however, even if not of the whole structure. Because of the flexibility of many civil engineering structures, ambient vibration may drown out any induced motion such as by a shaker or hammer and signal to noise ratio may therefore be unacceptable.

6. <u>Conclusion</u>

In this paper, mathematical and experimental considerations for the structural system identification of linear structures from modal information were summarized. Parallel developments in finite element calculations of eigenvalues and eigenvectors and their sensitivity and in vibration testing yielding modal information in the form of resonant frequencies, damping ratios, and mode shapes of vibration now make it feasible and practical to arrive at quantitative information regarding the structural integrity of existing structures. There is a need for evaluation of these techniques for civil engineering structures before they become commonplace such as is required in an inspection context.

<u>Acknowledgements</u>

This work was supported by the National Sciences and Engineering Council of Canada.

<u>References</u>

1. Beliveau, J-G, "First Order Formulation of Resonance Testing", Journal of Sound and Vibration, Vol. 65, 1979, pp. 319-327.

2. Beliveau, J-G, "Eigenrelations in Structural Dynamics", Journal of the American Institute of Aeronautics and Astronautics, Vol. 15, 1977, pp. 1039-1041.

3. Beliveau, J-G, Vigneron, F.R., Soucy, Y., and Draisey, G., "Modal Parameter Estimation from Base Excitation", Journal of Sound and Vibration, Vol. 107, 1986, pp. 435-449.

4. Dolhon, A.M., "Modal Sensitivity for Structural Parameter Estimation", Masters Thesis, University of Vermont, 1987.

5. Beliveau, J-G, Lemieux, P. and Soucy, Y. "Partial Solution of Large Symmetric Generalized Eigenvalue Problems by Nonlinear Optimization of a Modified Rayleigh Quotient", Computer & Structures, Vol. 21, No. 4, 1985, pp. 807-813.

6. Beliveau, J-G, "Identification of Viscous Damping in Structures from Modal Information", Journal of Applied Mechanics, Vol. 43, No. 2, 1976, pp 335-339.

7. Beliveau, J-G, "System Identification of Civil Engineering Structures", Canadian Journal of Civil Engineering, Vol. 14, No. 1, 1987, pp 7-18.

System Identification Using Nonlinear Structural Models

P. Jayakumar* and J. L. Beck**

Earthquake Engineering Research Laboratory
California Institute of Technology
Pasadena, California, U.S.A.

1.   Introduction

Analytical modeling of structures subjected to ground motions is an important aspect of fully dynamic earthquake-resistant design. In general, linear models are only sufficient to represent structural responses resulting from earthquake motions of small amplitudes. However, the response of structures during strong ground motions is highly nonlinear and hysteretic, and the development of appropriate models is an important and challenging problem.

System identification is an effective tool for developing analytical models from experimental data. Since testing of full-scale prototype structures remains the most realistic and reliable source of inelastic seismic response data, a quasi-static procedure, called pseudo-dynamic testing, has recently been developed to subject full-scale structures to simulated earthquake response. The present study deals with structural modeling and the determination of optimal nonlinear models by applying a system identification technique to inelastic pseudo-dynamic data from a full-scale, six-story steel structure.

The nonlinear hysteretic behavior of the structure during strong ground motions is represented by a general class of Masing models. A simple model belonging to this class is chosen with parameters which can be estimated theoretically, thereby making this type of model potentially useful for response predictions during design. The above model is identified from the experimental data and then its prediction capability and application in seismic design and analysis are examined.

---

* Graduate Student in Civil Engineering
** Associate Professor of Civil Engineering

2. A General Class of Masing Models

2.1 Masing's Hypothesis

Masing, in a paper titled "Self Stretching and Hardening for Brass" in 1926 [1], assumed that a metal body consists of a system of elasto-plastic elements each with the same elastic stiffness but different yield limits. Based on this model and on axial loading tests he performed, Masing asserted that if the load-deflection curve for the entire system at the virgin loading is symmetric about the origin and is given by

$$f(x, r) = 0 \tag{1}$$

where x is the displacement and r is the restoring force, then the unloading and reloading branches of the hysteresis loops for steady-state response are geometrically similar to the virgin loading curve and are described by the same basic equation except for a two-fold magnification. Applying this to a hysteretic loop describing cyclic loading between (x_0, r_0) and $(-x_0, -r_0)$ as shown in Fig. 1, each branch of the hysteresis loop is given by:

$$f\left(\frac{x - x^*}{2}, \frac{r - r^*}{2}\right) = 0 \quad , \tag{2}$$

where (x^*, r^*) is the load reversal point for that particular branch curve.

2.2 Masing's Rules Extended for Transient Response

For steady-state cyclic response or loading between fixed limits, Masing's hypothesis will suffice. However, for cases of transient loading or loading between variable limits, the hypothesis was considered to be of no help. However, Masing's original hypothesis can be extended to transient loading in a manner which is simple and has a physical basis, as follows [2]:

- **Rule 1: Incomplete Loops**

The equation of any hysteretic response curve, irrespective of steady-state or transient response, can be obtained simply by applying the original Masing's rules to the virgin loading curve using the latest point of loading reversal.

For example, let the virgin loading curve OA in Fig. 2 be characterized by Eq. 1. Applying Rule 1, the equation for the branch curve CD in Fig. 2 becomes:

$$f\left(\frac{x - x_c}{2}, \frac{r - r_c}{2}\right) = 0 \quad . \tag{3}$$

Based on Eq. 3, it is easy to show that if the reloading curve CD in Fig. 2 had been continued, it would have formed a closed hysteresis loop ABCDA.

- **Rule 2: Completed Loops**

The ultimate fate of an interior curve under continued loading or unloading can be determined by choosing one of the following two interpretations [3]:

    (i) Force-deflection values are given by a hysteretic curve originating from the point of most recent loading reversal until a specified upper or lower boundary is contacted. Thereafter, the force-deflection values are given by that boundary until the direction of loading is again reversed.

    (ii) If an interior curve crosses a curve described in a previous load cycle, the load-deformation curve follows that of the previous cycle.

The first interpretation is due to Jennings [4,5] and has also been used by Matzen and McNiven [3] in their system identification study using the Ramberg-Osgood model. Fan [6] and Prévost, et al. [7] have used the second interpretation to analyze earthquake response of steel frames and earth dams, respectively. Based on observations made from experimental hysteresis loops as explained below, we have chosen the second interpretation.

Özdemir [8] in testing mild steel for energy-absorbing devices under earthquake-type loading, obtained the hysteresis loops shown in Fig. 3. The closed hysteresis loops in Fig. 3 numbered as 4–5–7–8–4, 5–6–5, 8–9–8, 11–12–13–11 and 13–14–13 reinforce the implication of Rule 1, according to which every loop, irrespective of whether it corresponds to steady-state or transient response, forms a closed loop if continued long enough. More importantly, from the loops 5–6–5, 8–9–8, 4–5–7–8–4 and 13–14–13, it can be seen that they continue along 4–5–7, 7–8–4, 3–4–10 and 12–13–15, respectively, once they complete their respective inner loops. Hence it is assumed hereafter that an interior hysteresis curve at its completion of an inner loop starts to continue the load-deformation curve from a previous cycle just outside the completed loop (Rule 2).

2.3 Summary of A General Class of Masing Models

A general class of Masing models may, therefore, be defined which consists of all those hysteretic models with a virgin loading curve, or skeleton curve, defined in general by

$$f(x,r) = 0 \quad , \tag{1}$$

and for which any other hysteretic response curve is described by the following equation:

$$f\left(\frac{x - x^*}{2}, \frac{r - r^*}{2}\right) = 0 \quad , \tag{2}$$

where (x^*, r^*) is the load reversal point chosen appropriately using Rules 1 and 2.

If the initial loading curve in Eq. 1 is written in terms of the instantaneous stiffness:

$$\frac{dr}{dx} = g(x,r) \quad , \tag{4}$$

then any branch curve is given by:

$$\frac{dr}{dx} = g\left(\frac{x - x^*}{2}, \frac{r - r^*}{2}\right) \quad , \tag{5}$$

where (x^*, r^*) is the point of load reversal chosen appropriately using Rules 1 and 2.

It is important to note that to specify any particular model in this class, only its initial loading curve need be prescribed. These models give a complete description of hysteretic behavior for every possible loading type. It can be shown that this class of models provides a unifying framework which incorporates other previously proposed hysteretic models [2].

2.4 Comparison with Iwan's Model

Iwan [9,10] has also assumed that a general hysteretic system may be thought of as consisting of a large number of ideal elasto-plastic elements, each with the same elastic stiffness but having different yield levels. He examined the behavior of a system composed of a parallel arrangement of N elements in which each element consists of a linear spring with stiffness K/N in series with a coulomb or slip damper which has a maximum allowable force r_i^*/N. If the total number of elements N becomes very large, based on contributions from the elements which have yielded and those which have not as yet yielded, the force-deflection relation for the system on initial loading, may be written as:

$$r = \int_0^{Kx} r^* \phi(r^*) dr^* + Kx \int_{Kx}^{\infty} \phi(r^*) dr^* \qquad \dot{x} > 0 \tag{6}$$

86

where $\phi(r^*)dr^*$ is the fraction of the total number of elements with strengths in the range $r^* \leq r_i^* \leq r^* + dr^*$. The initial loading force-deflection curve of such a system will have a general shape similar to that of curve OA in Fig. 2.

When the direction of loading is reversed after initial loading, as along curve ABC of Fig. 2, the total force will result from three different groups of elements: Those elements which were in a positive yield state after initial loading and have now changed to a negative yield state; those elements which were in a positive yield state after the initial loading but have not yet changed to a negative yield state; and those elements which were unyielded on initial loading and are still unyielded. Thus the restoring force along the path ABC will be given by:

$$r = \int_0^{K(x_a-x)/2} -r^*\phi(r^*)dr^* + \int_{K(x_a-x)/2}^{Kx_a} \left[r^* - K\left(x_a - x\right) \right] \phi(r^*)\, dr^*$$

$$+ Kx \int_{Kx_a}^{\infty} \phi(r^*)\, dr^* \qquad \dot{x} < 0, \ -x_a \leq x \leq x_a \quad . \tag{7}$$

The above procedure could be carried on indefinitely by keeping track of the fraction of elements in each of the yielded or unyielded category after each reversal in the direction of loading. However, for a nested set of hysteresis loops, a large number of integral terms is required and their derivation and computation becomes cumbersome.

Jayakumar [2] has recently shown that Iwan's model obeys the two hysteresis rules given earlier. This conclusion has two important consequences. First, the two hysteresis rules for the class of Masing models can be viewed as a much simpler way of implementing Iwan's model. The yield-level distribution function $\phi(r^*)$ need only be used to determine the virgin loading curve as in Eq. 6, and then the two hysteresis rules can be applied to follow the behavior of the system. This obviates the need to generate the integral terms in Iwan's model which keep track of the behavior of the elasto-plastic elements in different states. On the other hand, Iwan's model provides additional support for the hysteresis rules for the Masing models by showing that they are consistent with the physics of a particular class of mechanical systems, namely those consisting of a collection of elasto-plastic elements connected in parallel. For example, one can conclude that the Masing models cannot exhibit any nonphysical behavior, such as self-generation of energy, under any loading history. The possibility of nonphysical behavior under some complicated loading history has been an area of concern in earlier formulations of hysteresis rules.

2.5 A Special Masing Model

A special Masing model is chosen in this study by defining the restoring force-deformation relation for the virgin loading by the differential equation:

$$\frac{dr}{dx} = K\left[1 - \left|\frac{r}{r_u}\right|^n\right] \ ,\tag{8}$$

where K, r_u and n are three model parameters which are sufficient to capture the essential features of the hysteretic behavior being modeled. The initial stiffness of the model is given by K, and the ultimate strength of the system is r_u, while the smoothness of the transition from elastic to plastic response of the force-displacement curve is controlled by n as seen in Fig. 4.

The force-deflection relation for any loading other than the virgin loading is defined by the differential equation:

$$\frac{dr}{dx} = K\left[1 - \left|\frac{r - r_0}{2r_u}\right|^n\right] \ ,\tag{9}$$

where r_0 is the restoring force at the point of load reversal chosen according to Rules 1 and 2.

For the special cases $n = 1$ and 2, Eq. 8 results in the following simple relationships when $r > 0$:

$$\frac{r}{r_u} = 1 - e^{-Kx/r_u} \qquad \text{and} \qquad \frac{r}{r_u} = \tanh\left(\frac{Kx}{r_u}\right) \ ,\tag{10}$$

respectively.

Eq. 8 for the virgin loading curve is similar to a special case of the equations used in several other models, including Wen's, Özdemir's and the Endochronic models [11,8,12]. The major difference is that in the latter models the corresponding equation is used to describe the complete force-deformation relationship, not just the virgin loading case. These models are not supplemented by hysteresis rules. However, it has been shown recently [2] that without the hysteresis rules, these models lead to inconsistent behavior in certain situations. For example, the hysteresis loops are not always closed under cyclic displacement loading and the loops drift continuously under cyclic force loading whereas the Masing models behave in a manner consistent with steel members under the same bending or axial loading histories.

3.   Pseudo-Dynamic Testing

3.1   Introduction

The pseudo-dynamic test method is a recently developed quasi-static procedure
[13] for subjecting full-scale structures to simulated earthquake response by means of
on-line computer control of hydraulic actuators. The inertial effects of the structure
are modeled in an on-line computer, but in contrast to the usual quasi-static test procedures, the relation between the interstory forces and deformations is not prescribed
prior to the test. Instead, feedback from displacement and load transducers is used to
force the appropriate earthquake behavior on the structure in an interactive manner as
the experiment proceeds. Hence, full-scale structures can be tested at strong-motion
amplitude levels without making any assumptions about the stiffness and damping characteristics of the structure. Also, it is relatively inexpensive to test full-scale structures
by the pseudo-dynamic method compared with the construction and instrumentation
of a big shaking table facility.

In the pseudo-dynamic method, a multi-story building structure is modeled as a
lumped-mass discrete system. The equation of motion of such a system when excited
by earthquake ground accelerations $\ddot{z}(t)$ is given by:

$$M\ddot{\underset{\sim}{x}} + C\dot{\underset{\sim}{x}} + \underset{\sim}{R} = \underset{\sim}{F}(t) = -M\ddot{z}(t)\,\underset{\sim}{1} \quad , \tag{11}$$

For implementation, Japanese researchers [13] chose to use the central-difference method for which Eq. 11 becomes:

$$[M + C\,\Delta t/2]\,\underset{\sim}{x}_{i+1} = (\Delta t)^2\,[\underset{\sim}{F}_i - \underset{\sim}{R}_i] + 2M\underset{\sim}{x}_i + [C\,\Delta t/2 - M]\,\underset{\sim}{x}_{i-1} \quad . \tag{12}$$

The mass matrix M is prescribed from the known mass distribution of the test
structure so that the on-line computer can simulate its inertial effects, and the viscous
damping matrix C is set equal to that derived from the preliminary free and forced
vibration tests of the structure at low amplitudes assuming Rayleigh damping. From
the knowledge of measured restoring forces and calculated displacements at the previous
time steps, the displacement at time step $(i+1)$ is calculated using Eq. 12 in an on-line
computer. Hydraulic actuators are then used to force the structure quasi-statically to
deflect to the calculated position (Fig. 5). When the desired displacement is achieved
within prescribed tolerances, load cells mounted on the actuators measure the restoring
forces and displacement transducers on the structure measure the final displacements
achieved. This information is fed back to the on-line data processing computer to
calculate the displacements to be imposed at the next time step.

3.2 BRI Testing Program

A six-story, two-bay, full-scale steel structure (Fig. 6) was tested by the pseudo-dynamic method at the Building Research Institute (BRI) in Tsukuba, Japan during November, 1983–March, 1984. This structure, which represented Phase II of the steel program under the U.S.—Japan Cooperative Earthquake Research Program Utilizing Large-Scale Testing Facilities, was designed to satisfy the requirements of both the 1979 U.S. Uniform Building Code (UBC) and the 1981 Architectural Institute of Japan code, using eccentric K-bracings [14]. It was 15 m×15 m in plan and 21.5 m high. The two exterior frames A and C were unbraced moment-resisting frames with one column in each oriented for weak-axis bending in order to increase the torsional stiffness, and the interior frame B was a braced moment-resisting frame with eccentric K-bracing in its north bay. All the girder-to-column connections were designed as moment connections in the loading direction and shear connections in the transverse direction. The floor system consisted of a formed metal decking with cast-in-place light-weight concrete acting compositely with the girders and floor beams. No non-structural component was attached to the frame system.

The BRI tests were performed at low amplitudes to give nominally elastic response and at larger amplitudes to excite the structure into the inelastic range. The uni-directional loading in the elastic and inelastic tests was produced by the Taft S21W component from the 1952 Kern County, California, earthquake scaled to peak accelerations of 6.5% g and 50% g, respectively. The elastic test data have been studied previously by the authors using linear models and single-input single-output and multiple-input multiple-output system identification methods [15,16]. In this paper, a study of the inelastic test data is reported.

4. System Identification Applied to Inelastic Pseudo-Dynamic Test Data

4.1 Simplified Structural Model

The equation of motion of a building modeled using N degrees of freedom can be written as:

$$M\underset{\sim}{\ddot{x}} + \underset{\sim}{R} = -M\ddot{z}(t)\underset{\sim}{1} \quad , \tag{13}$$

where $\underset{\sim}{R}$ is the vector of restoring forces at each degree of freedom. The restoring force R_i at the degree of freedom i stems from the interaction of that degree of freedom with all other degrees of freedom and with the ground. Therefore, the following relationship is assumed:

$$R_i = \sum_{\substack{j=0 \\ \neq i}}^{N} R_{ij} \quad , \tag{14}$$

in which R_{ij} $(j \neq 0)$ is the restoring force exerted at degree-of-freedom i by degree-of-freedom j due to the relative motion between the two, and R_{i0} is similar except it represents the interaction with the ground.

By introducing a shear-building approximation, it is assumed that

$$R_{ij} = 0 \qquad \forall\, i,\, j \ni |i - j| > 1 \quad . \tag{15}$$

From Eq. 14,

$$R_i = R_{i,\, i-1} + R_{i,\, i+1} \quad . \tag{16}$$

With the introduction of story shears as shown in Fig. 7 in which $R_{i,i-1} = r_i$, $R_{i,i+1} = -r_{i+1}$, Eq. 16 becomes:

$$R_i = r_i - r_{i+1} \quad . \tag{17}$$

The special Masing model is used to relate the story shear forces and story drifts (Fig. 7), so that, for example, from Eq. 8, the virgin loading curve for story i is:

$$\dot{r}_i = K_i \left(\dot{x}_i - \dot{x}_{i-1} \right) \left[1 - \left| \frac{r_i}{r_{u,\, i}} \right|^{n_i} \right] \tag{18}$$

In the above relationship, K_i is the initial stiffness, $r_{u,i}$ is the ultimate strength, and n_i is a model parameter for story i. This shear-building approximation is used in the modeling of the pseudo-dynamic test structure described previously.

4.2 Analytical Estimation of Model Parameters

The model parameters K_i and $r_{u,i}$ for each story can be estimated from material properties and structural plans. This makes the model potentially useful for predicting structural response during earthquake-resistant design. The analytical estimates of the parameters can also be used to provide initial estimates in the optimization algorithm used to estimate the parameters from the measured response of an existing structure.

Two methods were used to estimate analytically the story stiffnesses K for the shear-building model: the fundamental-mode approximation (FMA) method and Biggs' formula.

In the FMA method [17,18], the story stiffnesses are calculated using the fundamental frequency and modeshape of the structure being modeled. The modal properties may be from experiments in the case of an existing structure, or from a more detailed linear finite-element model of the structure. The modal equations corresponding to the fundamental mode of the shear-building model can be expressed as follows:

$$K \underset{\sim}{\phi}^{(1)} = \omega_1^2 M \underset{\sim}{\phi}^{(1)} \tag{19}$$

where K is the stiffness matrix of the shear building, M is the mass matrix, ω_1 is the fundamental frequency and $\underset{\sim}{\phi}^{(1)}$ is the fundamental modeshape vector. In this study, the properties of the fundamental mode of the test structure, as identified from the elastic pseudo-dynamic test, were used [16]. The system of linear equations in Eq. 19 was then solved for the story stiffnesses $K_1, K_2, \ldots, K_6$. Their variation with story level is plotted in Fig. 8 where the stiffnesses are marked at the mid-points of the stories. The values are in good agreement with the stiffnesses estimated from the overall slopes of the experimental hysteresis loops obtained from the inelastic pseudo-dynamic test. The latter values are tabulated in Table 1 under the appropriate column labelled "prior" and are plotted in Fig. 8.

Also plotted in Fig. 8 are the theoretical stiffnesses obtained by Biggs' formula, as presented by Anagnostopoulos [19] for braced structural frames. Biggs' formula is based on beam theory along with some simplifying assumptions about joint rotations and shears. In the present case, Biggs' formula gives story stiffnesses which are too large, although it should be noted that the results are based purely on theory, whereas the FMA results use modal properties derived from a test on the structure. Also, despite the high stiffnesses from Biggs' formula, the corresponding first three modal frequencies are within 10% of those estimated from the elastic pseudo-dynamic test data, and the corresponding theoretical modeshapes and the modeshapes identified from the elastic test data, labelled MI-MO in Fig. 9, are also in good agreement.

The ultimate strength $r_{u,i}$ of each story in the test structure can be estimated by assuming that the story has been transformed into a mechanism and then employing the principle of virtual work [2]. The results of the calculations are shown in Table 1 under the appropriate column labelled "prior."

4.3 Optimal Estimation of Model Parameters by System Identification

An output-error approach [20] for system identification is used in conjunction with a combination of the steepest-descent and modified Gauss-Newton methods to determine the optimal estimates of the parameters for the hysteretic model from experimental data. The approach is implemented in a computer program called HYSID [2].

The model parameters are:

$$\underset{\sim}{\theta} = [\ldots, \{K_i,\ r_{u,i},\ n_i\}, \ldots]^T \quad i = 1, 2, \ldots, N \tag{20}$$

where N is the number of stories in the structure, and K_i, $r_{u,i}$ and n_i are the story parameters described previously.

The optimal values for the model parameters are calculated by minimizing a measure-of-fit $J(\underset{\sim}{\theta})$ which describes the matching of the experimental story shears $r_{0,i}$ by the model:

$$J(\underset{\sim}{\theta}) = \frac{\sum_{i=1}^{N} \int_{t_s}^{t_e} \left[r_{0,i} - r_i(\underset{\sim}{\theta}) \right]^2 dt}{\sum_{i=1}^{N} \int_{t_s}^{t_e} r_{0,i}^2 dt}$$

4.4 Analysis of Inelastic Pseudo-Dynamic Test Data

The inelastic pseudo-dynamic test data are analyzed using the hysteretic system identification program HYSID, in order to determine the optimal estimates of the model parameters. The optimal estimates of story stiffnesses, story strengths and n values are plotted in Fig. 10 along with their respective mean values to examine the reliability of the identified parameter values from nine computer runs which start with different initial estimates. The mean values are also tabulated in Table 1 under the columns labelled "optimal."

It should be expected that all the runs would result in the same optimal estimates for the parameters. However, Fig. 10 shows that the parameters are not identified uniquely. This could be due to interaction between the parameters which can result in a fairly flat region containing all the "minimum" points.

In spite of the diversity of the initial starting estimates for the different runs of HYSID, the optimal estimates are close in all cases, except for the story strengths of the top three stories which show a large scatter about their mean values in Fig. 10. This was expected because the structure did not experience any significant inelastic deformation at the top three stories, and hence the determination of story strengths from these data is an ill-conditioned process since large changes in r_u make only small changes in the response. It is clear that a story strength cannot be estimated with any confidence if the story is not exercised well into its inelastic regime.

The mean values of the stiffnesses and strengths in Fig. 10 exhibit a decreasing trend with height, as expected, except for a lower first-story stiffness because of its greater height. Also, n takes values around 2.0, which corresponds to the hyperbolic tangent force-deformation relation (Eq. 10). However, the fifth-story stiffness, strength of the fourth story and n values for stories 4 and 5 appear anomalous. The strength and n values may be interacting during optimization because of the insignificant inelastic deformation of the top three stories during the test. The high stiffness of the fifth story may be due to the effect of the feedback errors during the pseudo-dynamic tests [16].

Hysteresis curves obtained from the pseudo-dynamic test are compared with the curves corresponding to the optimal hysteretic model in Fig. 11. The test roof displacement and base shear histories are compared with their counterparts predicted by the optimal model in Fig. 12.

These plots demonstrate that the hysteretic model developed in this study is able to capture the important features of the inelastic response of the test structure. This suggests that the special Masing model employed may be useful in predicting the response of steel structures to prescribed earthquake ground motions, although further tests of the model using structural response data are required. As described for the test structure, the story parameters corresponding to the initial stiffness and ultimate strenth, K_i and $r_{u,i}$, respectively, can be estimated theoretically from material properties and structural plans. For the model parameter n, based on this study, we tentatively propose the constant value n=2 for the inelastic, undamaged behavior of steel-frame structures, although it is recognized that this parameter may actually vary about this value from story to story, and from one structure to another.

5.   Conclusions

The improvement of earthquake-resistant design of structures requires a knowledge of the nonlinear response of structures. This, in turn, requires a method for describing the dynamic force-deflection relation of structural systems. A smooth, nonlinear hysteretic relation would be a realistic model for most applications.

A general class of Masing models is presented to represent this dynamic, hysteretic behavior. Masing's original hypothesis was extended to arbitrary transient loading by the introduction of two simple hysteresis rules. Any model within this general class of Masing models can be prescribed by giving its virgin loading curve. This class of models was shown, analytically and experimentally, to exhibit reasonable response behavior for arbitrary loading patterns.

Based on a study of some previous models, a simple virgin loading curve was chosen to prescribe a special Masing model for modeling the nonlinear hysteretic behavior of steel structures. This gives a force-deflection relation which is general enough to be potentially useful as a model for the hysteretic, dynamic behavior of a wide range of softening materials and structures. An important feature of the model is that two of its three parameters can be estimated theoretically based on material properties and structural plans. Also, by testing the model sufficiently with available experimental data, a set of feasible values for the parameter n for different materials and structures can be obtained. This is necessary if the model is to have the potential of predicting structural response prior to a structure being built and tested.

The inelastic pseudo-dynamic test data were analysed using a hysteretic system identification program, HYSID, in order to examine the applicability of the new hysteretic model to a real structure. The theoretical estimates of the initial stiffnesses and ultimate strengths were consistent with the optimal values determined for the six-story test structure, except for a couple of anomalies. The optimal estimation also gave an opportunity to examine suitable values for n, since this parameter cannot be determined theoretically. It was shown that the simple three-parameter model for each story shear-deformation relationship appears to be sufficient to capture the essential features of the nonlinear behavior of the steel frame test structure.

With these encouraging results, further exploration of the nonlinear model is planned, such as the generalization of the hysteretic model to problems in continuum mechanics using a multiple-yield-surface plasticity theory, application to damage detection in structures, further experimental verification using shaking-table tests, modeling of deterioration of material properties such as stiffness and strength during strong ground motions, and the development of a seismic design methodology using the hysteretic model.

REFERENCES

[1] Masing, G., "Eigenspannungen und Verfestigung beim Messing," *Proceedings of the 2nd International Congress for Applied Mechanics*, Zurich, Switzerland, 332–335, 1926. (German)

[2] Jayakumar, P., "Modeling and Identification in Structural Dynamics," Report No. EERL 87-01, Earthquake Engineering Research Laboratory, California Institute of Technology, Pasadena, California, May 1987.

[3] Matzen, V.C. and H.D. McNiven, "Investigation of the Inelastic Characteristics of a Single Story Steel Structure using System Identification and Shaking Table Experiments," Report No. EERC 76-20, Earthquake Engineering Research Center, University of California, Berkeley, California, August 1976.

[4] Jennings, P.C., "Response of Simple Yielding Structures to Earthquake Excita-

tion," Ph.D. Dissertation, California Institute of Technology, Pasadena, California, June 1963.

[5] Jennings, P.C., "Earthquake Response of a Yielding Structure," *Journal of Engineering Mechanics Division*, ASCE, Vol. 91(4), 41–68, August 1965.

[6] Fan, W.R.-S., "The Damping Properties and the Earthquake Response Spectrum of Steel Frames," Ph.D. Dissertation, University of Michigan, 1968.

[7] Prévost, J.-H., A.M. Abdel-Ghaffar and A.-W.M. Elgamal, "Nonlinear Hysteretic Dynamic Response of Soil Systems," *Journal of Engineering Mechanics*, ASCE, Vol. 111(5), 696–713, May 1985.

[8] Özdemir, H., "Nonlinear Transient Dynamic Analysis of Yielding Structures," Ph.D. Dissertation, Division of Structural Engineering and Structural Mechanics, Department of Civil Engineering, University of California, Berkeley, California, June 1976.

[9] Iwan, W.D., "A Distributed-Element Model for Hysteresis and Its Steady-State Dynamic Response," *Journal of Applied Mechanics*, ASME, Vol, 33(4), 893–900, December 1966.

[10] Iwan, W.D., "The Distributed-Element Concept of Hysteretic Modeling and Its Application to Transient Response Problems," *Proceedings of the 4th World Conference on Earthquake Engineering*, Vol. II, A-4, 45–57, Santiago, Chile, 1969.

[11] Wen, Y.-K., "Method for Random Vibration of Hysteretic Systems," *Journal of the Engineering Mechanics Division*, ASCE, Vol. 102(2), 249–263, April 1976.

[12] Sandler, I.S., "On the Uniqueness and Stability of Endochronic Theories of Material Behavior," *Journal of Applied Mechanics*, ASME, Vol. 45(2), 263–266, June 1978.

[13] Takanashi, K. et al., "Non-Linear Earthquake Response Analysis of Structures by a Computer-Actuator On-Line System," *Bulletin of Earthquake Resistant Structure Research Center*, No. 8, Institute of Industrial Science, University of Tokyo, Japan, 1–17, December 1974.

[14] Askar, G., S.J. Lee and L.-W. Lu, "Design Studies of Six Story Steel Test Building: U.S.–Japan Cooperative Earthquake Research Program," Report No. 467.3, Fritz Engineering Laboratory, Lehigh University, Bethlehem, Pennsylvania, June 1983.

[15] Beck, J.L. and P. Jayakumar, "Application of System Identification to Pseudo-Dynamic Test Data from a Full-Scale Six-Story Steel Structure," *Proceedings of the International Conference on Vibration Problems in Engineering*, Xian, China, June 1986.

[16] Beck, J.L. and P. Jayakumar, "System Identification Applied to Pseudo-Dynamic Test Data: A Treatment of Experimental Errors," *Proceedings of the 3rd ASCE Engineering Mechanics Specialty Conference on Dynamic Response of Structures*, University of California, Los Angeles, California, April 1986.

[17] Nielsen, N.N., "Dynamic Response of Multistory Buildings," Ph.D. Dissertation, California Institute of Technology, Pasadena California, June 1964.

[18] Lai, S.-S.P. and E.H. Vanmarcke, "Overall Safety Assessment of Multistory Steel Buildings Subjected to Earthquake Loads," Publication No. R80-26, Department of Civil Engineering, Massachusetts Institute of Technology, Cambridge, Massachusetts, June 1980.

[19] Anagnostopoulos, S.A., J.M. Roesset and J.M. Biggs, "Non-Linear Dynamic Response and Ductility Requirements of Building Structures Subjected to Earthquakes," Publication No. R72-54, Department of Civil Engineering, Massachusetts Institute of Technology, Cambridge, Massachusetts, September 1972.

[20] Beck, J.L.,"Determining Models of Structures from Earthquake Records," Report No. EERL 78-01, Earthquake Engineering Research Laboratory, California Institute of Technology, Pasadena, California, June 1978.

[21] Okamoto, S. et al., "Techniques for Large Scale Testing at BRI Large Scale Structure Test Laboratory," Research Paper No. 101, Building Research Institute, Ministry of Construction, Japan, May 1983.

Story	Elastic Stiffness (tonf/cm)		Strength (tonf)		n	
	Prior	Optimal	Prior	Optimal	Prior	Optimal
1	184.7	191.5	295.1	318.1	1.8	1.6
2	236.3	213.3	323.8	277.5	1.8	1.8
3	200.0	171.8	264.9	257.6	1.8	2.2
4	161.8	153.1	255.5	348.9	1.8	3.2
5	130.6	253.4	183.1	221.7	1.8	1.1
6	89.4	111.0	172.1	165.3	1.8	2.0

Table 1 Comparison of the Prior and Optimal Estimates of the Structural Parameters

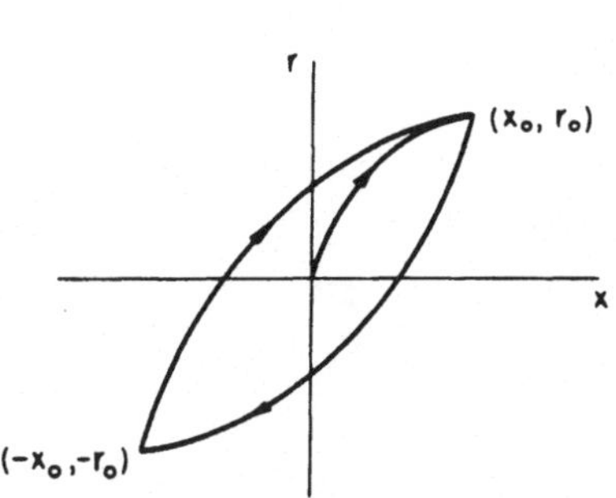

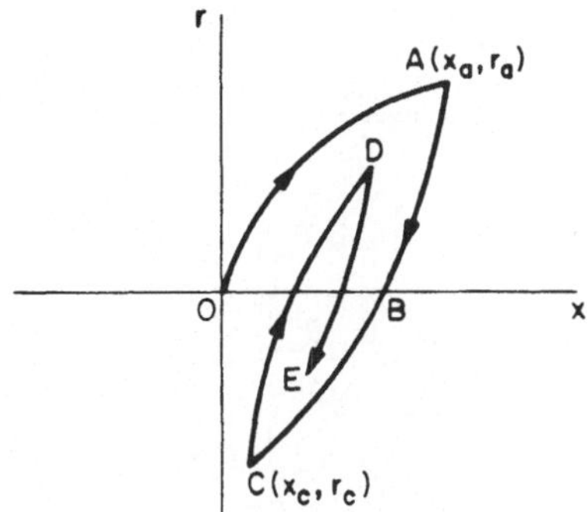

Fig. 1   Hysteretic Loop for Cyclic Loading   **Fig. 2**   Hysteretic Loop for Transient Loading

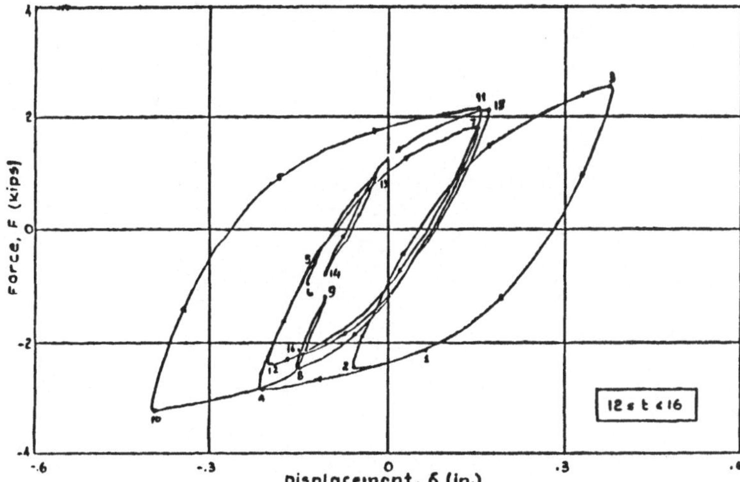

Fig. 3   Experimental Hysteresis Loops [8]

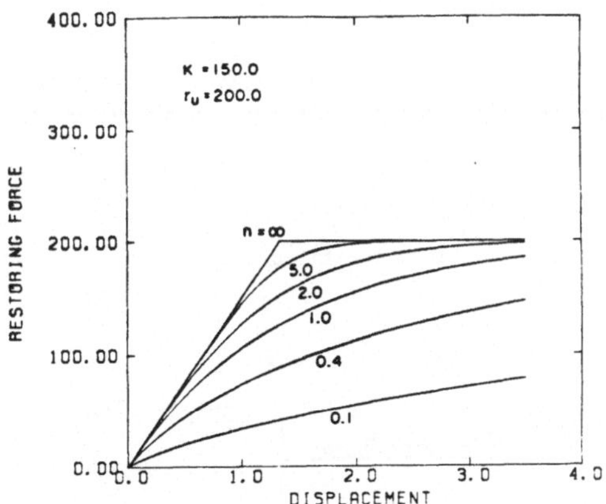

Fig. 4   Effect of n on the Force-Displacement Curve of Eq. 8

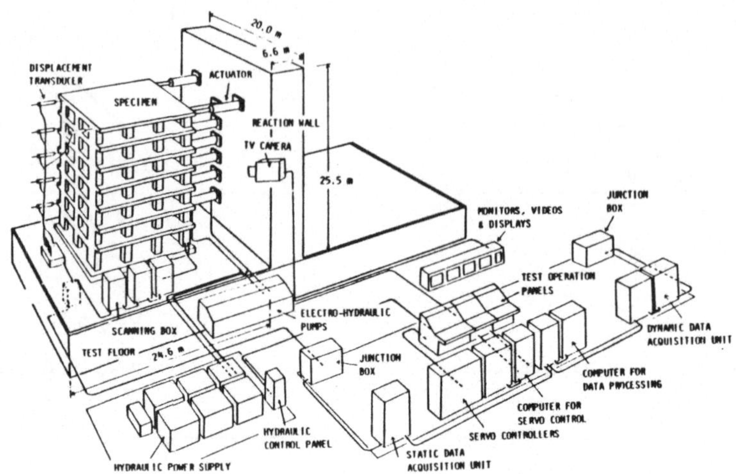

Fig. 5 Pseudo-Dynamic Testing Facility at the Building Research Institute in Tsukuba, Japan [21]

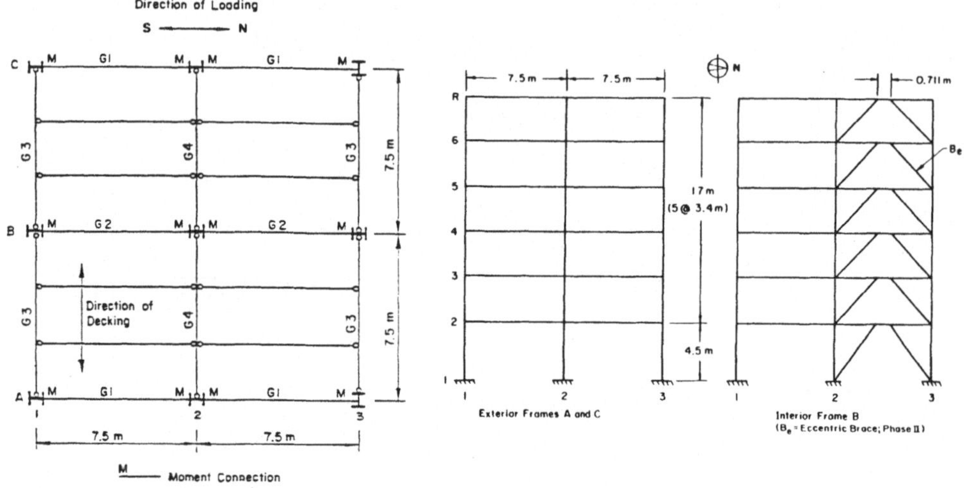

Fig. 6(a) Plan of Test Structure [14]

Fig. 6(b) Elevation Views of Exterior and Interior Frames of Test Structure [14]

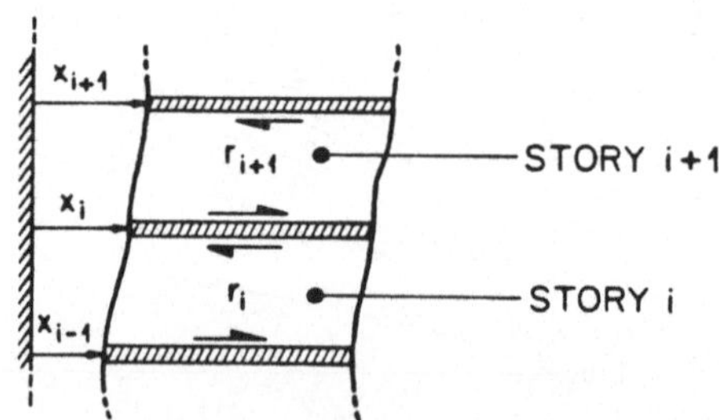

Fig. 7 Illustration of Story Shear Forces and Story Drifts

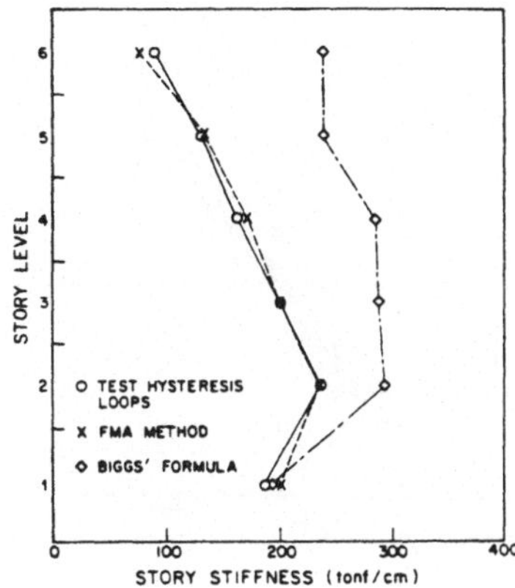

Fig. 8   Variation of Stiffness with Story
Level Estimated by Different Methods

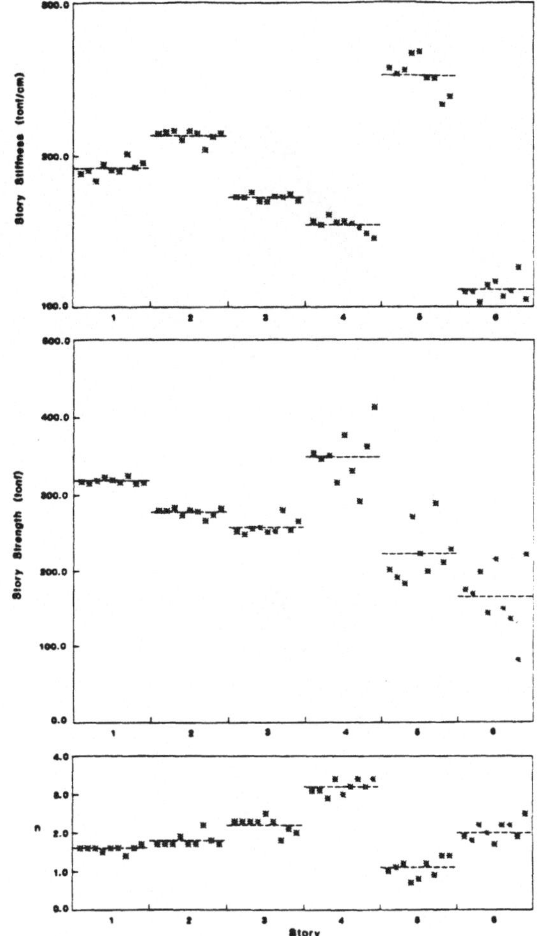

Fig. 10   Plots of Optimal Estimates and the Mean
Values of Structural Parameters

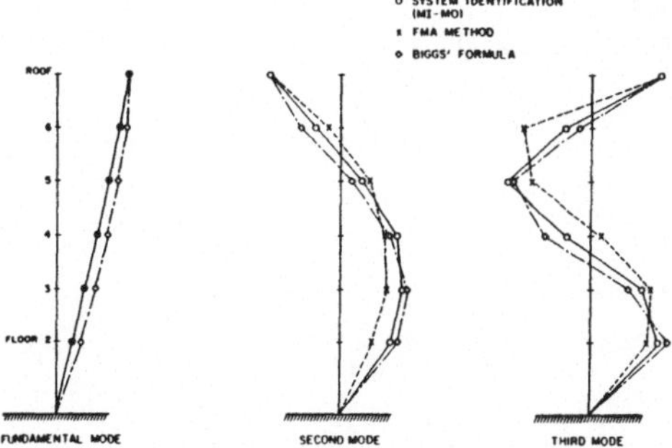

Fig. 9   Modeshapes Obtained by Different Methods

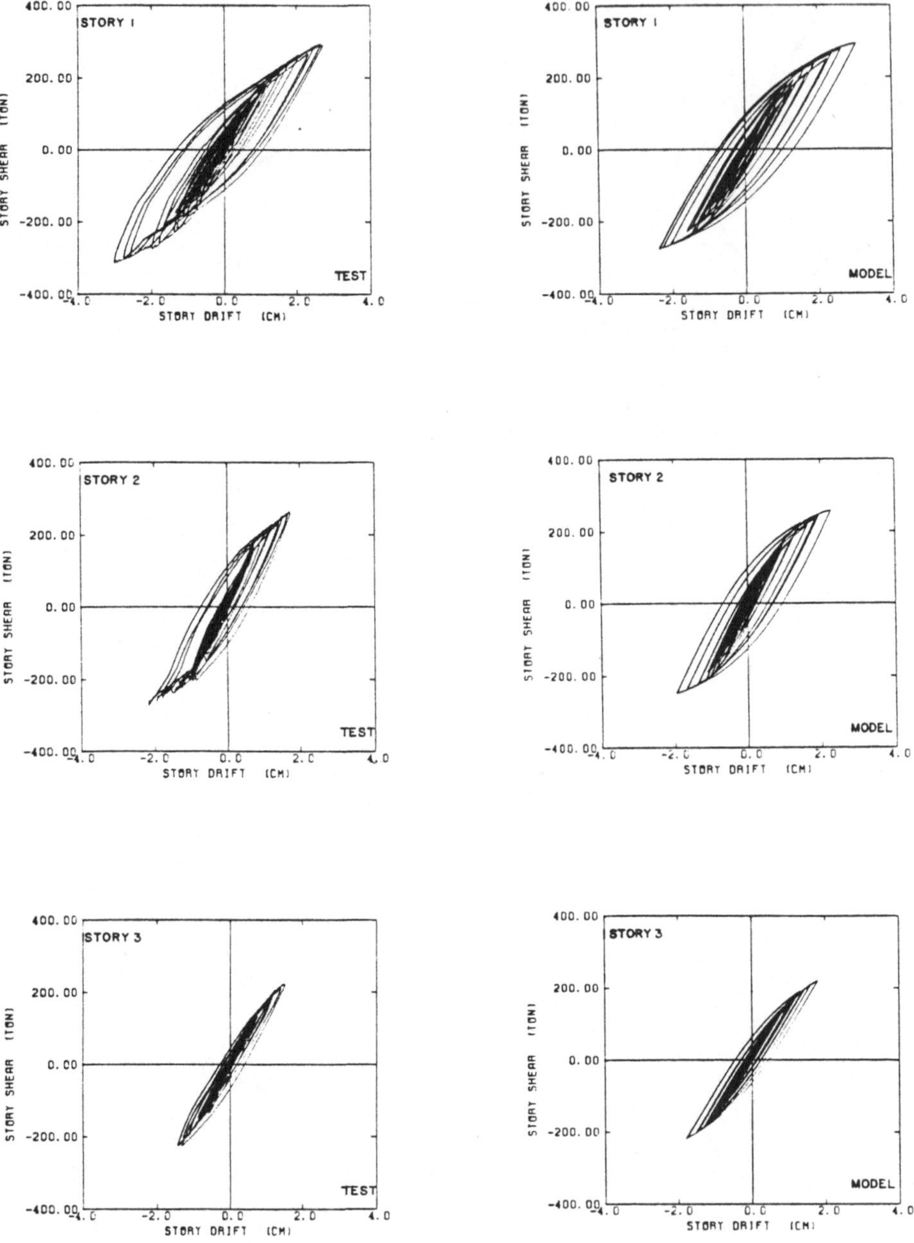

Fig. 11 Comparison of Hysteresis Behavior from the Inelastic Pseudo-Dynamic Test and the Optimal Hysteretic Model

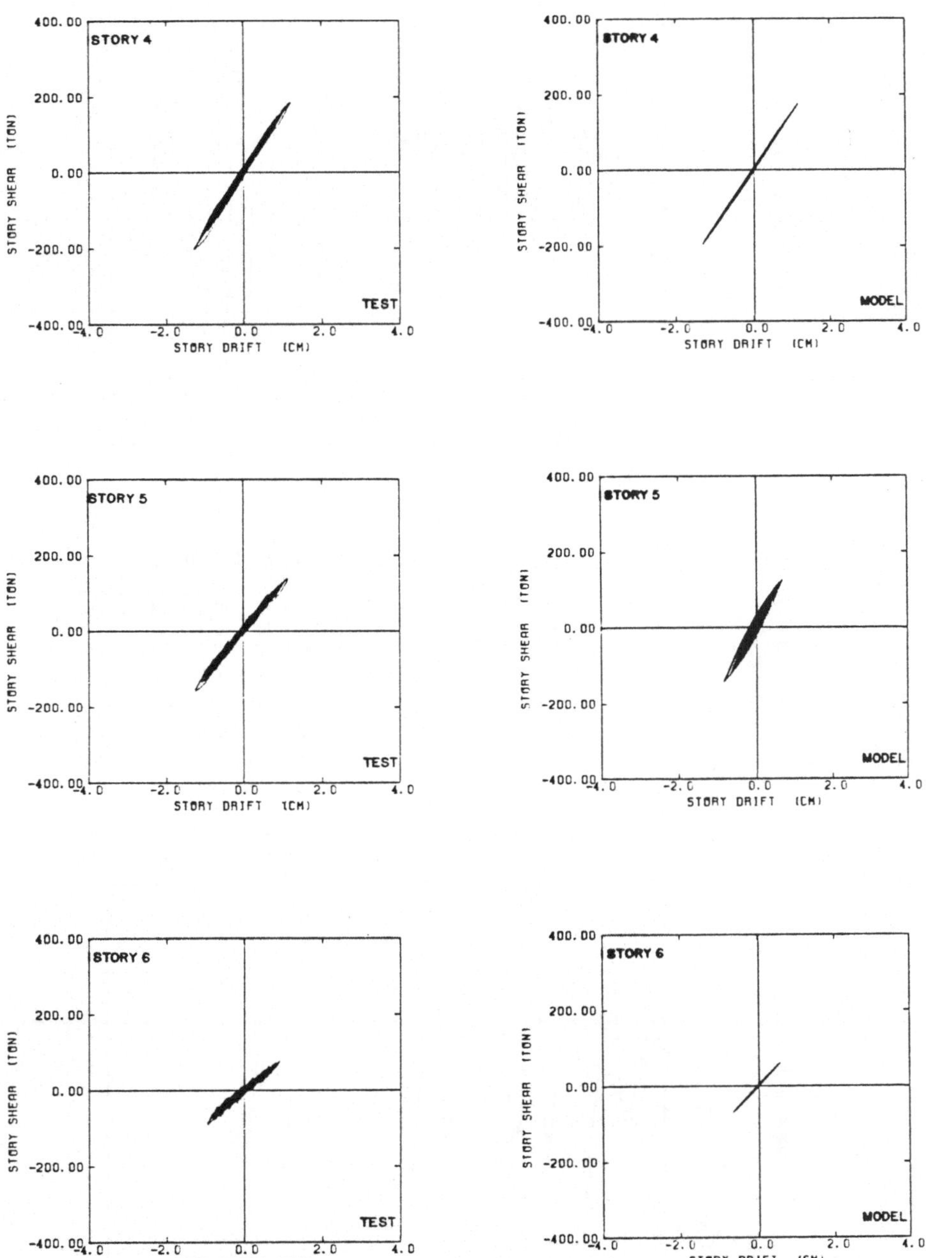

Fig. 11 (continued)

Comparison of Hysteresis Behavior from the Inelastic Pseudo-Dynamic Test and the Optimal Hysteretic Model

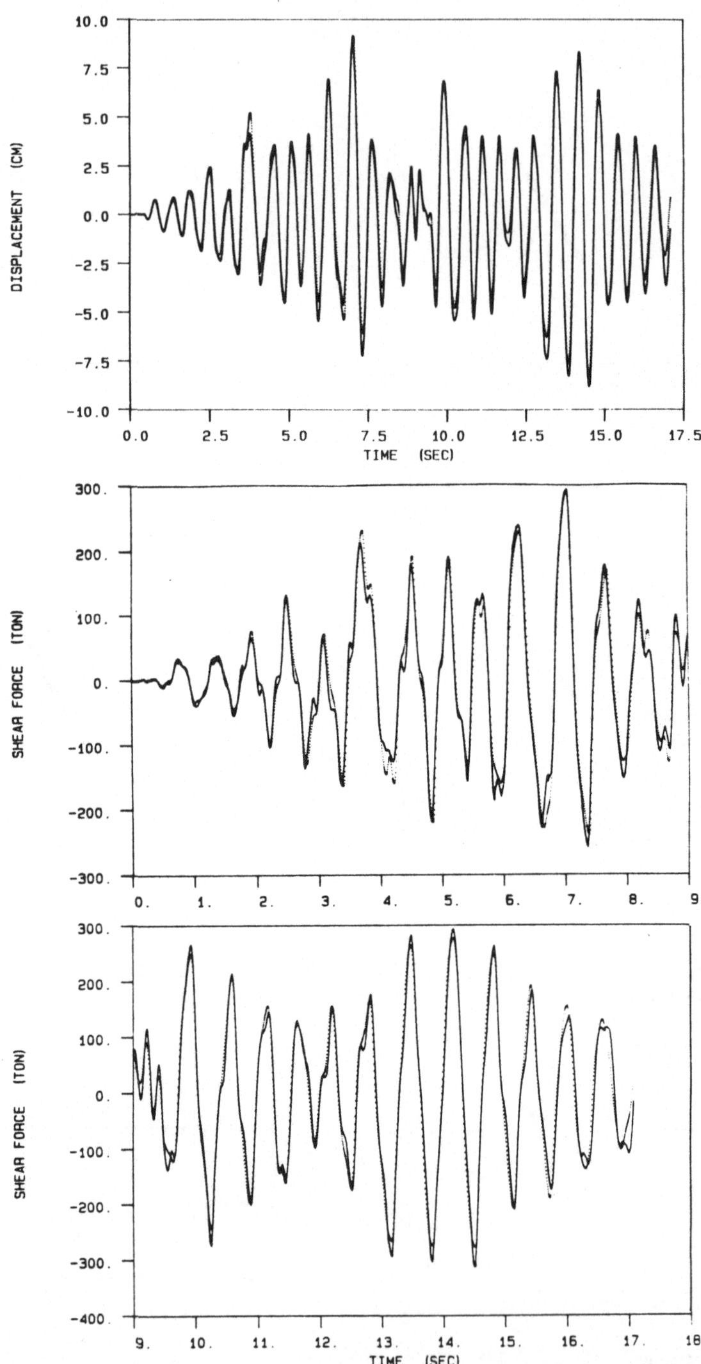

Fig. 12 Comparison of Roof Displacements and Base Shears (— *Inelastic* Pseudo-Dynamic Test; - - - Optimal Hysteretic Model)

Identification of Hysteretic Structural Behaviour from Strong Motion Accelerograms

C.-Y. Peng & W. D. Iwan

California Institute of Technology
Pasadena, California 91125   USA

Abstract

The problem of identifying the properties of hysteretic structures from strong motion earthquake records is examined. A method is presented for identifying the generalized restoring force diagram of different "modes" of the structural response. The method thereby offers a means for characterizing structural damage. The approach presented can be employed even when records are available from only a small number of locations in the structure. The method is applied to pseudo-dynamic test data from a full scale six-story steel-frame structure.

1.   Introduction

One of the best methods for studying the seismic behavior of an as-built structure is to record its response during an actual earthquake and then analyze the response records using system identification techniques. Strong motion acclerograph records provide one of the few sources of information concerning the large amplitude dynamic behavior of actual structures.

Two problems are common to all efforts to analyze structural response from earthquake records. First, the number of response measurements is usually small. Frequently, only two records are available, one at the base of the structure and the other near the top of the structure. Second, nonlinear behavior is observed for many cases of strong shaking. Thus, linear time-invariant models cannot be used successfully to treat the entire duration of response. The absence of a well-established analytical technique for determining nonlinear structural models from earthquake data has seriously limited the utility of this data.

This paper presents the results of research directed at solving some of these problems. It describes a new system identification method which can be used to extract hysteretic structural behavior from strong motion accelerograms.

2. Identification Method

Consider a nonlinear dynamic system governed by the equation of motion

$$M\ddot{y} + \underset{\sim}{f}(y,\dot{y}) = -M\underset{\sim}{1}\,\ddot{z}(t) \tag{1}$$

where M is the mass matrix of order n, y is the relative displacement vector with respect to the base, $\underset{\sim}{f}$ is the nonlinear restoring force vector and $\ddot{z}(t)$ is the base acceleration.

Let

$$\underset{\sim}{y}(t) = \Phi\,\underset{\sim}{u}(t) \tag{2}$$

where Φ is an $n \times m$ matrix whose columns are a set of appropriate orthonormal "modal vectors" for the system (1).

Furthermore, define

$$y_i^r(t) = \phi_{ir}\,u_r(t) \tag{3}$$

Then, it may be shown that

$$\ddot{y}_i^r + h_i^r\,(y_i^s,\dot{y}_i^s) = -\beta_i^r\,\ddot{z}(t)\;; \qquad s = 1, 2, \ldots, m \tag{4}$$

where y_i^r may be considered to be the generalized modal displacement, h_i^r the generalized modal restoring force and β_i^r is the effective modal participation factor.

The modal equations (4) may be rearranged as

$$h_i^r\,(y_i^s,\dot{y}_i^s) = -\beta_i^r\,\ddot{z}(t) - \ddot{y}_i^r \tag{5}$$

When the right hand side of equation (5) is specified, the identification problem is reduced to identifying the generalized restoring force $h_i^r\,(y_i^s,\dot{y}_i^s)$. The $\ddot{y}_i^r$ can be estimated from the frequency domain information by band-pass filtering. Thereby, coupling in h_i^r is effectively eliminated.

Nonparametric identification techniques are used initially to estimate h_i^r. Since the modal restoring force is known as a function of the modal displacement and velocity, no time integration of the equations of motion need be performed. Furthermore, the use of nonparametric identification avoids the need for nonlinear optimization. This results in considerable computational saving.

The only parameter left to be determined during the initial stage of the identification is the effective participation factor β_i^r. The optimal estimate of β_i^r may readily be obtained by any one-dimensional nonlinear optimization scheme.

After nonparametric identification has been used to obtain an initial nonhysteretic estimate of h_i^r and nonlinear optimization has been used to determine an optimal estimate of β_i^r, a parametric hysteretic model is employed to refine the final answer.

A variety of different error minimization criteria may be employed in both the nonparametric and parametric identification. However, it has been found that a simple criterion which minimizes the difference between the actual and model response at peaks only is adequate. This approach is motivated by the observation that peaks are generally the points of greatest significance in the response time history.

3. Restoring Force Models

In the case of linear systems with classical normal modes, the generalized restoring force will be of the form

$$h_i^r = 2\varsigma_r\, \omega_r\, \dot{y}_i^r + \omega_r^2\, y_i^r \tag{6}$$

where ς_r and ω_r are the modal damping ratio and frequency, respectively.

For a general nonlinear system, the analytical form of the generalized restoring force is unknown and can only be estimated. Let an initial estimate of h_i^r be given by $\widehat{h}_i^r$, where $\widehat{h}_i^r$ is a power series in y_i^r and $\dot{y}_i^r$. That is,

$$\widehat{h}_i^r = \sum_p \sum_q C_{pq}\, (y_i^r)^p\, (\dot{y}_i^r)^q \quad . \tag{7}$$

The coefficients C_{pq} may be determined by approximating h_i^r in the least squares sense of Chebyshev.

Two nonlinear restoring force models are employed herein in the generalized modal identification method. The first, a nonparametric model is used to obtain an initial estimate of the backbone of the hysteretic restoring force and the second, a parametric hysteretic model is used to obtain the final modal model.

3.1 Four-parameter "Nonparametric" Model

Initially, it is assumed that

$$\widehat{h}_i^r = a_1^r\, y_i^r + a_2^r\, (y_i^r)^3 + a_3^r\, \dot{y}_i^r + a_4^r\, (\dot{y}_i^r)^3 \tag{8}$$

where a_1 and a_3 control the small amplitude behavior of the modal response and a_2 and a_4 control the large amplitude behavior. This is a truncated form of the general nonparametric model (7) obtained by eliminating those terms which are "non-physical". This model may be considered the simplest extension of the linear model (6).

3.2 Two-parameter Distributed Element Model

In order to better model the hysteretic behavior of the structure, a parametric distributed-element model is used to complete the structural identification. The model used is shown in Figure 1. The backbone of the hysteresis curves is assumed to be of the form

$$
\begin{aligned}
h_i^r &= b_1^r\, y_i^r + b_2^r\, (y_i^r)^3 \quad ; \quad y_i^r \le \sqrt{-b_1^r/3b_2^r} \\
&= (2b_1^r/3)\, \sqrt{-b_1^r/3b_2^r}\; ; \quad y_i^r > \sqrt{-b_1^r/3b_2^r}
\end{aligned}
\tag{9}
$$

where the initial estimate of the coefficients is $b_1^r = a_1^r$ and $b_2^r = a_2^r$. The stiffness distribution of sub-elements of the distributed element model may be determined in a straightforward manner from the backbone curve. In this study, the hysteretic behavior of the model is generated from the elasto-plastic behavior of each sub-element and no mathematical rules are needed.

Since the model parameters identified during the initial stage nonparametric identification are generally very close to the optimal parameters of the parametric model, error minimization for the final stage of the identification process is very direct and efficient. No convergence problems have been encountered.

4. Verification with Simulated Data

Generalized modal identification incorporating the approach outlined above has been verified using simulated data generated for the hysteretic structure shown in Figure 2(a). The characteristics of the structural model were chosen to approximate a structure tested on the shaking table at The University of California, Berkeley, Figure 2(b).

Two different earthquake accelerograms were used as a base excitation to generate response data for the three-degree-of-freedom system. The first accelerogram, El Centro, 1940, S00E, was used to identify the structure and the second accelerogram, Taft, 1952, S69E, was used to study the prediction capability of the method.

Both earthquakes were scaled to a peak acceleration of 0.5g to assure significant nonlinear response behavior. The base input and top floor accelerations only were used in this analysis. The actual inter-story restoring force diagrams for the Taft excitation are shown in Figure 3.

Figure 4 shows the identified and predicted displacement response using the initial stage nonparametric model only. It is seen that this model gives a very good frequency and amplitude match of response data but cannot reproduce the permanent or drift displacement observed at the end of the response. The failure to identify permanent displacement is to be expected since the nonparametric model has no mechanism with which to describe hysteretic drift behavior. The good match of frequency and amplitude response implies that the backbone of the generalized restoring force is well identified and can be used as a good estimate for the backbone of the distributed-element hysteretic model.

The identification and prediction of the displacement response are greatly improved when the parametric distributed-element model is used, as shown in Figure 5. These results appear to validate the proposed method.

5. Application to Pseudo-Dynamic Test Data

The identification method was applied to pseudo-dynamic test data from a full-scale six-story steel-frame structure tested in Japan as part of a US–Japan coorporative research program. The input and roof response data only were used in the analysis. Identification results using the nonparametric model are shown in Figures 6 and 7. Figure 8 shows the final identification results for the modal restoring force using the distributed element model. The time histories of the response of the actual structure and the identified model are shown in Figure 9.

It is believed that the results obtained are quite good, especially considering the fact that only one input record and one response record were used in the system identification algorithm.

6. Conclusions

An efficient system identification technique has been presented for use in analyzing strong motion accelerograph data from structures. The validity of the approach has been verified using simulated data. Application of the method to pseudo-dynamic test data shows that it is capable of providing an accurate representation of the hysteretic nature the structural response. The method therby offers a promising means for characterizing structural damage.

Acknowledgement

This paper describes research conducted under grants from the US National Science Foundation. Any conclusion or opinions presented are those of the authors and do not necessarily reflect those of the Foundation. The authors wish to thank J. Beck and P. Jayakumar for allowing use of their version of the US–Japan pseudo-dynamic test data.

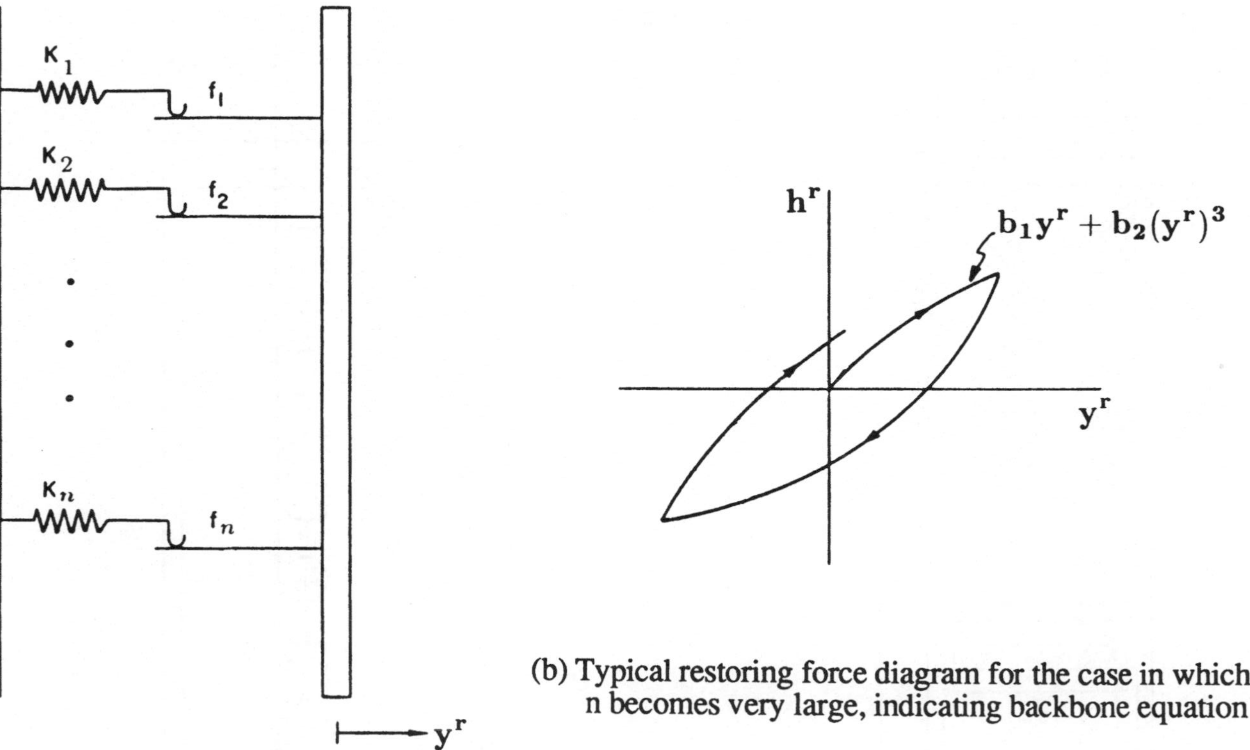

(a) Physical representation of model

(b) Typical restoring force diagram for the case in which
n becomes very large, indicating backbone equation

Figure 1 Distributed Element Hysteresis Model

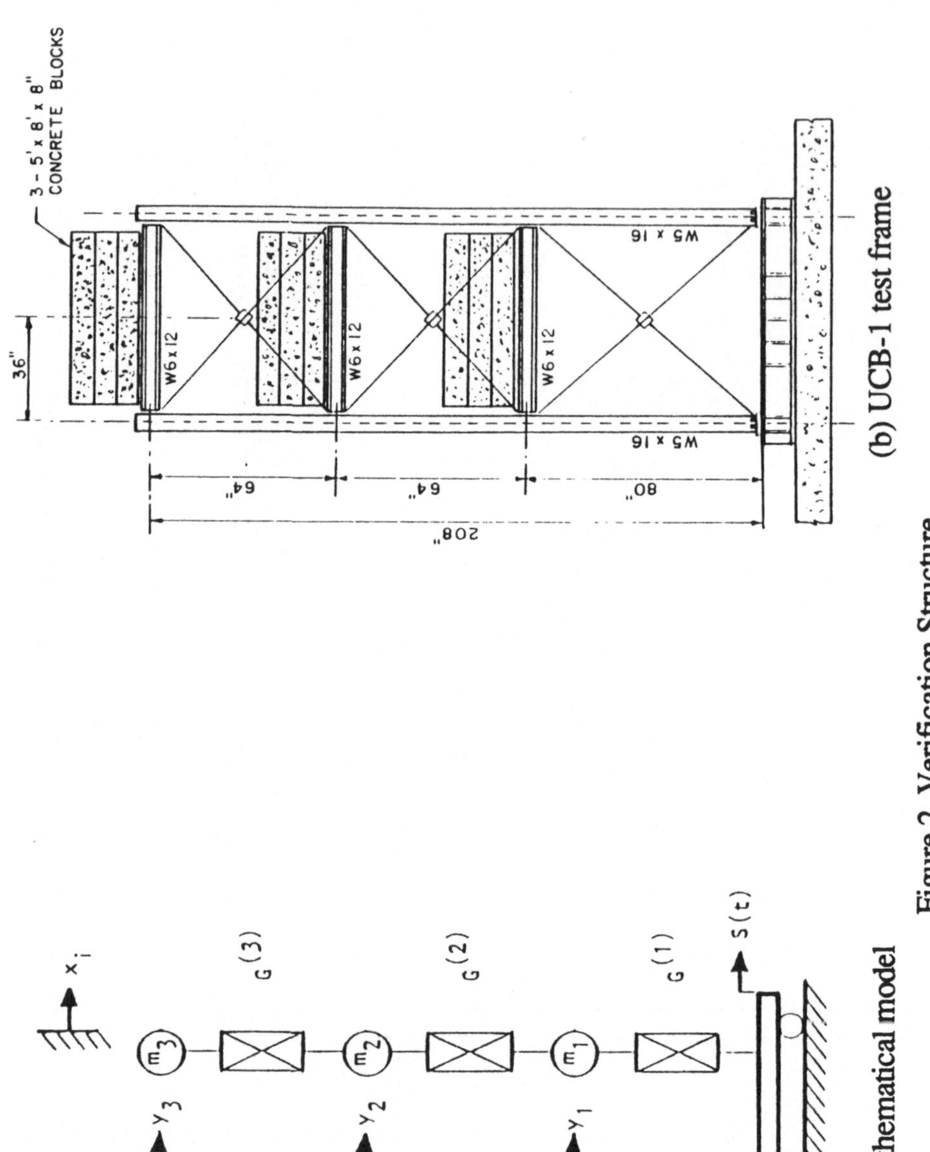

(a) Mathematical model

(b) UCB-1 test frame

Figure 2 Verification Structure

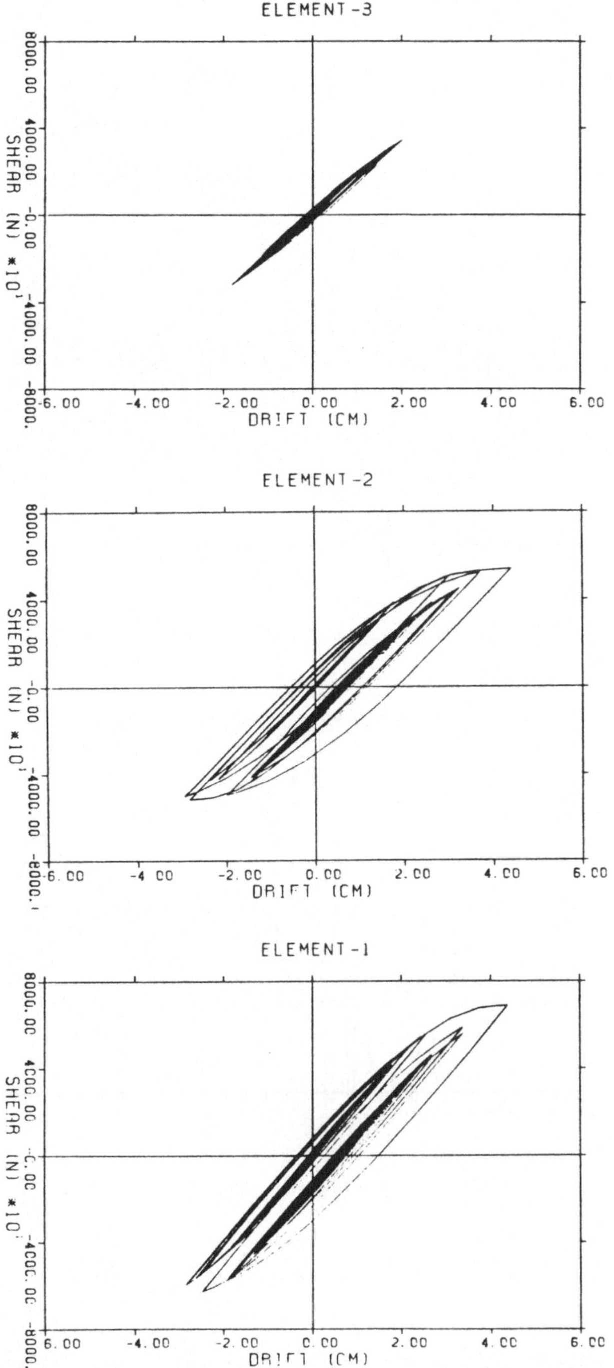

Figure 3 Interstory Restoring Force Diagrams (Taft Earthquake); Verification Structure

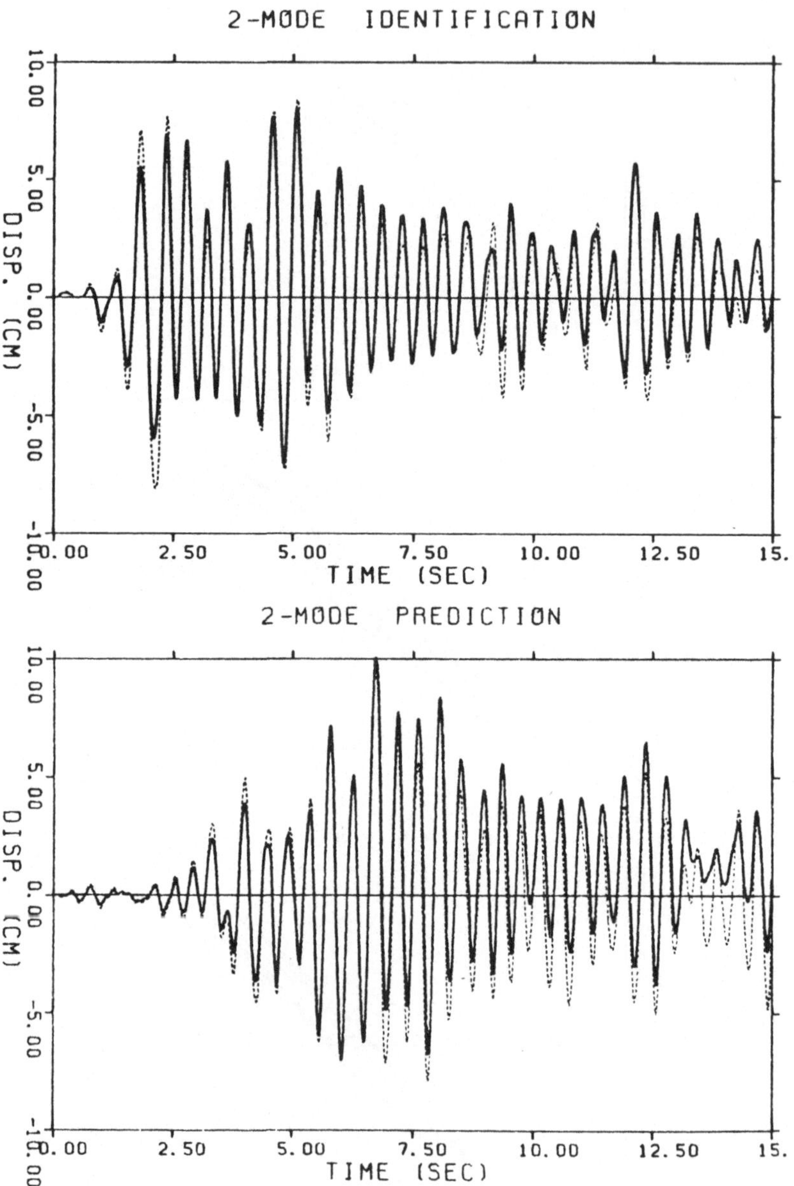

Figure 4 Identification and Prediction Using Nonparametric Model;
Verification Structure

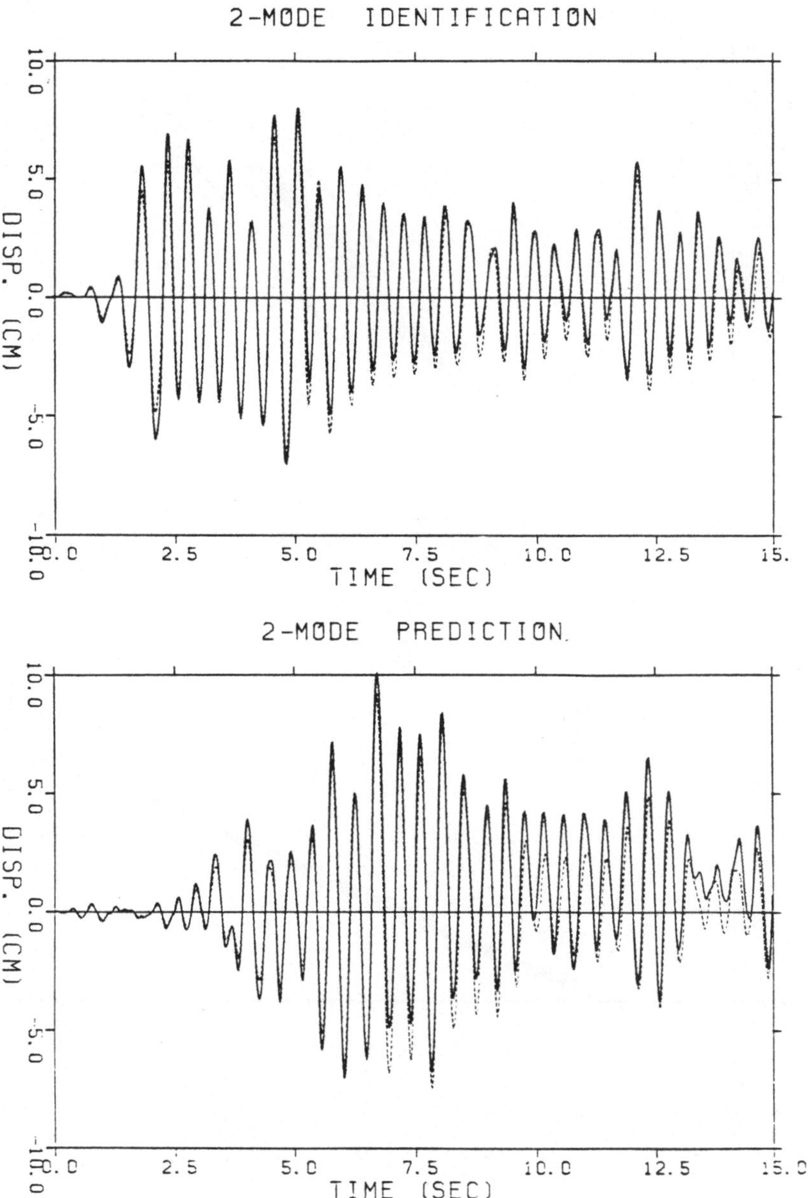

Figure 5 Identification and Prediction Using Distributed Element Model;
Verification Structure

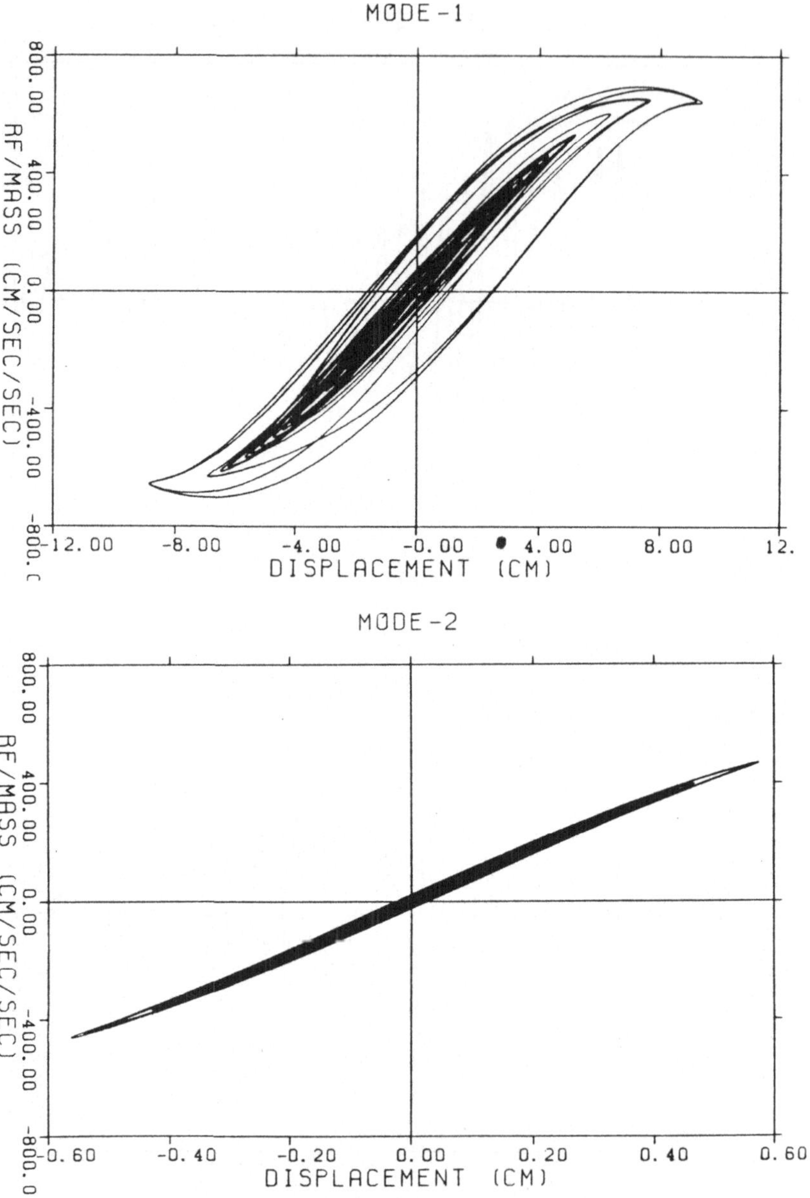

Figure 6 Generalized Modal Restoring Force Diagrams
Using Nonparametric Model; US-Japan Test Data

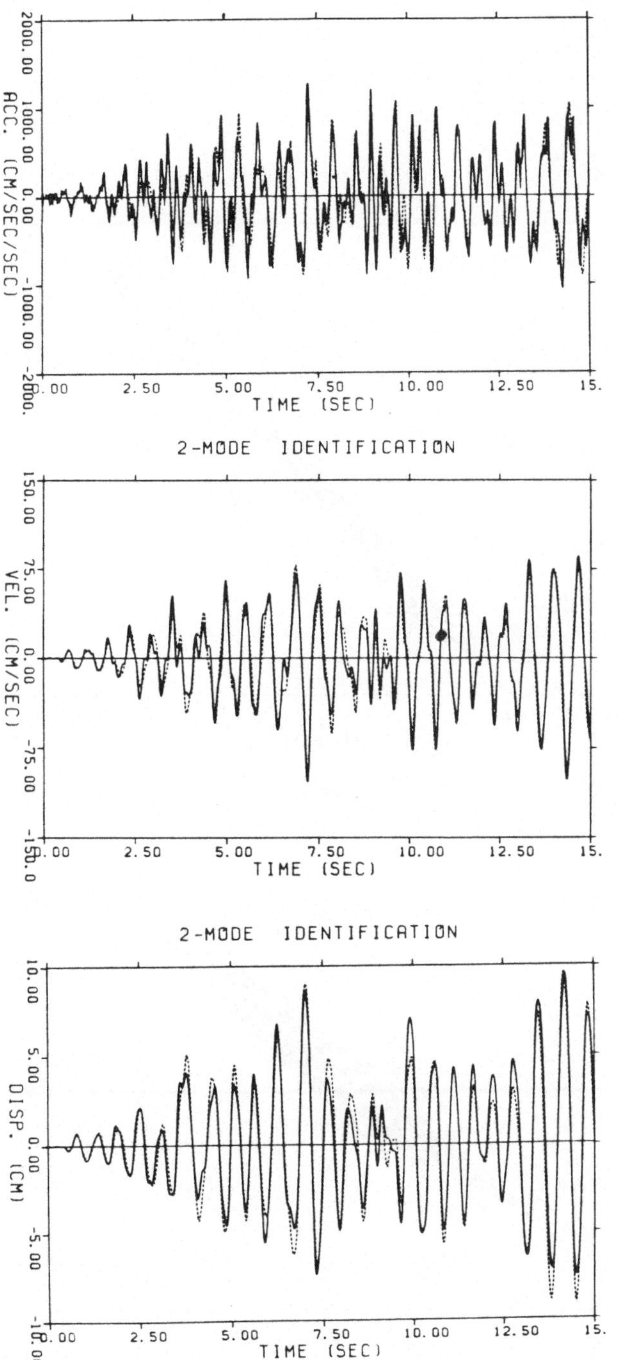

Figure 7 Identification Using Nonparametric Model;
US-Japan Test Data

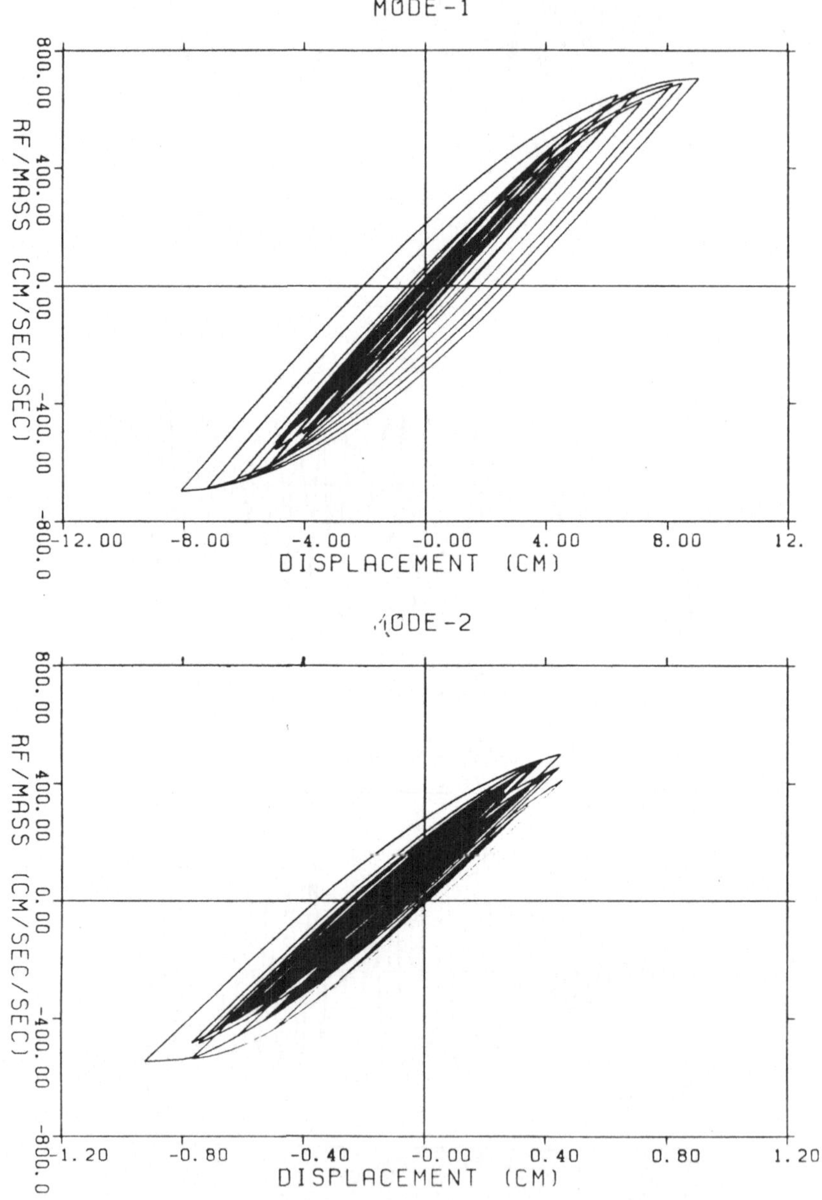

Figure 8 Generalized Modal Restoring Force Diagrams
Using Distributed Element Model; US-Japan Test Data

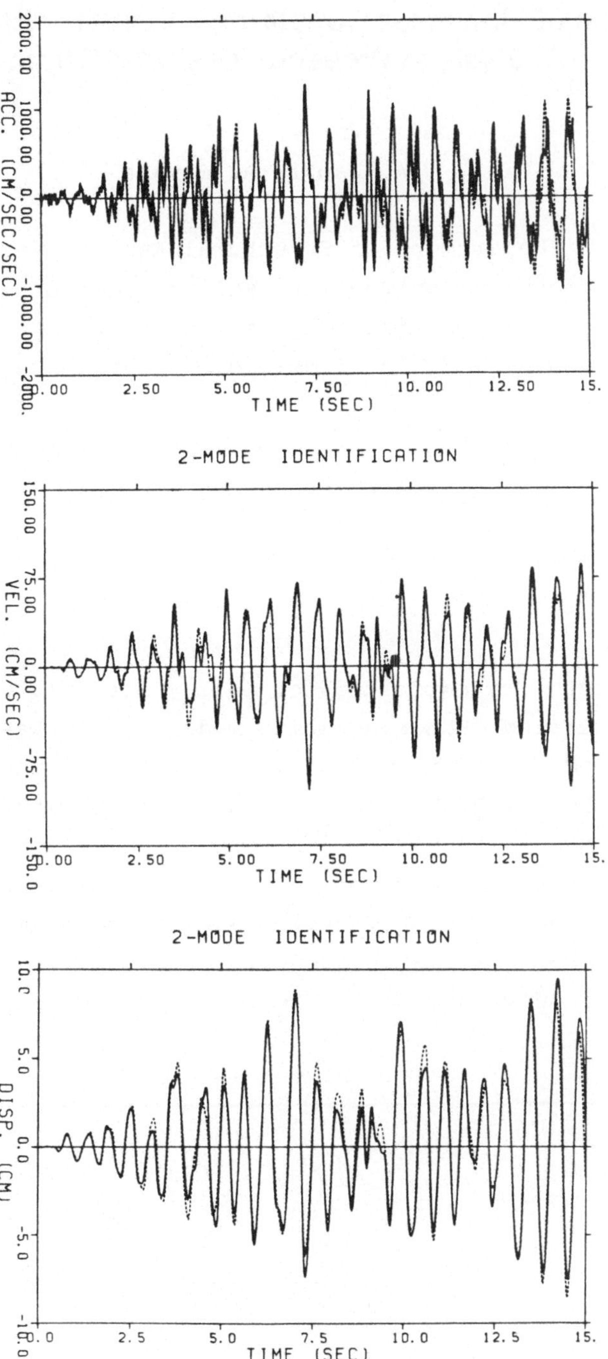

Figure 9 Identification Using Distributed Element Model;
US-Japan Test Data

Calibration of Nonlinear Constitutive Laws for Elastic-Plastic Analysis in Presence of Creep Strains

A. Nappi, A. Gavazzi

Department of Structural Engineering,
Technical University (Politecnico) of Milan
Piazza L. Da Vinci, 32
20133 Milano - Italy

<u>Abstract</u>

Parameters are estimated for Bailey-Norton's creep law and for a simplified ORNL plasticity model on the basis of Gauss-Newton's and complex methods. Creep and plastic strains as given by the calibrated models are compared with experimental data and good agreement is found (particularly for creep strains). Finally, a numerical analysis has been carried out by considering a simple system subject to plastic and creep strains. The relevant results are reported and show satisfactory agreement with experimental data.

Introduction

Parameter estimation techniques can advantageously lead to the optimal values of parameters which are part of constitutive models to be employed in structural analysis. A correct estimate of such parameters is clearly necessary in order to exploit mathematical modelling properly. In fact, discrepancy between experimental data and numerical computations is generally caused by poor knowledge of parameters rather than by solution algorithms.

Typical fields where system identification can be used are found in geotechnics, since 'in situ' measurements tend to provide more reliable information than laboratory tests [1,2]. In structural engineering potential advantages are also to be expected, since non-inspectable (or hardly inspectable) failures can be detected by proper measures of the overall response of the system. For instance, the response of offshore platforms is affected by cracks, which modify stiffness and even mass (when water is allowed to fill structural elements). Further applications in structural engineering may consist in detecting non-homogeneous zones, plastic strain distributions and other peculiarities inside a continuum medium by means of external measurements (along the surface).

It may be noted that system identification has been widely used and extensively studied in structural dynamics [3-8], while research activities concerned with statics and relevant applications are rather lacking. However, as happens, for instance, in geotechnics, there is a limited but important class of quasi-static problems where parameter estimation coupled with structural analysis provides significant advantages. Apparently, this is the case of the structural identification problem which will be discussed in the present contribution.

An elastic-viscoplastic constitutive model for the uniaxial behaviour of stainless steel at high temperatures has been calibrated on the basis of experimental data concerning plastic and creep deformations, separately at various temperatures. The parameters to identify are those which characterize Bailey-Norton's creep law and a plastic model allowing for a combination of kinematic and isotropic hardening. In order to minimize the overall discrepancy between measured and computed material data, a Gauss-Newton approach and a direct search algorithm have been employed: the former for Bailey-Norton's parameters, the latter for the plastic model. Then, the calibrated models have been used for the analysis of a simple structural system subjected to cyclic thermal loading. Such analysis has been carried out in the framework of EEC-sponsored benchmark calculations for high temperature of nuclear components in the creep range [9].

Gauss-Newton method

Let $\underline{D}(\underline{p})$ be the q-vector of discrepancies between measured and computed quantities for a given vector $\underline{p}$, which collects p independent parameters. If a linear Taylor expansion is taken around an initial vector $\underline{p}_0$, we obtain:

$$\hat{\underline{D}}(\underline{p}) = \underline{D}(p_0) + \frac{\partial \underline{D}}{\partial \underline{p}^T} (\underline{p} - \underline{p}_0) \tag{1}$$

Then we define an objective function

$$f(\underline{p}) = \tfrac{1}{2} \hat{\underline{D}}^T(\underline{p}) \, \hat{\underline{D}}(\underline{p}) \tag{2}$$

which is to be minimized with respect to $\underline{p}$ in order to obtain those parameters which make the model response as close as possible to the true response. By substituting Eq. (1) into Eq. (2) and by taking the first derivative with respect to $\underline{p}$, the following expression is obtained:

$$\frac{\partial f}{\partial \underline{p}^T} = \underline{D}^T(\underline{p}_0) \, \underline{S} + (\underline{p}^T - \underline{p}_0^T) \, \underline{S}^T \, \underline{S} \tag{3}$$

where $\underline{S} = \partial \underline{D}(\underline{p})/\partial \underline{p}^T$ is the Jacobian matrix of $\underline{D}(\underline{p})$ computed at $\underline{p} = \underline{p}_o$. By setting $\Delta \underline{p} = \underline{p} - \underline{p}_o$, it is easy to solve the linear system of p equations

$$\left[\underline{S}^T \underline{S}\right] \{\Delta \underline{p}\} = \{- \underline{S}^T \underline{D}(\underline{p}_o)\} \tag{4}$$

and compute the new estimate

$$\hat{\underline{p}} = \underline{p}_o + \Delta \underline{p} \tag{5}$$

When the model is nonlinear, iterations are required in order to find the optimal value for $\underline{p}$: $\underline{p}_o$ is set equal to $\hat{\underline{p}}$ and system (3) is solved again until $\Delta \underline{p}$ does not exceed a given tolerance. As usual, a possible criterion to establish when the iterative process should stop may consist in the constraint

$$\frac{\|\Delta \underline{p}\|}{\|\underline{p}_o\|} \leq \eta \tag{6}$$

where η is a convenient upper limit for the ratio between the norms of $\Delta \underline{p}$ and $\underline{p}_o$.

It is of some interest to notice that a Taylor expansion of $\underline{D}(\underline{p})$ around $\underline{p}_o$ including terms of the second order, yields to a linear system such as (4). However, the matrix of coefficients becomes the matrix typical of Newton's method and has the following pattern:

$$\underline{M} = \underline{S}^T \underline{S} + \sum_{1}^{q} D_i(\underline{p}) \underline{D}_i''(\underline{p}) = \underline{S}^T \underline{S} + \underline{S}^*(\underline{p}) \tag{7}$$

rather than $\left[\underline{S}^T \underline{S}\right]$. In Eq. (7) $D_i(\underline{p})$ and $\underline{D}_i''(\underline{p})$ represent the i-th entry of $\underline{D}(\underline{p})$ and the Hessian matrix of $D_i(\underline{p})$, respectively, computed at $\underline{p} = \underline{p}_o$. Simple matrix manipulations show that $\underline{M}$ is the Hessian matrix of the function $\{\frac{1}{2} \underline{D}^T(\underline{p}) \underline{D}(\underline{p})\}$, while $[\underline{S}^T\underline{S}]$ is the Hessian matrix as obtained after a linear expansion.

On the basis of Eq. (7), some comments are now possible about local convergence for Gauss-Newton method [10]. In fact, when $\underline{S}^*(\underline{p}^*) = \underline{0}$ at the optimal point $\underline{p}^*$, we can find the same convergence properties of Newton's method, since Newton's and Gauss-Newton's method coincide. On the other hand, when $\underline{S}^*(\underline{p}^*) \neq \underline{0}$ and $\underline{M}$ is positive definite, local convergence takes place only when residuals are 'small enough or linear enough' [10] and the first derivatives of $\underline{D}(\underline{p})$ are Lipschitz continuous. This condition is met for a function $\phi(\cdot)$ at a point P, if three arbitrary constants A, C, α exist such that [11]

$$|\phi(Q) - \phi(P)| \leqq A\, r^\alpha \tag{8a}$$

$$0 < \alpha \leqq 1 \tag{8b}$$

for any point Q such that

$$r = \overline{QP} \leqq C \tag{8c}$$

Application of Gauss-Newton method to the Bailey-Norton model

When creep of metal materials takes place, a convenient constitutive law which gives inelastic strains associated to creep as functions of stress, temperature and time is the one proposed by Bailey and Norton:

$$\varepsilon(\sigma,\ T,\ t) = B\, \sigma^n\, t^m \tag{9}$$

where B, n, m are parameters which depend upon temperature. Hence, when experimental material data at different stress levels are available for a given temperature, the discrepancy between computed and measured values becomes:

$$\varepsilon_{jk}^C - \varepsilon_{jk}^M = B\, \sigma_j^n\, t_k^m - \varepsilon_{jk}^M \tag{10}$$

Terms such as $(\varepsilon_{jk}^{C} - \varepsilon_{jk}^{M})$ can be collected in the q-vector $\underline{D}(\underline{p})$ defined in the previous Section. In this case, q denotes the total pairs of stresses σ_j and times t_k, at which strains are measured.

The sensitivity matrix $\underline{S}$ in Eq. (3) is easily computed, since derivatives of each term of $\underline{D}(\underline{p})$ with respect to $\underline{p}$ read

$$\sigma_j^n \, t_k^m \, , \quad B \, \sigma_j^n \, \lg \sigma_j \, t_k^m \, , \quad B \, \sigma_j^n \, t_k^m \, \lg t_k \qquad \text{(11a,b,c)}$$

as $\underline{p}^T = [B \quad n \quad m]$.

Numerical tests have been carried out by simulating experiments where measured values are the strains given by a viscous law, which is able to reproduce experimental data concerning 316 stainless steel. Namely, the following constitutive law has been considered:

$$\varepsilon(\sigma, T, t) = A \, e^{-\frac{P}{T+273}} \, \sigma^4 (1 - e^{-\alpha t^\beta}) +$$
$$+ \, t \, \dot{\varepsilon}^S(\sigma, T) \qquad \text{(12a)}$$

where

$$\dot{\varepsilon}^S(\sigma, T) = e^{(-\gamma - \frac{Q}{T+273})} \, \sigma^4 \qquad \text{(12b)}$$

$$\text{if } \sigma \leq \hat{\sigma}(T) = e^{-\frac{1}{\delta(T+273)}(\xi - \frac{R}{T+273})} \qquad \text{(12c)}$$

$$\dot{\varepsilon}^S(\sigma, T) = e^{(-\nu - \frac{S}{T+273})} \, \sigma^{[4+\delta(T-366)]} \qquad \text{(12d)}$$
$$\text{if } \sigma \geq \hat{\sigma}(T)$$

The above constants have been given the following values:

$A = 2.202 \ 10^2$; $P = 2.709 \ 10^4$; $Q = 1.97 \ 10^4$; $S = 1.43 \ 10^5$; $R = 15.4 \ 10^4$; $\delta = 0.05$; $\xi = 238.93$; $\nu = 251.1$. A set of measured strains has been determined by considering 8 stress levels (10, 46, 82, ..., 190 MPa) and 36 times (1, 25, 49, ..., 865 hours) at temperatures $T = 590°C$ and $593°C$. Gauss-Newton's method has been applied by assuming $B_o = 10 \cdot 10^{-13}$, $n_o = 5$, $m_o = 0.78$ as initial estimates. A tolerance $\eta = 10^{-10}$ as defined in the previous Section has been set. In order to avoid numerical problems due to figures of different magnitude, measured strains have been multiplied by 10^{13}, so that the initial and final values of B during the estimation process was equal to the actual current parameter multiplied by 10^{13}.

The logarithm of the error function for $T = 593°C$ is reported in Fig. 1 vs. the number of iterations. Optimal values are obtained quite soon. In fact the final estimates are found after a few steps, while most iterations are required only to satisfy the strict tolerance $\eta = 10^{-10}$. However, the objective function is not monotonically decreasing (as evidenced by the logarithmic scale of Fig. 1). Faster convergence could be obtained, e.g., by a modified Newton method where convenient search lines and step-sizes are determined at each iteration [12]. Details of the estimation procedure are given in Fig. 2, where parameters are reported as functions of the iteration steps. As pointed out before, few iterations are required for a correct estimate of the parameters and no significant change occurs after the dashed line in Fig. 2. The performance of the method can be appreciated by observing Fig. 3. Here creep strains given by the calibrated Bailey-Norton law are compared with the ones provided by Eqs. (12). Similar plots can be drawn for different stress levels. Therefore, it appears that the law discussed in this Section and governed by three parameters is capable to represent (with good accuracy) the creep behaviour described by a more complicated law.

A further test has been carried out by adopting a two-step identification procedure. First, the parameters have been estimated by considering a subset of experimental material

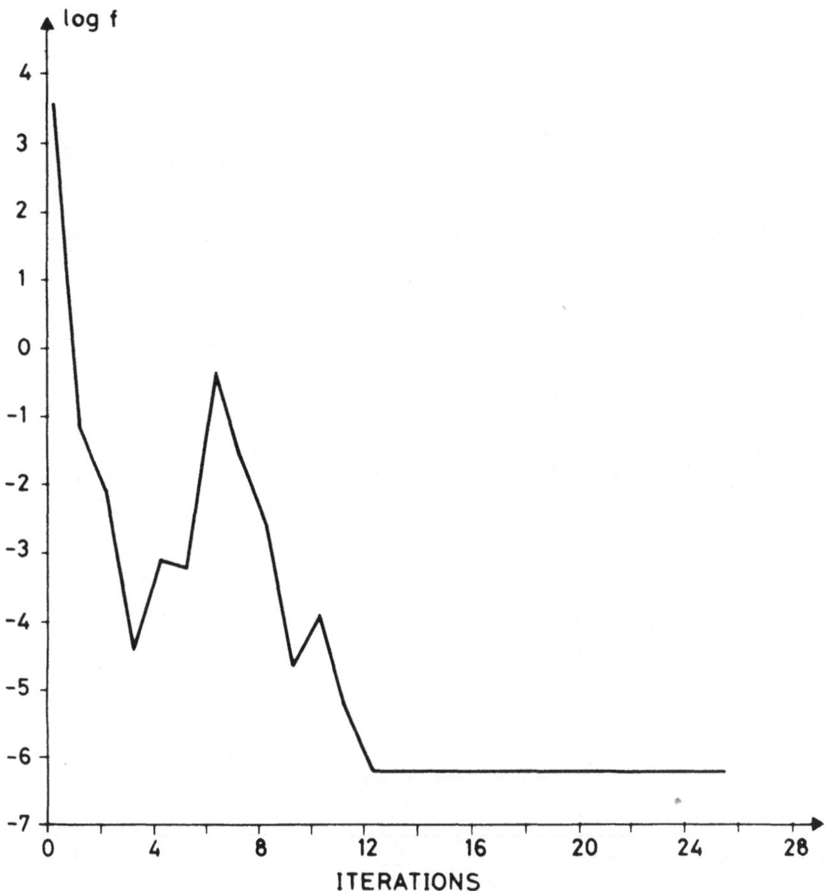

Fig. 1 - Logarithm of error function f vs. number of iterations at T=593°C for the estimation of parameters concerning Bailey-Norton's law

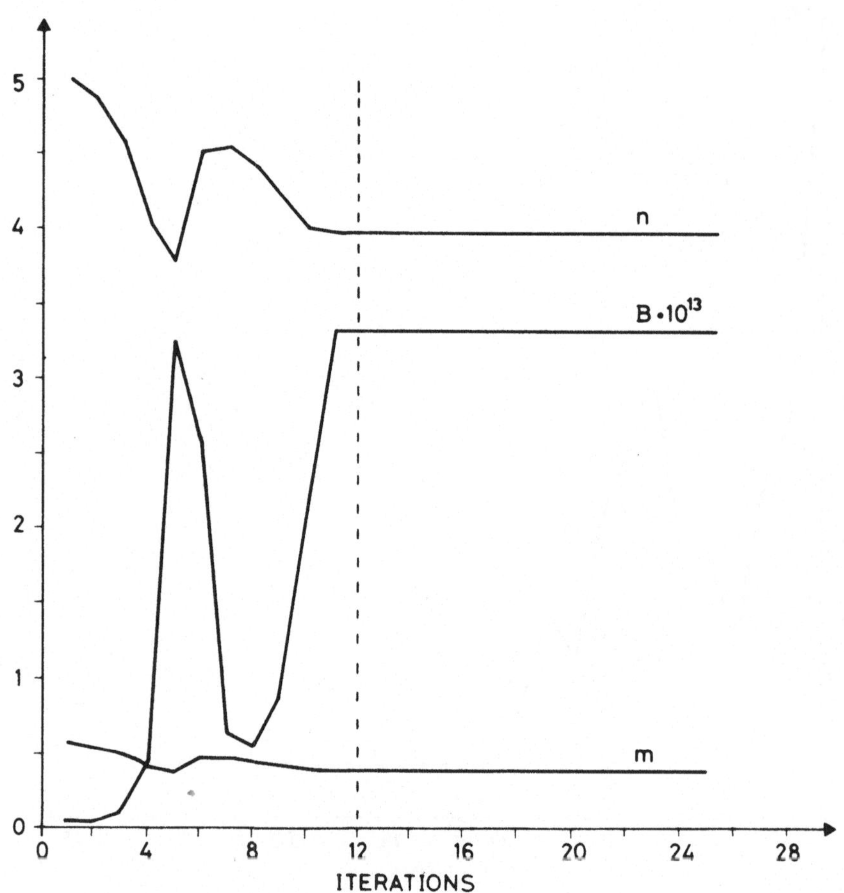

Fig. 2 - Parameters of Bailey-Norton's law vs. number of iterations at T=593°C

data. Then, m has been held constant (m = 0.46), while B and n have been estimated on the basis of the whole set of measurements. This way, it has been possible to represent both the identification path over the plane B-n and isovalue contours for m = 0.46 (Fig. 4). Clearly, the results obtained by means of this procedure are expected to be less accurate than before. This fact can be easily checked by comparing plots similar to the one reported in Fig. 3. Here this comparison is skipped for sake of brevity, but less agreement with experimental data is found when the new parameters are used. In addition, the optimal value of the objective function at the end of the two-step procedure (Fig. 4) is about twice than before.

The same parameters have been estimated also on the basis of experimental data at T = 500°C. The performance of the method has been again satisfactory. Starting from the same initial values as before, the following estimates have been found: B = 8.93 10^{-15}, m = 0.417, n = 4.

The Complex method

Optimization problems can be solved by means of a strategy which tends to decrease (or increase) an objective function on the basis of values attained at properly selected points. A technique which makes use of this strategy represents what is termed as a 'direct search' method. Probably, the Simplex method is the most famous technique of this kind and a possible improvement is offered by the Complex method.

If we consider the version presented by Ref. 13, it can be summarized as follows:

(a) A polihedron of 2n vertices is generated in the n-dimensional space where the objective function is defined (n being the number of independent variables). This operation is performed by computing the objective function at 2n points which represent the vertices. If inequality

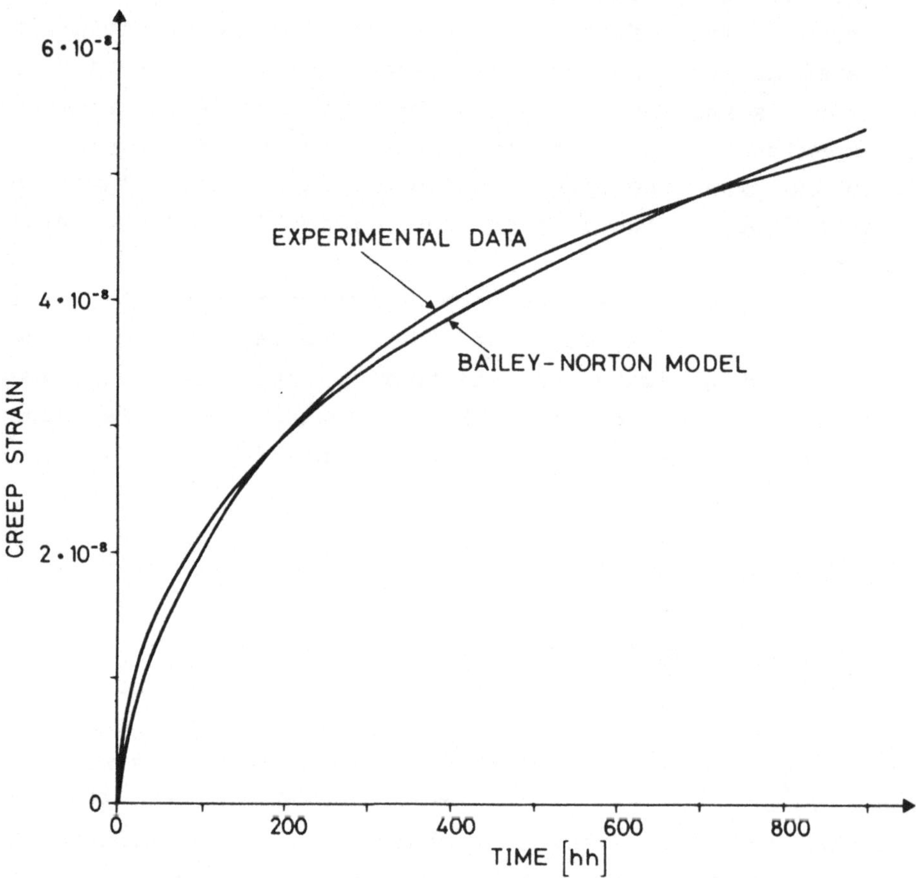

Fig. 3 - Comparison between measured creep strains and creep strains given by Bailey-Norton's law (T=593°C, σ=10 MPa)

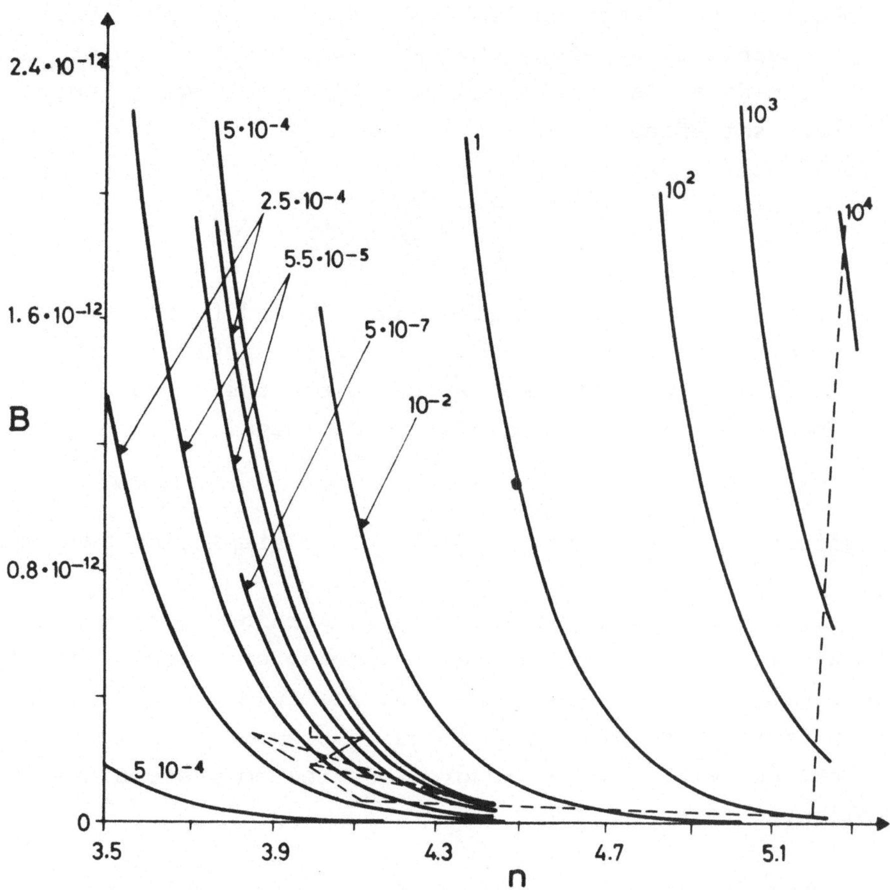

Fig. 4 - Isovalue contours for m=0.46 showing the pattern of the objective function (solid lines) and path followed for the estimation of parameters B and n at T=593°C (dashed line)

constraints and lower or upper bounds are given, the 2n selected points must be feasible.

(b) The vertex corresponding to the worst value of the objective function is substituted on the basis of a convenient strategy by a new feasible point and the procedure is continued until the following inequalities are satisfied:

$$|1 - f_b/f_w| < \eta \qquad , \qquad \sum_{1}^{n} {}_i \frac{1}{n} |x_b^i - x_w^i| < \eta \qquad (13a,b)$$

where η is a given tolerance, f_b and f_w are the best and the worst values of the objective function at the current vertices, while x_b^i and x_w^i represent the coordinates of the points at which f_b and f_w are computed.

Application of the Complex method to a simplified ORNL model

In order to account for plastic strains, a simplified ORNL (Oak Ridge National Laboratory) model utilized in the ABAQUS code [14] has been considered and the relevant parameters have been estimated by means of the Complex method. This model is concerned with single degree of freedom systems and has the following features:

(a) There exist an initial yield limit $\sigma_o^*(T)$ and a hardened yield limit $\sigma_{10}^*(T)$ which depend on temperature T. The latter corresponds to the yield limit at the tenth cycle.

(b) After yielding, linear kinematic hardening always takes place and is described by an internal variable α, which gives the centroid of the current yield points.

(c) The yield limit can be updated only once and becomes $\sigma_{10}^*(T)$ when $\alpha \, \dot{\varepsilon}^p < 0$ or $\varepsilon^H > 0.002$, where ε^H is defined as 'effective creep strain'. Namely, if ε_i^c represents the creep strain at the i-th cycle and $\varepsilon^+ = \min_i \{\varepsilon_i^C\}$, $\varepsilon^- = \max_i \{\varepsilon_i^C\}$, then $\varepsilon^H \equiv \varepsilon^C - \varepsilon^+$ for $\sigma > 0$ and $\varepsilon^H = \varepsilon^C - \varepsilon^-$ for $\sigma < 0$.

(d) As shown in Fig. 5, the model is piecewise-linear in plane $\sigma - \varepsilon$ and four parameters are needed to define it: the elastic limits σ_o^*, σ_{10}^* and the slopes E, E^T. Incidentally, it may be noted that the model implies a constant slope E^T (at the first and at the tenth cycle, both for positive and for negative stresses).

For identification purposes, the elastic modulus E has been assumed as known ($E = 0.1998 \cdot 10^6 - 0.7518 \cdot 10^2$ T, where E is expressed in MPa and T in centigrade degrees). Therefore only three parameters have been estimated. This task has been accomplished on the basis of measurements at three different temperatures. In order to calibrate the model, experimental data similar to the ones shown in Fig. 6 have been available. Namely, the strain ranges and the cycles given in Tab. 1 have been considered. An initial vertex has been supplied (see Tab. 2). The remaining five initial vertices have been determined by means of a random number generator which finds points between the upper and the lower bounds. The final estimates are shown in Tab. 2. Further information about the identification process is given in Figs. 7-9, which show the objective function and the parameters vs. the number of iterations. It should be noted that here by 'iteration' we mean the evaluation of a new vertex, which does not imply an improvement of the current best value of the objective function. This fact explains why the objective function often remains constant (Fig. 7), as well as the parameters corresponding to the current best point (Figs. 8,9). In Fig . 10 a typical comparison between an experimental cycle and the estimated model is reported.

The interested reader may find a detailed discussion of ORNL constitutive equations in Ref. 15.

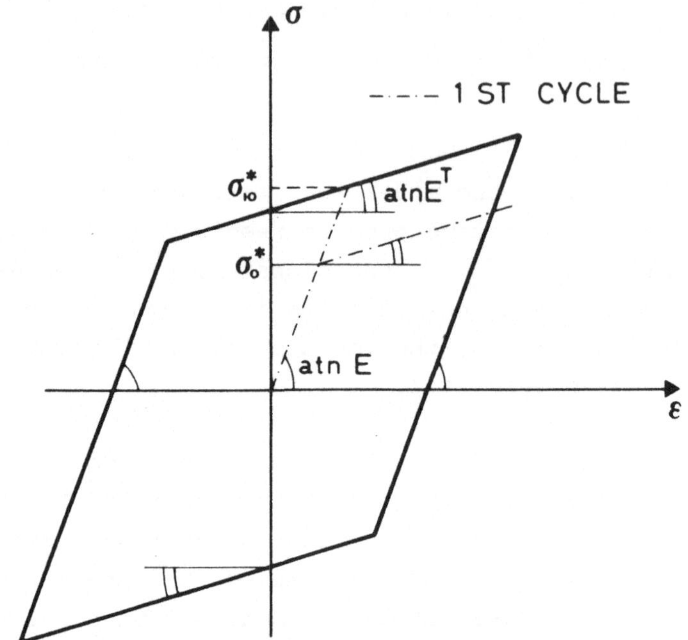

Fig. 5 - Sketch of the simplified ORNL model for plasticity

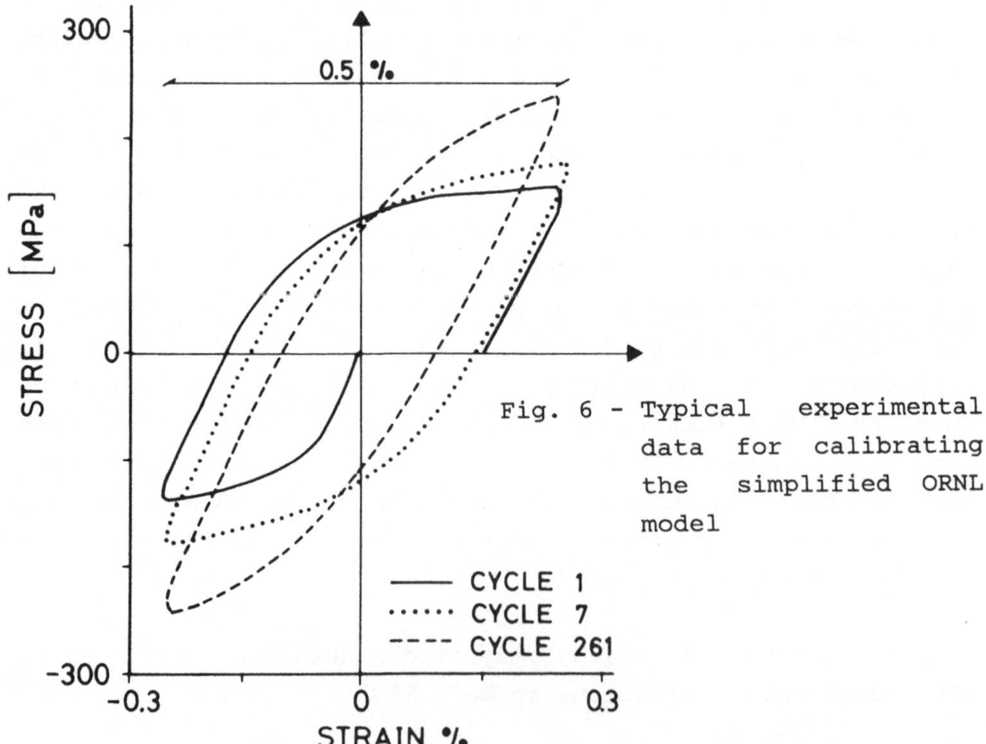

Fig. 6 - Typical experimental data for calibrating the simplified ORNL model

STRAIN RANGE	400°C	550°C	625°C
0.25 %			1,7,261
0.30 %		1,20	
0.50 %	1,13	1,20,300	10,120
0.75 %	12	12,577	12,485
1.50 %	1,8,48		

Tab. 1 - Experimental cycles at which measures have been used for identification

T [°C]	σ_o^* [MPa]	σ_{10}^* [MPa]	E^T [MPa]
400	119 (140)	145 (255)	28893 (30000)
550	103 (140)	204 (255)	25700 (30000)
625	144 (140)	218 (255)	27317 (30000)

Tab. 2 - Estimated parameters for simplified ORNL model and initial values (in brackets)

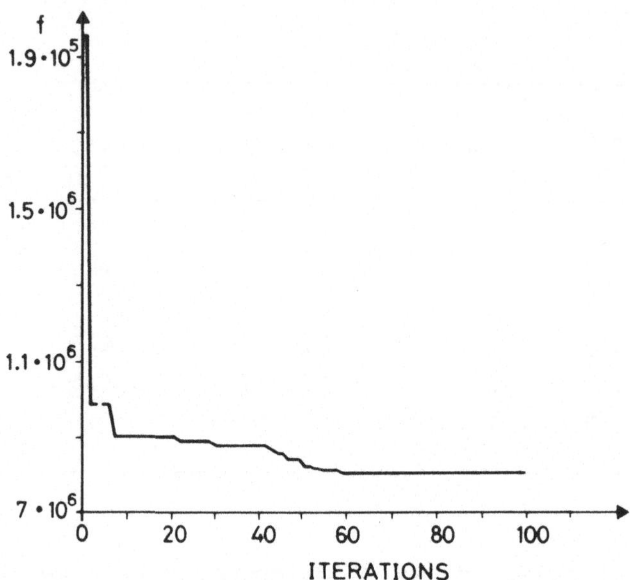

Fig. 7 - Objective function vs. number of iterations for the estimate of parameters concerning the simplified ORNL model

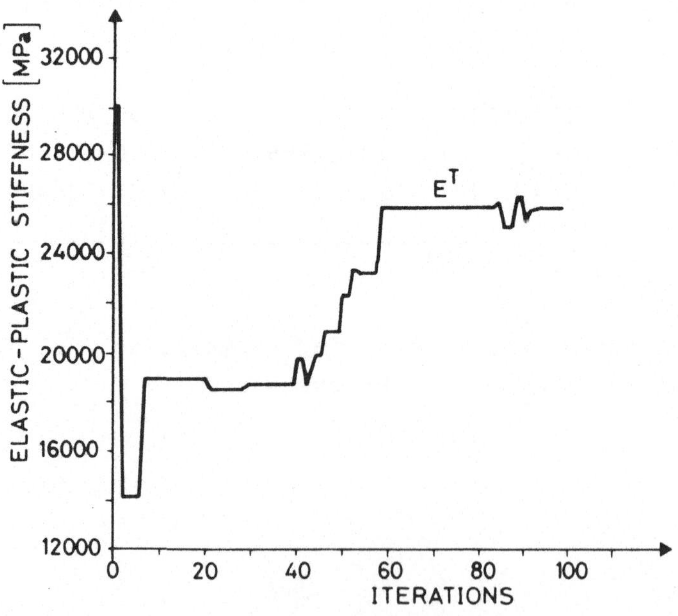

Fig. 8 - Parameter E^T vs. number of iterations

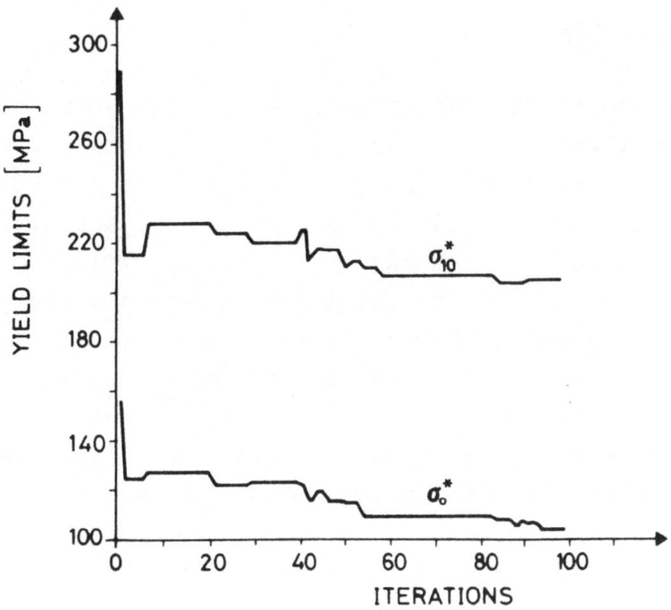

Fig. 9 - Parameters σ_0^* and σ_{10}^* vs. number of iterations

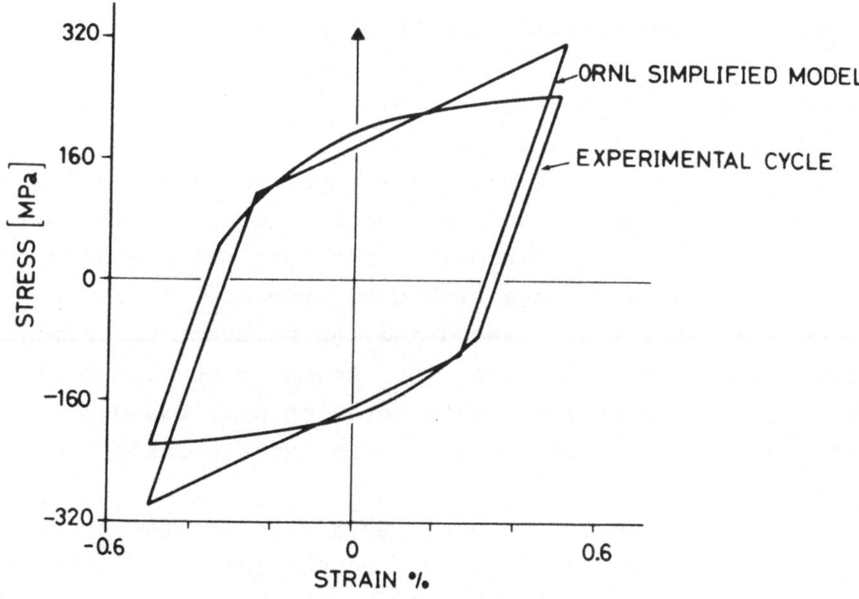

Fig. 10 - Comparison between experimental 10th cycle and
strains given by the calibrated simplified ORNL
model (T=593°C)

Analysis of a simple system

The parameters determined as shown in the above Sections have been utilized for the analysis of the simple system of Fig. 11, already discussed in Ref. 16. Two bars are subjected to equal strains and to loads which remain constant over given time intervals. During each day there are thermal cycles. The whole test is supposed to last 36 days and the loading conditions are given in Tab. 3. Results in terms of stresses are reported in Tab. 4 and compared with experimental data. It can be noted that both mean values and ranges are fairly accurate, which is a good outcome of this numerical experiment, since stress ranges are closely related to internal damage. The output of the analysis was also concerned with strains: computed strain ranges (generally about 0.3 %) turned out to be nearly identical to the experimental ones. However, significant discrepancies between measured and computed mean values were found, particularly during the second half of the test (last 18 days). This fact seems to be caused by the material data selected to estimate the parameters of the simplified ORNL model. In fact, these data were referred to several strain ranges (often much higher than the ones found experimentally _and_ given by the analysis). This is shown by Fig. 12, where all experimental cycles at 625°C considered for model calibration are reported. The same figure also shows the calibrated piecewise-linear model, which appears to be a good compromise, but does not reproduce the cycles at law strain range with good accuracy. Therefore, if more accurate results are needed and the relevant experimental material data are available, it seems to be reasonable to proceed with a further model calibration only based on those material data which concern the actual strain range (in this case about 0.3 %).

The above remark is supported by a parallel investigation due to P.S. White (private communication). In this case parameters have been determined on the basis material data concerning ± 0.2 % and ± 0.3 % strain ranges. Then the relevant analysis of the same system of Fig. 11 has led to accurate results, including strains very close to the experimental ones.

Closing remarks

The work discussed in the paper suggests the following comments:

a) The primary and secondary phase of creep are fairly well described by Bailey-Norton's law, which can be easily calibrated since only three parameters are needed at each temperature.

b) The direct search technique has turned out to be a convenient, suitable method for the estimation of the parameters which characterize the simplified ORNL model. Although the objective function is generally non-convex, satisfactory results have always been obtained and the calibrated model has always represented a good compromise with experimental cycles.

c) When constitutive models are used to study a real system, results of the analysis should be handled with care. Particular attention is to be paid to the actual strain range. A further analysis may be required in order to estimate parameters only on the basis of those experimental data which refer to the actual strain range.

d) Good estimates of strain ranges and stress ranges have been obtained also by means of the analysis based on a set of experimental material data, which refer to several strain ranges (selected in a random way and generally much different from the strain range of the system under test). The results reported in the previous Section appear to be particularly interesting and meaningful, since stress ranges (which are sufficiently accurate) are closely related to internal damage.

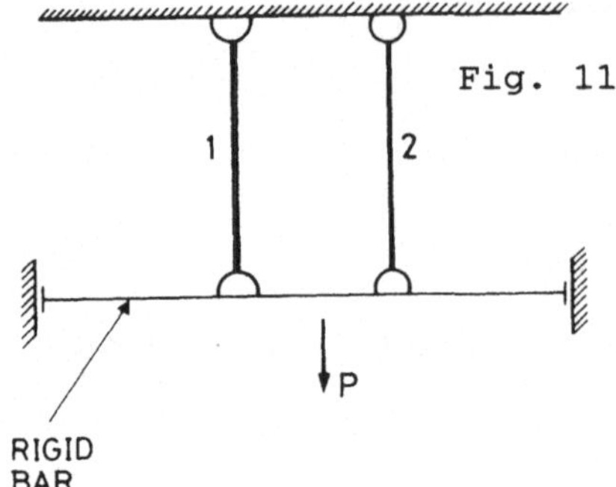

Fig. 11 - System for elastic plastic analysis in presence of creep strains. Cross sectional areas: 5.064 cm^2 for bar 1, 1.266 cm^2 for bar 2

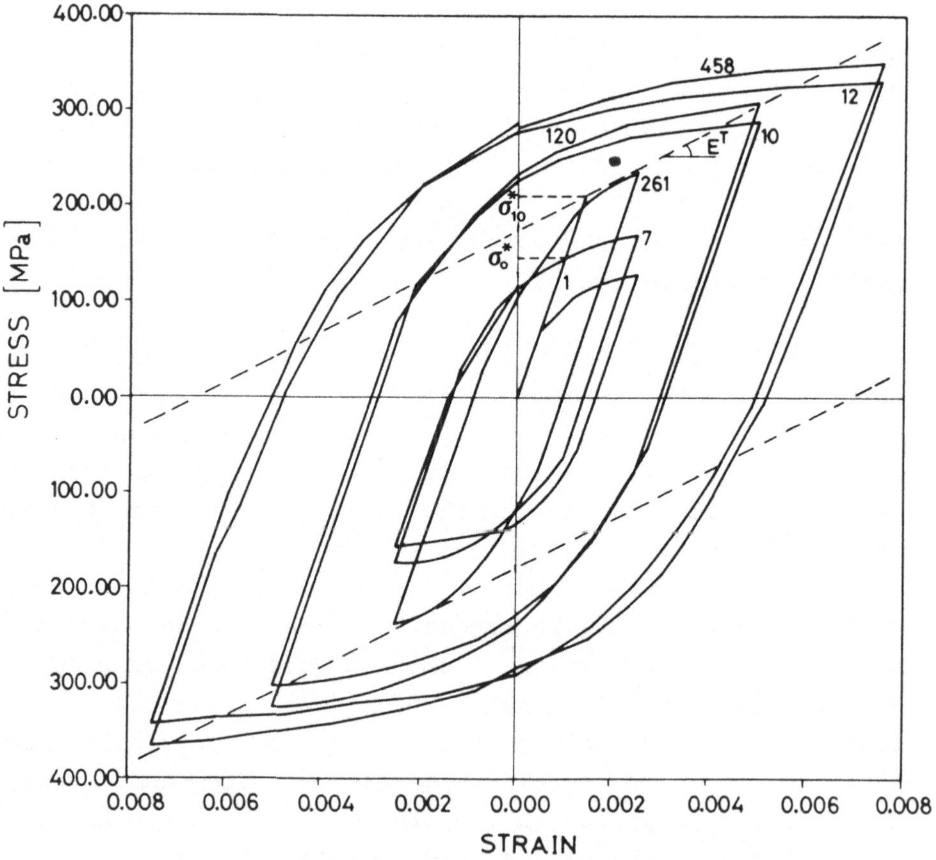

Fig. 12 - Experimental cycles accounted for at T=625°C (solid lines) and calibrated elastic-plastic branches after hardening (dashed lines)

DAY	P [MN]		MINUTE	TEMPERATURE [°C] BAR 1	BAR 2
1	0.02993		1 TO 350	600	300
2 TO 7	0.03525		360 TO 410	500	200
8 TO 15	0.04054		420 TO 1370	600	300
16 TO 26	0.05080		1380 TO 1430	500	200
27 TO 36	0.05891		1440	600	300

Tab. 3 - Loads applied to the system of Fig. 11 during 36 days
(left side) and thermal cycles (right side). Thermal
cycles are periodically repeated (one per day)

STRESSES IN BAR 1   [MPa]

DAY	EXPERIMENTAL RESULTS			RESULTS BASED ON ESTIMATED PARAMETERS		
	MIN	MAX	RANGE	MIN	MAX	RANGE
6	22	102	80	-2	94	96
12	30	112	82	6	102	96
18	48	136	88	26	120	94
24	46	138	92	25	124	99
30	60	150	90	36	136	100
36	60	152	92	35	136	101

Tab. 4 - Stresses in bar 1 at intervals of six days

Acknowledgements

The first author is indebted to Prof. H.G. Natke for his hospitality in Germany and for covering travel expenses from Italy, when the paper was presented at the 'Workshop on structural safety evaluation based on system identification approaches' held at Lambrecht, 29 June - 1 July, 1987.

Both authors are grateful to Prof. G. Maier for his suggestions and to ENEA for the possibility of utilizing information gathered in the context of a subcontract with the Polytechnic of Milan.

A grant from CNR is also acknowledged.

References

[1] Isenberg, J., Collins, J.D., Kavarna, J., "Statistical estimations of geotechnical material model parameters from in situ test data", Proc. ASCE Spec. Conf. on Probabilistic Mechanics and Structural Reliability, Tucson, 1979, 348-352.

[2] Gioda, G., Maier, G., "Direct search solution of an inverse problem in elastoplasticity: Identification of cohesion friction angle and 'in situ' stress by pressure tunnel tests", Int. J. Num. Meth. Eng., 15, 1980, 1823-1848.

[3] Ibañez, P., "Identification of dynamic parameters of linear and nonlinear structural Models from Experimental Data", Nucl. Eng. Design, 1972, 25-30.

[4] Beliveau, J.G., "Identification of viscous damping in structures from modal information", J. Appl. Mech. 98, 2, 1976, 335-339.

[5] Hart, G.C., Torkamani, M.A.M., "Structural system identification", in Stochastic problems in mechanics, Eds. Ariaratuam S.T., Leipholz M.M.E., Univ of Waterloo Canada, 1977, 207-228.

[6] Hart, G.C., Yao, J.T.P., "System identification in structural dynamics", J. Eng. Mech. Div., Proc. ASCE, 103, 6, 1977, 1089-1104.

[7] Natke, H.G., "Die Korrectur des Rechnenmodelles eines Elastomechanischen Systems mittels gemessener erzungener Schwingungen", Ing. Arch., 46, 1977, 169.

[8] Yun, C.B., Shinozuka, M., "Identification of non-linear structural dynamic systems", J. Struct. Mech., 8, 2, 1980, 187-203.

[9] Tonarelli, F., Corsi, F., "Benchmark calculation programme - Step 2, Phase 4 - Leicester 2 bar test - Final report" (CEE Study Contract)

[10] Dennis, J.E., "A user's guide to nonlinear optimization algorithms", Proc. of IEEE, Vol.72, N.12, 1984, 1765-1776.

[11] Smirnov, V.I., "A course of higher mathematics", Vol.3, Pergamon Press, Oxford, 1964.

[12] Dempster, M.A.H., "Elements of optimization", Chapman & Hall, London, 1975

[13] Box, M.J., "A new method of constraint optimization and comparison with other methods", Computer Journal, 1965, 8, 42

[14] Hibbit, Karlsson, Sorensen, Inc., Abaqus theory manual, Version 4.5, Providence, Rhode Island, 1984

[15] White, P.S., "An account of the ORNL constitutive equations", GEC Internal Report, Mechanical Engineering Laboratory, Whetstone, Leicester, UK

[16] Megahed, M., Ponter, A.R.S., Morrison, C.J., "An experimental and theoretical investigation into the creep properties of a simple structure of 316 stainless steel", Int. J. Mech. Sci., Vol. 26, 1984, 149-164

Inelastic Modeling and System Identification

Y. K. Wen[*] and A. H-S. Ang[*]

Abstract

Structures under severe loads often go to inelastic range and show nonlinear behavior. Proper modeling of restoring force such that the physical properties can be realistically represented is essential in study of performance and safety of structures under natural or man-made hazards. Also, the estimation of the inelastic structural system parameters and their uncertainties should be based on experimental or field evidence of actual structures, therefore requires methods of system identification and regression analysis. Recent developments in the above areas are summarized with emphasis on restoring force hysteresis, degradation, biaxial interaction and capacity against damage. A simple time domain least-square technique for practical evaluation of restoring force parameters is also presented. Numerical examples are given and applications to damage prediction and damage limiting design are mentioned.

1   Introduction

Structures generally become nonlinear and inelastic before damage or collapse occurs. Proper understanding and analytical modeling of nonlinear structural behavior in the inelastic range is therefore essential in structural safety evaluation. When structures go into inelastic range, the restoring force becomes hereditary, i.e., displacement time history dependent; it may deteriorate in stiffness, or strength, or both, causing progressive type of failure. Also, biaxial interaction may become important; namely, the loading in one direction may affect the stiffness and load carrying capacity in the perpendicular direction. Traditionally, these behaviors are described by empirical hysteresis rules which allow one to model approximately the restoring force in a step-by-step time history response analysis. However, these rules are difficult to express in any mathematically tractable form such that analytical solution of the response may be obtained. An entirely theoretical approach, say based on

---

[*]Professor of Civil Engineering, University of Illinois at Urbana-Champaign, Urbana, Illinois, USA.

plasticity theory, on the other, would be analytically and computationally impractical for study of structural systems. Recently a restoring force model based on nonlinear differential equations has been developed which seem to be able to capture most of the important characteristics of inelastic systems, such as hereditary behavior, deterioration, and biaxial interaction. Although these models are phenomenological in nature, they correspond closely to the rate-type constitutive equations in plasticity, therefore do have a sound theoretical basis. A summary of the development of this model and the identification of the system parameters from data are given in the following. The application of this method to damage and performance evaluation of structural and geotechnical systems under seismic excitation are also mentioned.

2   Modeling of Restoring Force

The hereditary behavior of a hysteretic restoring force indicates that the force-displacement relationship can no longer be algebraic, i.e., restoring force cannot be expressed in terms of only the instantaneous displacement and velocity. Consider first a single-degree-of-freedom system under uniaxial load, one can add a hysteretic part and model the restoring force as

$$Q(x,\dot{x},t) = g(x,\dot{x}) + h(x) \tag{1}$$

in which g = a nonhysteretic component, an algebraic function of the instantaneous x and $\dot{x}$. h = a hysteretic component, a function of the time history of x. As an example, the restoring force of a nearly elasto-plastic system may be modeled by

$$Q(x,t) = \alpha ku + (1 - \alpha)kz \tag{2}$$

in which k = the pre-yielding stiffness; α = ratio of post-yielding stiffness to pre-yielding stiffness; and $(1 - \alpha)kz$ = the hysteretic part of the restoring force in which z is described by the following nonlinear differential equation (1)

$$\dot{z} = \frac{1}{\eta}\left[A\dot{x} - \nu\left(\beta|\dot{x}||z|^{n-1}z - \gamma\dot{x}|z|^{n}\right)\right] \tag{3}$$

with the force parameters A, β, γ, η, ν, and n governing the amplitude, shape of the hysteretic loop, and the smoothness of transition from elastic to inelastic ranges. Proper choices of parameters give various softening as well as hardening systems. The absolute value signs play the role of specifying different force-displacement relationships during loading and unloading, analogous to the constitutive equations in plasticity theory. However, the relationship here is given in a mathematically convenient form.

Degradation of the restoring force can be included by prescribing the parameters to be functions of the severity of the response, such as amplitude of the response and total hysteretic energy dissipation. The hysteretic energy dissipation which is a measure of the cumulative effect of severe response and repeated oscillations, is given by (1),

$$\varepsilon_T(t) = (1 - \alpha)k \int_0^t z(\tau)\dot{x}(\tau)d\tau \tag{4}$$

For example, both stiffness and strength degradation can be introduced by prescribing A as a decreasing function of ε_T; i.e.,

$$A(t) = A_o - \delta_A \varepsilon_T(t) \tag{5}$$

in which δ_A is the deterioration rate. Similarly, strength degradation can be introduced by

$$\nu(t) = \nu_o + \delta_\nu \varepsilon_T(t) \tag{6}$$

and stiffness degradation by

$$\eta(t) = \eta_o + \delta_\eta \varepsilon_T(t) \tag{7}$$

The restoring forces of a large number of structural and geotechnical systems can be modeled closely with proper choices of the parameters. A response amplitude dependant deterioration model is given by (2),

$$\eta = A \frac{\left(u_{P_i} - u_{P_{i-1}}\right)}{\left(z_{P_i} - z_{P_{i-1}}\right)} \tag{8}$$

in which u_{P_i} and z_{P_i} are the displacement and hysteresis amplitude in the i-th half cycle. It reproduces well reinforced concrete member and system behavior under cyclic loads. The above model has been extended to include the pinching of the hysteresis loops as well (3,8).

For two-dimensional structures under biaxial excitations, the interaction of the restoring forces in the two directions may significantly alter the response behavior. The above restoring force has been extended to include such interaction by requiring that the hysteretic components in the two directions z_x and z_y satisfy the following coupled differential equations (4)

$$\dot{z}_x = A\dot{u}_x - \beta|\dot{u}_x z_x|z_x - \gamma\dot{u}_x z_x^2 - \beta|\dot{u}_y z_y|z_x - \gamma\dot{u}_y z_x z_y$$

$$\dot{z}_y = A\dot{u}_y - \beta|\dot{u}_y z_y|z_y - \gamma\dot{u}_y z_y^2 - \beta|\dot{u}_x z_x|z_y - \gamma\dot{u}_x z_x z_y \tag{9}$$

in which u_x and u_y are, respectively, the displacement in the x and y direction. Note that Eq. 9 gives an isotropic system, for an orthotropic system whose stiffness and strength in the two orthogonal directions are different, one can introduce a simple transformation (scaling) of response variables and still use the same equations (4). As in the uniaxial model, deterioration can be introduced by letting parameters A, β, and γ be function of time depending on the severity of the response; e.g., maximum response amplitude or hysteretic energy dissipation

$$\varepsilon(t) = \int_0^t [z_x(\tau)\dot{u}_x(\tau) + z_y(\tau)\dot{u}_y(\tau)]d\tau \tag{10}$$

or both.

The accuracy and capability of this method of modeling is indicated by comparison of the biaxial force-displacement relationship with those based on analytical and experimental studies. Shown in Figs. 1 and 2 is the restoring force of a nondegrading system under different displacement paths (Fig. 3), according to Eq. 9. Whereas Fig. 4 gives the results of the corresponding analytical model based on plasticity theory by Powell and Chen (5). Figure 5a shows the experimental results of a R.C. column under biaxial load (6) with a nearly square loading path. Figure 5b shows the corresponding degrading system based on the proposed model in which β, γ, and α are function of energy dissipation and maximum displacement (4).

An important advantage of this method of modeling is that it is amenable to analytical solution and can be applied to systems of considerable complexity and under random excitation. Accurate response statistics can be obtained via an equivalent linearization method including those associated with the hysteretic energy dissipation which is a good measure of the cumulative damage under stress reversals. This method has been applied to damage prediction of reinforced concrete buildings (7), masonry buildings (8), and soil deposits (9) based on a damage measure depending on both statistics of maximum response and hysteretic energy dissipation. In these studies, the capacity of the systems against damage are also studied in terms of the foregoing damage measure by extensive analysis of experimental results. The uncertainties of this capacity are also investigated. The predictions of damage based on this method for systems in recent earthquakes have been compared with field evidence and the results are found to be quite satisfactory. Details can be found in Refs. 7, 8, and 9.

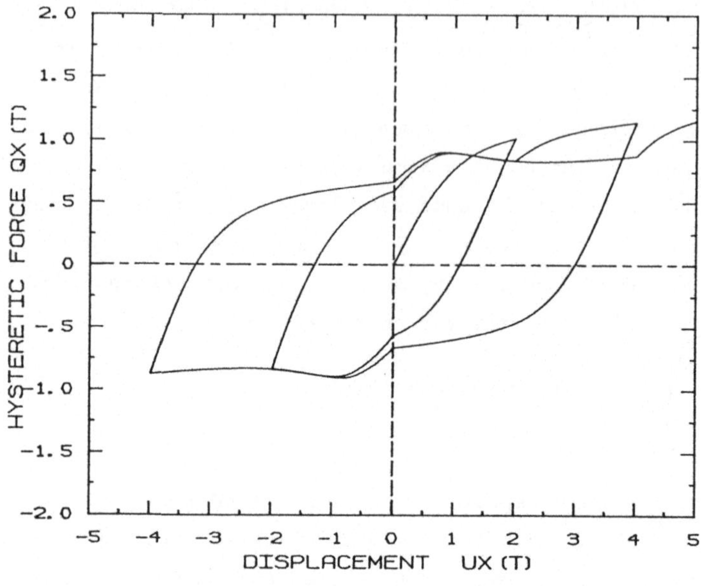

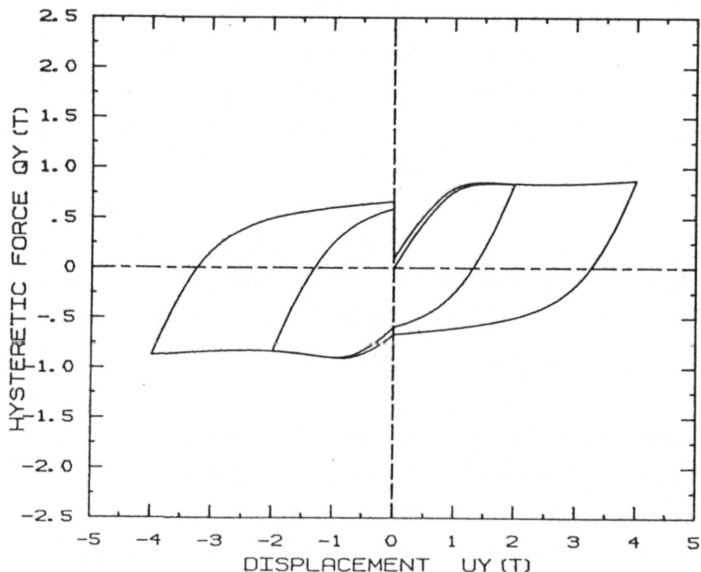

Fig. 1   Force-Displacement Relationship of A Nondegrading
Biaxial System with Diamond-Shaped Loading Path

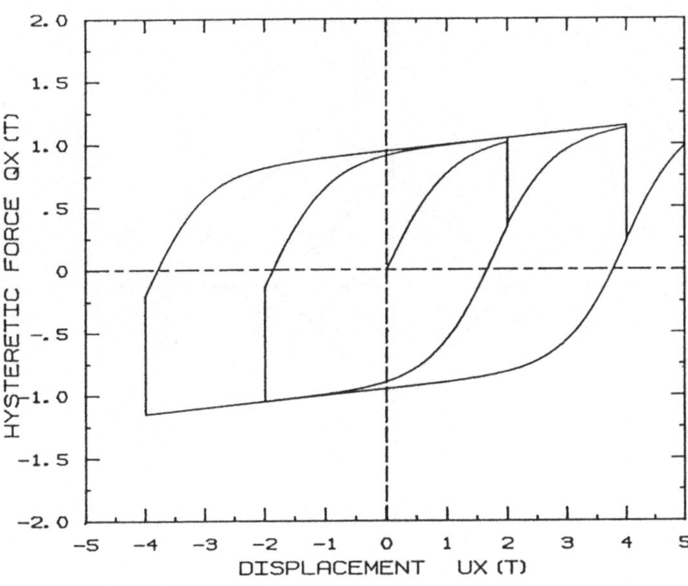

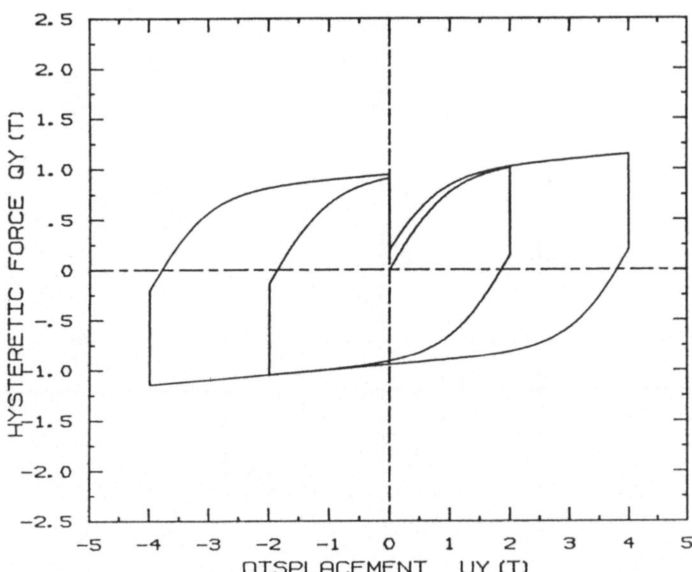

Fig. 2 Force-Displacement Relationship of A Nondegrading
Biaxial System with Square Loading Path

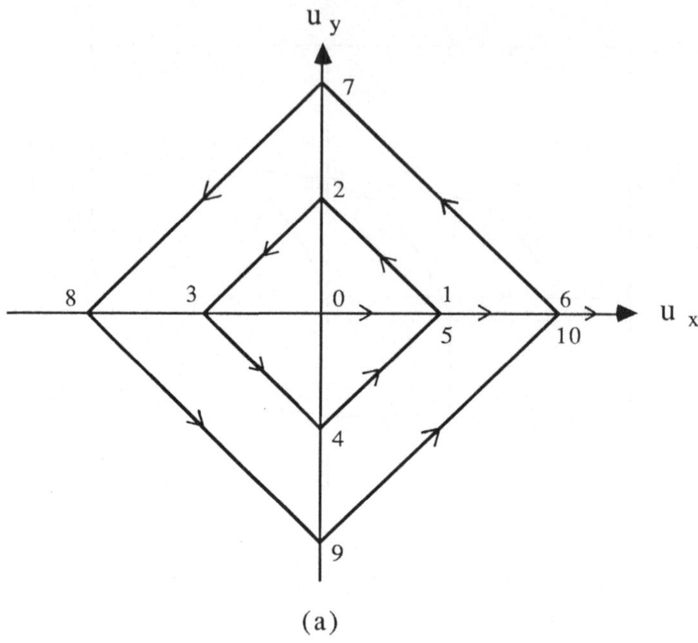

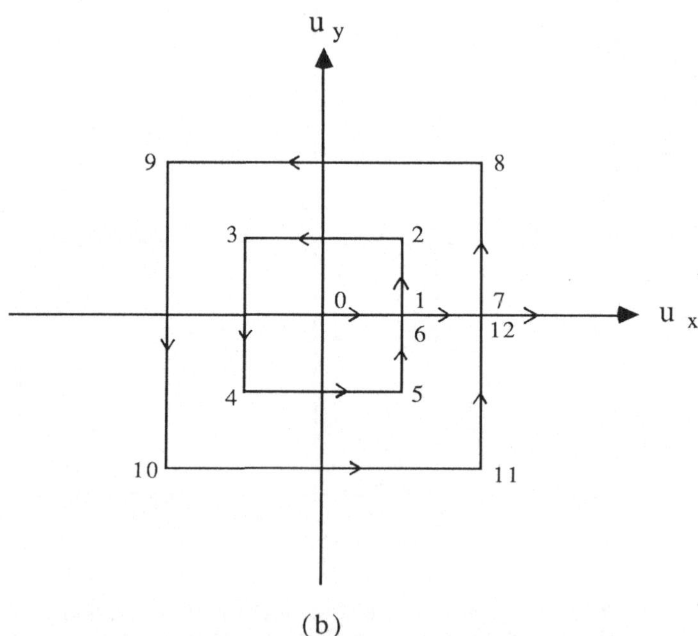

Fig. 3 Bi-Axial Displacement Loading Path
(a) Diamond (b) Square

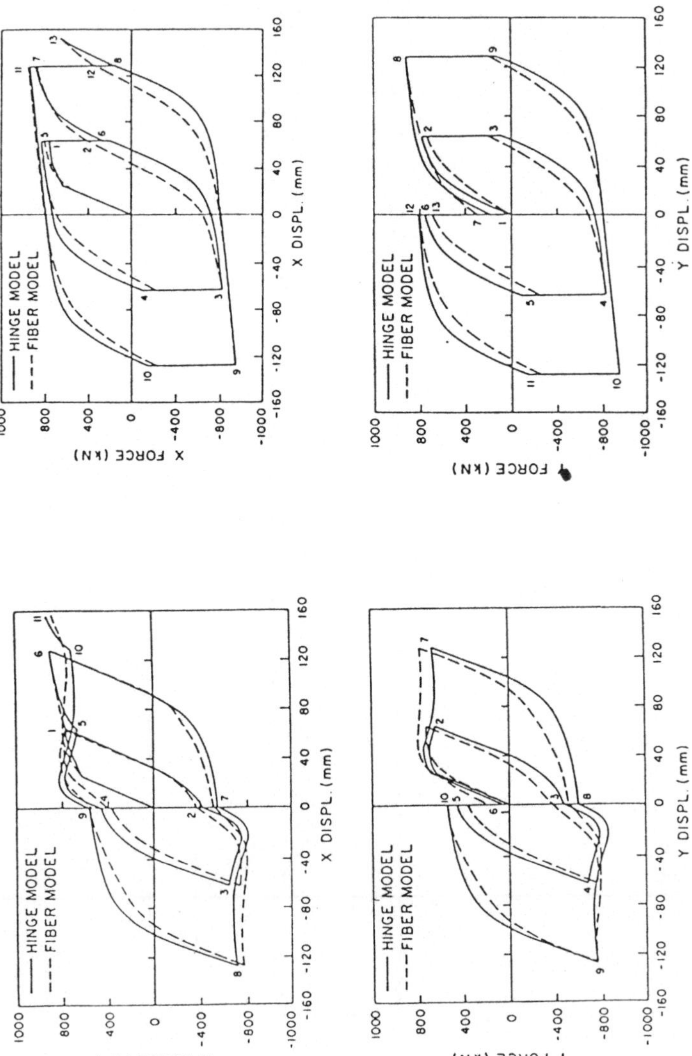

Fig. 4 Force-Displacement Relationship of Analytical Model by Powell and Chen (1986)

150

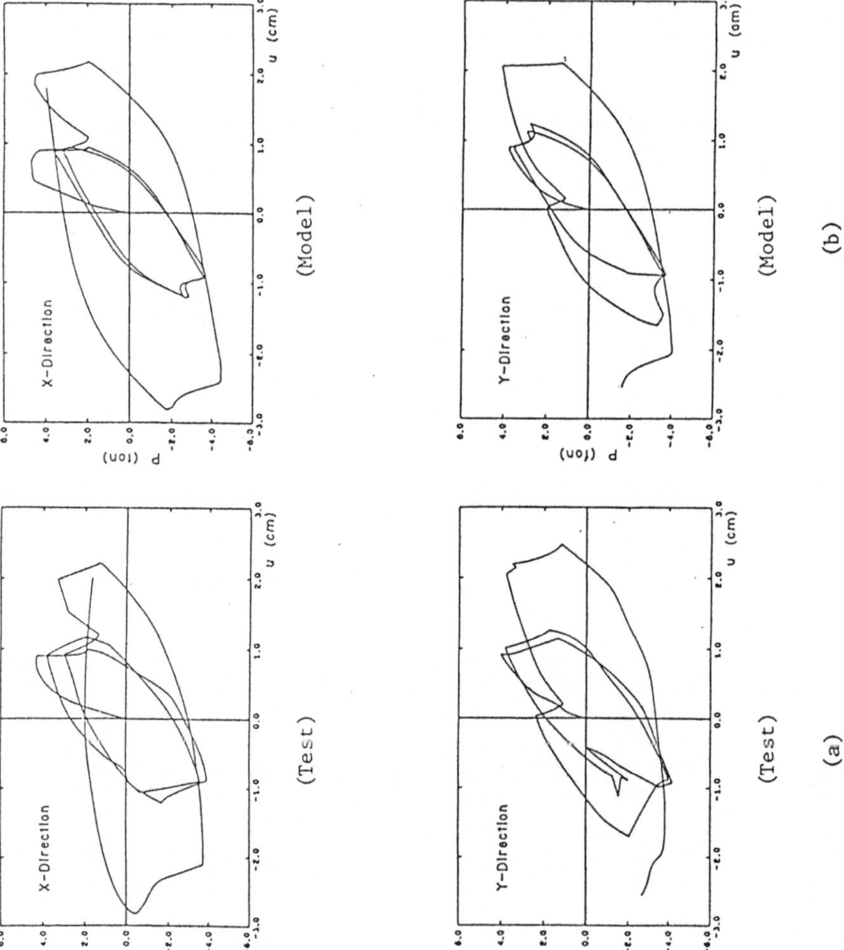

Fig. 5 Comparison of Test (after Takizawa and Aoyama, 1976) and Model Results

3   Identification of Force Parameters

To properly model the restoring force of actual structures, the force parameters need to be determined according to experimental or field evidence of inelastic structural behavior, therefore a system identification procedure is required for this purpose. Since the system is nonlinear, time domain approaches are more appropriate. A technique based on an invariant imbedding filter (10) was tried. It required the solution of a large number of first-order nonlinear differential equations. For example, for a single-degree-of-freedom, nondegrading system under uniaxial load, twenty are required. Although this method has certain advantages, the numerical effort required seems inordinately large and results were less than satisfactory when the response history used was highly irregular, as in the case of earthquake response data. For these reasons, a simple technique based on least square error minimization was developed (11,12) for uniaxial systems.

The method was tested using results based on solution of Eq. 3 as well as actual experimental data. The results are found to be quite satisfactory. For example, a comparison is given of the force-displacement curves from tests (Fig. 6a) and by Eqs. 2, 3, and 8 in which system coefficients are determined by the proposed method (Fig. 6b). The extension of this method to biaxial system is summarized in the following. Details of this method for uniaxial systems can be found in Ref. 12. It is noted, however, that a Kalman filter method has been applied to this restoring force model with some success by Hoshiya and Maruyama (13). A modification of the restoring force model has been used by Beck and Jayakumar (14) in their work on model identification based on minimization of output error function with good results.

3.1   Least Square Method for Biaxial System

Consider first a nondegrading biaxial hysteretic system which can be modeled by Eq. 9. From laboratory or field observations the force-displacement relationship such as those in Figs. 1 and 2 are obtained. The objective is to determine the required system parameters A, β, and γ.

Integrating Eq. 9 with respect to time, one obtains in the x direction, for given force-displacement data points $(z_x^i, z_y^i, u_x^i, u_y^i)$ in which i refers to observation at $t = t^i$,

(a)

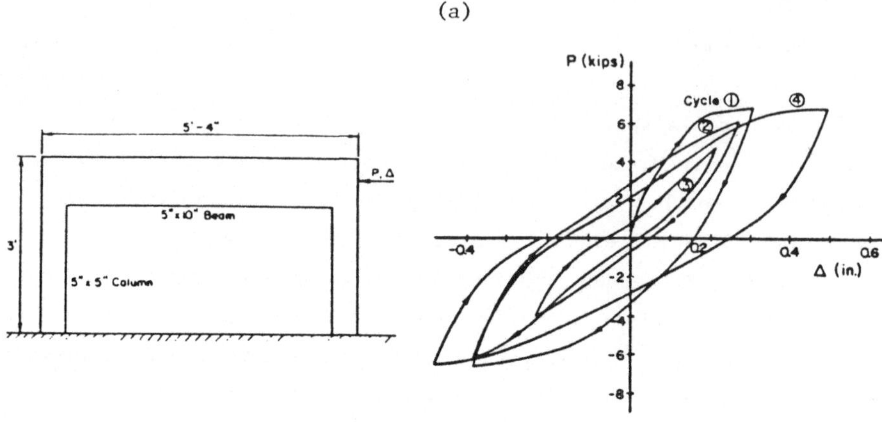

(b)

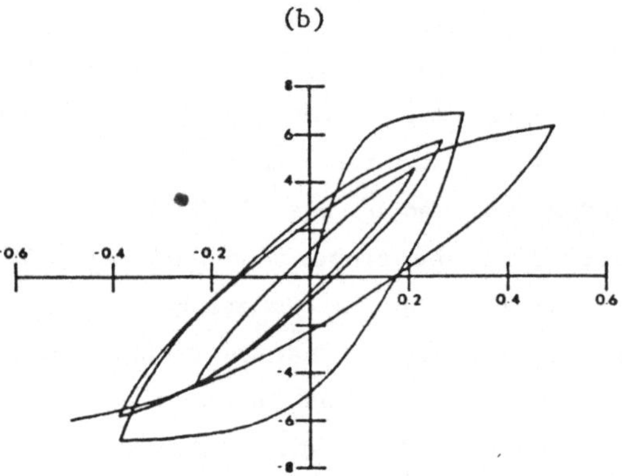

Fig. 6 Comparison of Test (a) (after Gulkan and Sozen, 1971) and
Model Results (b)

$$\bar{z}_x^i = z_x^o + AI_{1x}^i - \beta I_{2x}^i - \gamma I_{3x}^i \tag{11}$$

in which

$$I_{1x}^i = \int_{u_x^o}^{u_x^i} du_x = u_x^i - u_x^o$$

$$I_{2x}^i = \int_{u_x^o}^{u_x^i} \frac{|\dot{u}_x z_x|}{\dot{u}_x} z_x \, du_x + \int_{u_y^o}^{u_y^i} \frac{|\dot{u}_y z_y| z_x}{\dot{u}_y} \, du_y$$

$$I_{3x}^i = \int_{u_x^o}^{u_x^i} z_x^2 \, du_x + \int_{u_y^o}^{u_y^i} z_x z_y \, du_y$$

I_{1x}^i, I_{2x}^i, and I_{3x}^i can be calculated directly and may be considered as observable quantities. The same is done in the y direction and one obtains

$$\bar{z}_y^i = z_y^o + AI_{1y}^i - \beta I_{2y}^i - \gamma I_{3y}^i \tag{12}$$

where I_{1y}^i, I_{2y}^i, and I_{3y}^i are similarly defined. Now define the total error as

$$E = \sum_{i=1}^{n} \left(z_x^i - \bar{z}_x^i \right)^2 + \left(z_y^i - \bar{z}_y^i \right)^2 \tag{13}$$

minimization of E w.r.t. the system parameter A, β, and γ gives $\partial E/\partial A = 0$, $\partial E/\partial \beta = 0$, and $\partial E/\partial \gamma = 0$, and the required three linear equations for the parameters. The method is tested by using force-displacement solutions of Eq. 9 for several different loading paths as data points with A = 1.00 and $\beta = \gamma = 0.5$. The method recovers these parameters with excellent accuracy. The results are shown in Table 1. For systems under random excitation, a time history of a biaxial system under random ground excitation is generated. The horizontal components of the ground motion are modeled by two Gaussian processes with a power spectral density function of the Kanai-Tajimi form and assumed to be independent for simplicity. The force-displacement relationship is shown in Fig. 7. The same set of hysteretic parameters is used and the method gives A = 1.00, β = .482, and γ = .519.

For degrading systems, A, γ, and β, etc. are functions of time depending on the severity of the response. For example, if A is modeled

Table 1

Parameters of Nondegrading Biaxial System

Parameters	Loading Path
$A = 1.00$ $\beta = .499$ $\gamma = .502$	
$A = .999$ $\beta = .500$ $\gamma = .499$	Diamond (see Fig. 3)
$A = 100$ $\beta = .502$ $\gamma = .497$	Square (see Fig. 3)

by Eqs. 5 and 10, the system will deteriorate in strength and stiffness in both directions. In place of A, system parameter A_o, which is the initial value of A, and δ_A, which controls the rate of the deterioration, need to be identified.

Replacing A in Eq. 9 by $A_o - \delta_A \varepsilon_T$ and integrating with respect to time one obtains, after some algebra, in the x direction,

$$\bar{z}_x^i = A_o I_{1x}^i - \delta_A I_{2x}^i - \beta I_{3x}^i - \delta I_{4x}^i \tag{14}$$

in which

$$I_{1x}^i = u_x^i$$

$$I_{2x}^i = \int_{u_x^o}^{u_x^i} \varepsilon_T \, du_x$$

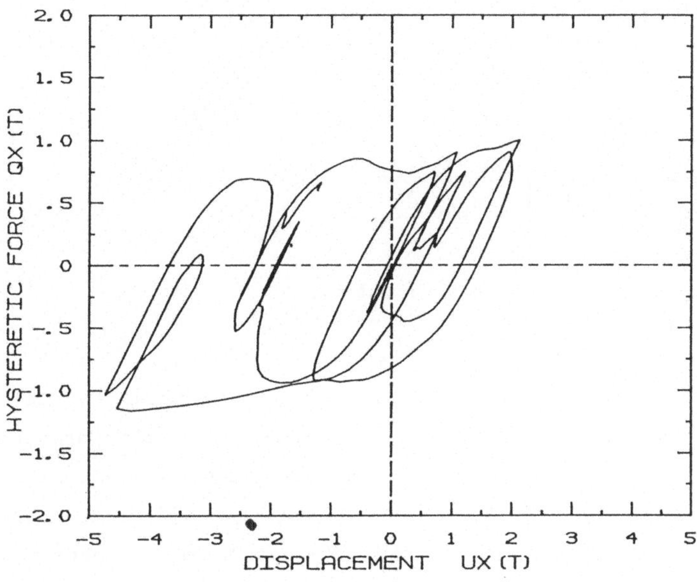

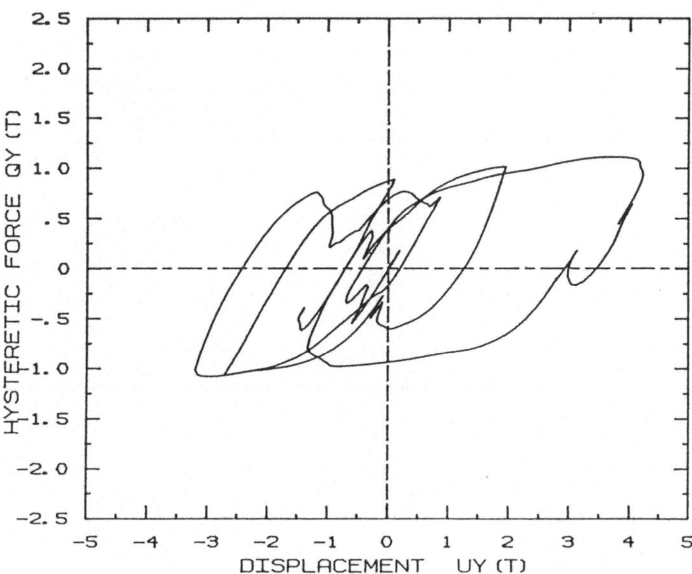

Fig. 7 Force-Displacement Relationship of A Nondegrading
Biaxial System under Random Excitation

156

$$I^i_{3x} = \int_{u^o_x}^{u^i_x} \frac{|\dot{u}_x z_x|}{\dot{u}_x} z_x \, du_x + \int_{u^o_y}^{u^i_y} \frac{|\dot{u}_y z_y|}{\dot{u}_y} z_y \, du_y$$

$$I^i_{4x} = \int_{u^o_x}^{u^i_x} z^2_x \, du_x + \int_{u^o_y}^{u^i_y} z_x z_y \, du_y$$

Similarly one obtains, in the y direction

$$\bar{z}^i_y = A_o I^i_{1y} - \delta_A I^i_{2y} - \beta I^i_{3y} - \gamma I^i_{4y} \tag{15}$$

Substituting Eqs. 14 and 15 into Eq. 13 and minimizing the total error, one obtains the required four linear equations for the parameters. For $A_o = 1.0$, $\delta_A = 0.02$, $\beta = \gamma = 0.5$, the force-displacement relations are generated by solving Eqs. 9, 5, and 10 for two different loading paths (Figs. 8 and 9). The method again recovers these parameters with very good accuracy. The results are given in Table 2.

Table 2

Parameter of Degrading Biaxial System

Parameters	Loading Path
$A_o = 1.00$	
$\beta = 0.503$	Diamond
$\gamma = 0.496$	(see Fig. 3)
$\delta_A = 0.020$	
$A_o = 1.00$	
$\beta = 0.508$	Square
$\gamma = 0.491$	(see Fig. 3)
$\delta_A = 0.020$	

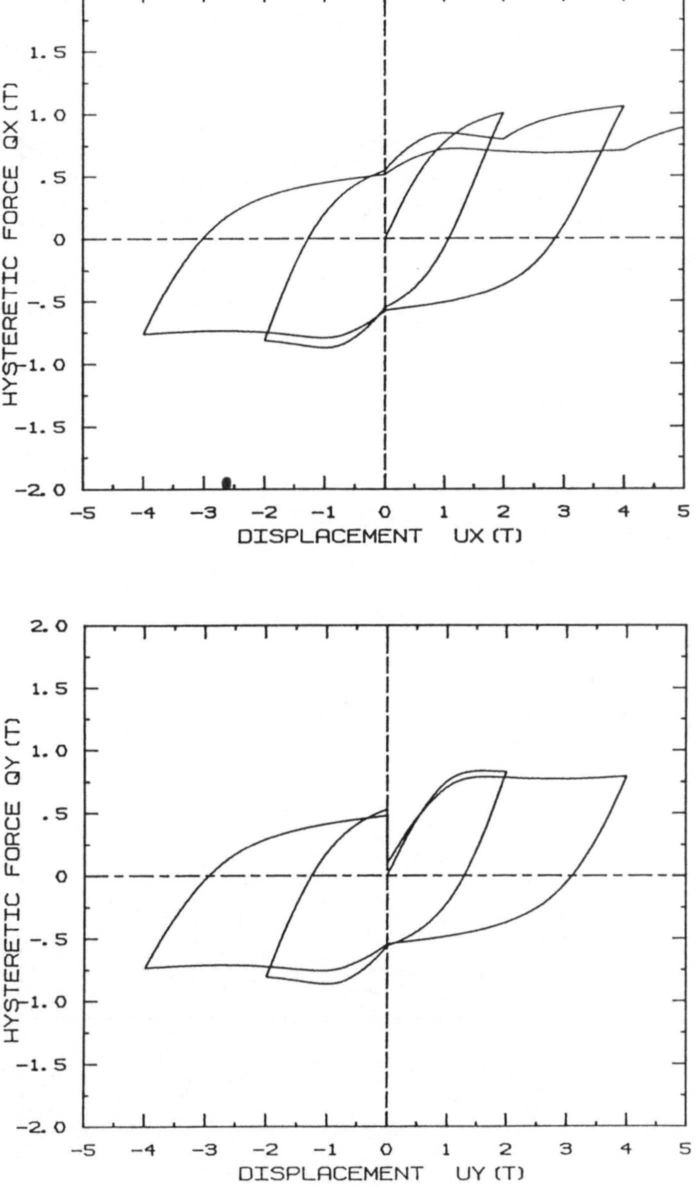

Fig. 8 Force-Displacement Relationship of A Degrading
System with Diamond-Shaped Loading Path

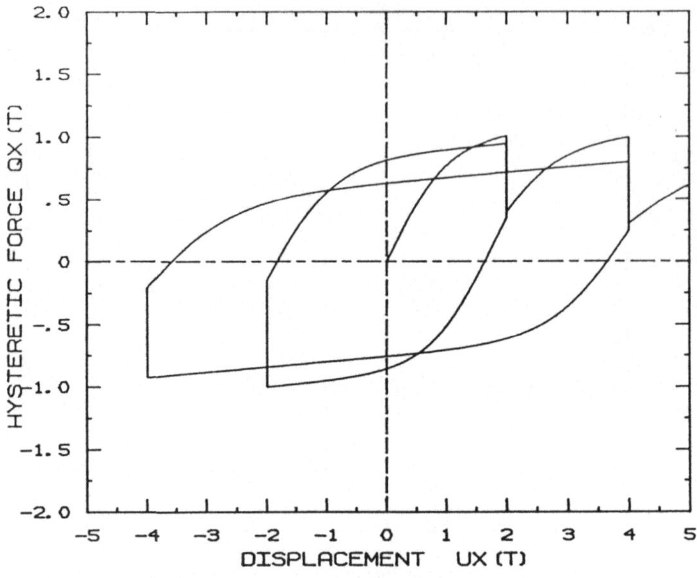

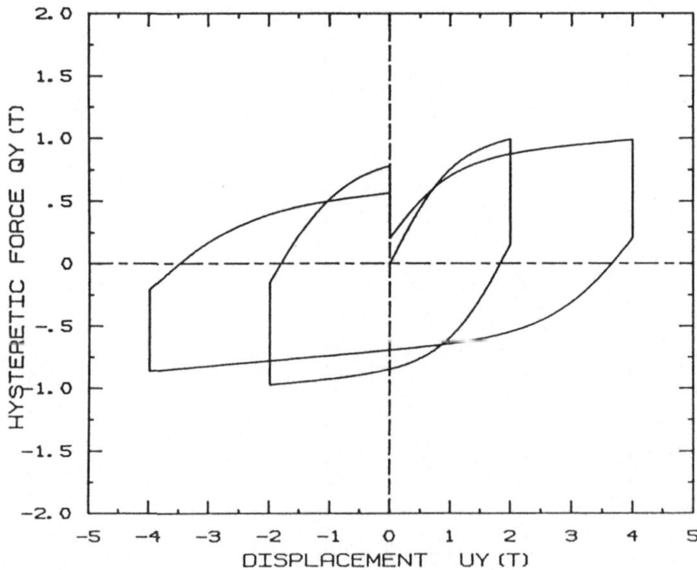

Fig. 9 Force-Displacement Relationship of A Degrading
System with Square Loading Path

4 Summary and Conclusion

Recent developments of a method of modeling inelastic structures and the identification of the system parameters are summarized. The emphasis is on realistic representation of restoring force hereditary behavior, deterioration and biaxial interaction. It is shown that the proposed differential equation model is accurate and efficient in reproducing most of the important characteristics of inelastic systems and the associated system parameters can be determined without difficulty. The method is a useful and practical tool for predicting structural performance (including damage) in severe natural hazards. Applications to damage prediction of structural and geotechnical systems are mentioned.

Acknowledgment

This research is part of an ongoing program on safety of structures under seismic excitation at the University of Illinois supported by the National Science Foundation (NSF ECE 85-11972). The support is gratefully acknowledged. Former and current graduate students are thanked for their contribution.

References

1. Baber, T. T. and Wen, Y. K. "Random Vibration of Hysteretic Degrading Systems," _Journal of Engineering Mechanics Division_, ASCE, December 1981, pp. 1069-1087.

2. Sues, R. H.; Wen, Y. K.; and Ang, A. H-S. "Stochastic Evaluation of Seismic Structural Performance," _Journal of Structural Engineering_, ASCE, Vol. 111, No. 6, June 1985, pp. 1204-1218.

3. Baber, T. T. and Noori, M. N. "Random Vibration of Pinching Hysteretic Systems," Report No. UUA/526378/CE84/102, University of Virginia, September 1983.

4. Park, Y. J.; Wen, Y. K.; and Ang, A. H-S. "Two-Dimensional Random Vibration of Hysteretic Structures," _Journal of Earthquake Engineering and Structural Dynamics_, Vol. 14, 1986, pp. 543-557.

5. Powell, G. H. and Chen, P. F-S. "3D Beam-Column Element with Generalized Plastic Hinges," _Journal of Engineering Mechanics_, ASCE, Vol. 112, No. 76, July 1986.

6. Takizawa, H. and Aoyama, H. "Biaxial Effects in Modeling Earthquake Response of R/C Structures," _Earthquake Engineering and Structural Dynamics_, Vol. 4, 1976, pp. 523-552.

7. Park, Y. J.; Ang, A. H-S.; and Wen, Y. K. "Seismic Damage Analysis of Reinforced Concrete Buildings," _Journal of Structural Engineering_, ASCE, Vol. 111, No. 4, April 1985, pp. 740-757.

8. Kwok, Y. H. "Seismic Damage Analysis and Design of Unreinforced Masonry Buildings," Ph.D. Thesis, University of Illinois at Urbana-Champaign, Urbana, Illinois, March 1987.

9. Pires, J.E.A.; Wen, Y. K.; and Ang, A. H-S. "Probabilistic Analysis of Seismic Safety Against Liquefaction," _Proceedings, 8th World Conference on Earthquake Engineering_, Vol. III, San Francisco, California, July 1984, pp. 159-166.

10. Distefano, N. and Pena-Pardo, B. "System Identification of Frames Under Seismic Loads," _Journal of Engineering Mechanics_, ASCE, Vol. 102, No. EM2, pp. 313-330, April 1976.

11. Sues, R. H.; Wen, Y. K.; and Ang, A. H-S. "Stochastic Seismic Performance Evaluation of Buildings," Civil Engineering Studies, Structural Research Series No. 506, UILU-ENG 83-2008, University of Illinois at Urbana-Champaign, Urbana, Illinois, May 1983.

12. Sues, R. H.; Mau, T. T.; and Wen, Y. K. "System Identification of Degrading Hysteretic Restoring Forces," accepted for publication, Journal of Engineering Mechanics, ASCE, 1987.

13. Hoshiya, M. and Maruyama, Osamu. "Identification of Nonlinear Structural Systems," _Proc., ICASP 5, 1987_, University of British Columbia, Vancouver, May 1987, pp. 182-189.

14. Beck, J. L. and Jayakumar, P. "Pseudo-Dynamic Testing and Model Identification," Research Report, Caltech, 1986.

Identification of Equivalent Linear Systems

by

C. Paliou[1], A.M. ASCE
M. Shinozuka[2], M. ASCE

[1] Postdoctoral Research Scientist, Department of Civil Engineering and Engineering Mechanics, 610 S.W. Mudd, Columbia University, New York, NY 10027.
[2] Renwick Professor of Civil Engineering, Department of Civil Engineering and Engineering Mechanics, 610 S.W. Mudd, Columbia University, New York, NY 10027.

Introduction

Civil engineering structures are designed, analyzed and constructed in order to fulfil specific needs and they are expected to perform their functions accordingly. However, civil engineering structures are subjected to random disturbances which may result in a random system performance. Furthermore, dynamic characteristics such as mass, stiffness and damping, used for design and analysis, involve idealization, approximation and uncertainty to the extent that their performance even under prescribed disturbances must be considered non-deterministic.

In a study by Paliou et al. (1986), for example, the dynamic response of an offshore tower to wind induced random wave forces was analyzed for the purpose of evaluating its fatigue performance. The analysis was performed by means of equivalent linearization techniques so that the original nonlinear equations of motion can be solved in the frequency domain in approximation. The results have been confirmed with the aid of the time domain (Monte Carlo) analysis in which the original nonlinear equations of motion were solved directly. The major analytical difference between these two methods lies in the following. The frequency domain technique is an approximate method and produces, under the assumption of stationarity, the variance of the response without a time history analysis. On the other hand, the time domain analysis provides exact results, at least in principle, by means of Monte Carlo simulation regardless of the severity of nonlinearity that the structural system exhibits. This is particularly important when the response is considered for high wind velocities under which the offshore tower oscillates in a more than moderate nonlinear range (Shinozuka et al., 1977).

Both methods assumed exact knowledge of the structural parameters involved in the analysis. In real life, however, the question often arises whether or not the structure actually possesses dynamic characteristics used in the analysis and design in the first place. In this respect, before any judgment is made on the adequacy of the structural performance or any course of action is taken to remedy apparent inadequacy in the performance, the dynamic characteristics of the structures must be known as precisely as possible. Indeed, this is the problem of system identification which usually reduces to that of identification of the structural parameters in the mathematical models, a situation known as the "gray box" problem. In this case, identification implies "estimation" of the parameters on the basis of observation of the excitation and response time histories.

To accomplish this estimation, Kalman filtering techniques are applied in the time domain under the assumption that a) all structural parameters of the original nonlinear equations of motion are unknown, b) the nonlinear system can be approximated by a linear system and c) the excitation and response time histories of the actual nonlinear system are available.

In the present study, the excitation time histories are numerically generated, and at the same time, appropriate values are assigned to the structural parameters of the original nonlinear system to generate response time histories. Estimation of the structural parameters of the corresponding linear system will be made in some optimal sense by means of Kalman filtering techniques, as mentioned above. When these optimally estimated structural parameters are used in the equations of motion of the linear system, this system is referred to as an "equivalent linear system". To examine the validity of this equivalent linearization technique, the exact response time histories derived from the time domain analysis of the original nonlinear equations of motion are compared with those obtained from the equivalent linear system. An equivalent linear system can also be developed on the basis of more conventional methods (Malhotra & Penzien, 1970; Paliou et al., 1986; Shinozuka et al., 1977), thus providing a second measure to check the validity of the results which will be obtained by the equivalent linear system developed by the Extended Kalman filtering procedures just mentioned.

Structural Models and Equations of Motion

The equation of motion of an offshore structure idealized as a discretized mass system can be written as (Malhotra & Penzien, 1970; Shinozuka et al., 1977; Paliou et al., 1986)

$$\mathbf{M}\ddot{\xi} + \mathbf{C}\dot{\xi} + \mathbf{K}\xi = \mathbf{C}_M(\ddot{v} - \ddot{\xi}) + \mathbf{C}_D(\dot{v} - \dot{\xi})|\dot{v} - \dot{\xi}| \tag{1}$$

where ξ, $\dot{\xi}$, $\ddot{\xi}$ = nodal displacement, velocity and acceleration vectors, $\dot{v}$, $\ddot{v}$ =

wave particle velocity and acceleration vectors, $\mathbf{M}$ = diagonal matrix of the lumped masses, $\mathbf{C}$ = structural damping matrix in the air, $\mathbf{K}$ = structural stiffness matrix, and $\mathbf{C}_M$, $\mathbf{C}_D$ = diagonal matrices containing the inertia and drag coefficients associated with the wave forces acting on the structure. Thus the nonlinearity of the system is due to drag forces arising from wave- structure interaction.

Using the notation $\bar{\mathbf{M}} = \mathbf{M} + \mathbf{C}_M$ for the effective mass matrix of the system and premultiplying both sides of Eq. 1 by $\bar{\mathbf{M}}^{-1}$, the equation of motion can be written as

$$\ddot{\xi} + \{\bar{\mathbf{M}}^{-1} \cdot \mathbf{C}\}\dot{\xi} + \{\bar{\mathbf{M}}^{-1} \cdot \mathbf{K}\}\xi = \{\bar{\mathbf{M}}^{-1} \cdot \mathbf{C}_M\}\ddot{v} + \{\bar{\mathbf{M}}^{-1} \cdot \mathbf{C}_D\}(\dot{v} - \dot{\xi})|\dot{v} - \dot{\xi}| \quad (2)$$

Replacing $\mathbf{C}_D(\dot{v} - \dot{\xi})|\dot{v} - \dot{\xi}|$ with $\bar{\mathbf{C}}(\dot{v} - \dot{\xi})$, in Eq. 2, one obtains

$$\ddot{\xi} + \{\bar{\mathbf{M}}^{-1} \cdot \mathbf{C}\}\dot{\xi} + \{\bar{\mathbf{M}}^{-1} \cdot \mathbf{K}\}\xi = \{\bar{\mathbf{M}}^{-1} \cdot \mathbf{C}_M\}\ddot{v} + \{\bar{\mathbf{M}}^{-1} \cdot \mathbf{C}_D\}(\dot{v} - \dot{\xi}) \quad (3)$$

Introducing the following matrices to Eq. 3

$$\mathbf{C}^* = \bar{\mathbf{M}}^{-1} \cdot \mathbf{C} \quad (4)$$

$$\mathbf{K}^* = \bar{\mathbf{M}}^{-1} \cdot \mathbf{K} \quad (5)$$

$$\mathbf{M}^* = \bar{\mathbf{M}}^{-1} \cdot \mathbf{C}_M \quad (6)$$

$$\mathbf{D}^* = \bar{\mathbf{M}}^{-1} \cdot \bar{\mathbf{C}} \quad (7)$$

the equation of motion of the equivalent linear system is finally given by

$$\ddot{\xi} + \mathbf{C}^*\dot{\xi} + \mathbf{K}^*\xi = \mathbf{M}^*\ddot{v} + \mathbf{D}^*(\dot{v} - \dot{\xi}) \quad (8)$$

The current identification problem consists of finding the optimal estimates of the unknown coefficient matrices $\mathbf{C}^*$, $\mathbf{K}^*$, $\mathbf{M}^*$ and $\mathbf{D}^*$. As mentioned earlier in the present study, once these "optimal" estimates are computed, Eq. 8 is considered to represent the equation of motion of a linear system "equivalent" to the original nonlinear system.

System Model and Measurement Model

In order to proceed with the identification, the first task is to model the system dynamics; Selecting ξ and $\dot{\xi}$ as the state variables and regarding the unknown coefficient matrices $\mathbf{C}^*$, $\mathbf{K}^*$, $\mathbf{M}^*$ and $\mathbf{D}^*$ as augmented state variables, the state vector $\mathbf{x}$ becomes

$$\mathbf{x}^T = [\xi_1\xi_2...\xi_n\dot{\xi}_1...\dot{\xi}_nC_{11}^*...C_{nn}^*K_{11}^*...K_{nn}^*$$

$$M_{11}^*...M_{nn}^*D_{11}^*...D_{nn}^*]^T \tag{9}$$

where C_{ij}^*, K_{ij}^*, M_{ij}^*, $D_{ij}^* =$ the ij elements of matrices $\mathbf{C}^*$, $\mathbf{K}^*$, $\mathbf{M}^*$ and $\mathbf{D}^*$ respectively. Eq. 9 indicates a $[2n + 4n^2]$ dimensional state vector.

The observations of the structural response $\xi(t)$ are assumed to be corrupted by noise $\eta(t)$ and the measurements are assumed to be made at discrete time instants with equal intervals so that

$$\mathbf{Y}(t_k) = \xi(t_k) + \eta(t_k) \tag{10}$$

for $k = 1, 2, ..., n$, where

$$t_k = k \cdot \triangle t \tag{11}$$

Furthermore, it is assumed that the excitation consisting of the wave particle velocities $\dot{\mathbf{v}}(t)$ and acceleration $\ddot{\mathbf{v}}(t)$ can be also observed without error at the same discrete time instants, i.e. $\dot{\mathbf{v}}(t_k)$, $\ddot{\mathbf{v}}(t_k)$, k = 1,2,...,n where n=$\frac{T}{\triangle t}$ and T= duration of observations. With the aid of Eqs 9 and 10, the system can be generally described by a set of linear stochastic differential equations.

$$\dot{\mathbf{x}}(t) = \mathbf{f}\left(\mathbf{x}(t), \dot{\mathbf{v}}(t), \ddot{\mathbf{v}}(t), t\right) \tag{12}$$

$$\mathbf{Y}(t_k) = \mathbf{h}\left(\mathbf{x}(t_k), t_k\right) + \eta(t_k) \tag{13}$$

For the present problem, it can be shown that the function $\mathbf{f}$ in Eq. 12 is a $[2n+4n^2]$-dimensional vector with

$$f_i = 0 \tag{14}$$

for $i = 2n + 1, ..., (2n + 4n^2)$. Since response observations are available for mass displacements only, Eq. 13 can also be reduced to

$$\mathbf{Y}_k = \mathbf{A}\mathbf{x}_k + \eta_k \tag{15}$$

$$\mathbf{A} = [\,\mathbf{I} \mid \mathbf{0}\,] \tag{16}$$

where $\mathbf{Y}_k$ is written for $\mathbf{Y}(t_k)$ for simplicity (henceforth, similar notation is used for other quantities) and $\mathbf{A}$ is an $n \times [2n + 4n^2]$-dimensional matrix. $\mathbf{I}$ indicates the $n \times n$ identity matrix and $\mathbf{0}$ an $n \times [n + 4n^2]$ null matrix.

The estimation of the unknown state vector is accomplished with the aid of linear filtering techniques. The response of the structure ξ and $\dot{\xi}$ as well as the unknown coefficient matrices $\mathbf{C}^*$, $\mathbf{K}^*$, $\mathbf{M}^*$ and $\mathbf{D}^*$ are identified simultaneously on the basis of the observations on the excitations $\dot{\mathbf{v}}_k$ and $\ddot{\mathbf{v}}_k$ and the noisy response $\mathbf{y}_k$. For the identification, the algorithm of the Extended Kalman filtering (EKF) technique is used in this study.

Numerical Example

System identification is performed for the offshore structure shown in Fig. 1. The structure is a two-degree-of-freedom system with both masses being under the water surface. The dynamics of the system are described by Eq. 1 and are to be approximated by the dynamics of an equivalent linear structure for which the equations of motion are given by Eq. 8.

In order to obtain the observation vector (Eq. 10) the time histories of the mass displacements are needed. In order to demonstrate the numerical procedure of the proposed equivalent linearization method, the numerical solution of the original nonlinear equation of motion is used in lieu of actual measurements. The solution is obtained in the time domain and a white noise component is added to simulate the measurement error. The values of the matrices $\bar{\mathbf{M}}^{-1} \cdot \mathbf{C}$, $\bar{\mathbf{M}}^{-1} \cdot \mathbf{K}$, $\bar{\mathbf{M}}^{-1} \cdot \mathbf{C}_M$ and $\bar{\mathbf{M}}^{-1} \cdot \mathbf{C}_D$, used for this purpose, are given in Table 1b following Yun & Shinozuka, (1980) and Shinozuka et al., (1982) .

The structure is assumed to be subjected to wind-induced random wave forces. The waves are modeled as stationary Gaussian random processes with zero mean (Malhotra & Penzien, 1970; Shinozuka et al., 1977; Paliou et al., 1986). Thus the wave particle velocities and accelerations are also stationary Gaussian random processes. In this study, as in Paliou & Shinozuka (1985) and Paliou et al. (1986), the Pierson-Moskowitz one-sided spectrum has been used to characterize the spectral content of the wave height;

167

$$S_{hh}(\omega) = \frac{\alpha g^2}{\omega^5} \cdot exp\left[-\beta\left(\frac{g}{\omega W}\right)^4\right] \tag{17}$$

where $\alpha = 8.1 \cdot 10^{-3}$, $\beta = 0.74$ and W is the average storm wind velocity at 64 ft (19.5 m) above the water surface. Two different values of W have been considered in this analysis: W = 25 ft/sec (7.62 m/sec) and W = 75 ft/sec (22.86 m/sec).

The power spectral density function $S_{\dot{v}\dot{v}}(\omega)$ and $S_{\ddot{v}\ddot{v}}(\omega)$ of the horizontal components of the water particle velocity and acceleration respectively, are functions of $S_{hh}(\omega)$ (e.g., Paliou & Shinozuka, 1985)

$$S_{\dot{v}\dot{v}}(\omega) = \omega^2 \cdot S_{hh}(\omega) \cdot exp\left[\frac{2\omega^2}{g} \cdot z\right] \tag{18}$$

$$S_{\ddot{v}\ddot{v}}(\omega) = \omega^2 \cdot S_{\dot{v}\dot{v}}(\omega) \tag{19}$$

where z = water depth

In order to proceed in the time domain analysis, $\dot{v}(t)$ and $\ddot{v}(t)$ are simulated and generated using the Fast Fourier Transform (FFT) algorithm in the following form (Shinozuka, 1974)

$$\dot{v}_l(k\triangle t) = Re \sum_{m=0}^{M_1-1} \left(\sqrt{2S_{\dot{v}\dot{v}}(\omega)\triangle\omega} \cdot e^{i\phi_m}\right) \cdot exp[2\pi i \cdot \frac{mk}{M_1}] \quad k = 0,..,M_1-1 \tag{20}$$

$$\ddot{v}_l(k\triangle t) = -Im \sum_{m=0}^{M_1-1} \left(\sqrt{2S_{\ddot{v}\ddot{v}}(\omega)\triangle\omega} \cdot e^{i\phi_m}\right) \cdot exp[2\pi i \cdot \frac{mk}{M_1}] \quad k = 0,..,M_1-1 \tag{21}$$

where

$$\triangle\omega = \frac{\omega_u}{N} \tag{22}$$

$$\triangle t = \frac{2\pi}{M_1\triangle\omega} \tag{23}$$

and ϕ_m are random phase angles uniformly distributed between 0 and 2π. The numerical values used for the generation of $\dot{v}(t)$ and $\ddot{v}(t)$ are N = 512, $M_1 = 2048$, $\omega_u = 16$ rad/sec, $z_1 = 10$ ft (3.05 m) and $z_2 = 40$ ft (12.2 m).

The observation vector, defined by Eq. 10, is created by adding a Gaussian white noise $\eta(t)$ to the response $\xi(t)$. The noise vector is simulated by

$$\eta_l(k\triangle t) = -Re \sum_{m=0}^{M_1-1} \left(\sqrt{2S_{\eta_l\eta_l}(\omega)\triangle\omega} \cdot e^{i\psi_m}\right) \cdot exp[2\pi i \cdot \frac{mk}{M_1}] \quad k = 0,..,M_1-1 \tag{24}$$

where $S_{\eta_l \eta_l}(\omega)$ is the one-sided spectral density of a band limited white noise and ψ_m are random phase angles uniformly distributed between 0 and 2π. The intensity of the observation noise is assumed to be 5% of the structural response $\xi(t)$ in terms of root mean square (RMS) value.

Following the analysis of the previous section, the 20-dimensional state vector $\mathbf{x}$ is

$$\mathbf{x}^T = [\xi_1 \xi_2 \dot{\xi}_1 \dot{\xi}_2 C^*_{11} C^*_{21} C^*_{12} C^*_{22} D^*_{11} D^*_{21} D^*_{12} D^*_{22}$$
$$K^*_{11} K^*_{21} K^*_{12} K^*_{22} M^*_{11} M^*_{21} M^*_{12} M^*_{22}]^T \tag{25}$$

With the aid of Eq. 25, Eq. 8 can be written in matrix form

$$\begin{pmatrix} \dot{x}_3 \\ \dot{x}_4 \end{pmatrix} + \begin{pmatrix} x_5 & x_7 \\ x_6 & x_8 \end{pmatrix} \cdot \begin{pmatrix} x_3 \\ x_4 \end{pmatrix} + \begin{pmatrix} x_{13} & x_{15} \\ x_{14} & x_{16} \end{pmatrix} \cdot \begin{pmatrix} x_1 \\ x_2 \end{pmatrix} =$$
$$\begin{pmatrix} x_{17} & x_{19} \\ x_{18} & x_{20} \end{pmatrix} \cdot \begin{pmatrix} \ddot{v}_1 \\ \ddot{v}_2 \end{pmatrix} + \begin{pmatrix} x_9 & x_{11} \\ x_{10} & x_{12} \end{pmatrix} \cdot \begin{pmatrix} \dot{v}_1 - x_3 \\ \dot{v}_2 - x_4 \end{pmatrix} \tag{26}$$

Introducing $\bar{x}_3 = \dot{v}_1 - x_3$ and $\bar{x}_4 = \dot{v}_2 - x_4$, the state equation of the system is given by

$$\begin{pmatrix} \dot{x}_1 \\ \dot{x}_2 \\ \dot{x}_3 \\ \dot{x}_4 \\ \dot{x}_5 \\ \vdots \\ \dot{x}_{20} \end{pmatrix} = \begin{pmatrix} x_3 \\ x_4 \\ -x_{13}x_1 - x_{15}x_2 - x_5 x_3 - x_7 x_4 + x_{17}\ddot{v}_1 + x_{19}\ddot{v}_2 + x_9 \bar{x}_3 + x_{11}\bar{x}_4 \\ -x_{14}x_1 - x_{16}x_2 - x_6 x_3 - x_8 x_4 + x_{18}\ddot{v}_1 + x_{20}\ddot{v}_2 + x_{10}\bar{x}_3 + x_{12}\bar{x}_4 \\ 0 \\ \vdots \\ 0 \end{pmatrix} \tag{27}$$

The measurement equation can be written, similarly, as

$$\begin{pmatrix} y_1 \\ y_2 \end{pmatrix} = \begin{pmatrix} 1 & 0 & 0 & \cdots & 0 \\ 0 & 1 & 0 & \cdots & 0 \end{pmatrix} \cdot \begin{pmatrix} x_1 \\ x_2 \\ x_3 \\ \vdots \\ x_{20} \end{pmatrix} + \begin{pmatrix} \eta_1 \\ \eta_2 \end{pmatrix} \tag{28}$$

According to the Extended Kalman filtering (EKF) algorithm, the identification, which consists of the optimal estimation of the state vector $\hat{\mathbf{x}}$ is accomplished by making use of the following equations:

$$\hat{\mathbf{x}}(t_{k+1}|t_{k+1}) = \hat{\mathbf{x}}(t_{k+1}|t_k) + \mathbf{K}\{t_{k+1}; \hat{\mathbf{x}}(t_{k+1}|t_k)\}\left[\mathbf{y}_{t_{k+1}} - \mathbf{h}\left(\hat{\mathbf{x}}(t_{k+1}|t_k), t_{k+1}\right)\right] \tag{29}$$

169

$$\mathbf{P}(t_{k+1}|t_{k+1}) = [\mathbf{I} - \mathbf{K}\{t_{k+1};\hat{\mathbf{x}}(t_{k+1}|t_k)\}\mathbf{M}\{t_{k+1};\hat{\mathbf{x}}(t_{k+1}|t_k)\}]\mathbf{P}(t_{k+1}|t_k)\times$$

$$\times [\mathbf{I} - \mathbf{K}\{t_{k+1};\hat{\mathbf{x}}(t_{k+1}|t_k)\}\mathbf{M}\{t_{k+1};\hat{\mathbf{x}}(t_{k+1}|t_k)\}]^T + \tag{30}$$

$$+\mathbf{K}\{t_{k+1};\hat{\mathbf{x}}(t_{k+1}|t_k)\}\mathbf{R}(k+1)\mathbf{K}^T\{t_{k+1};\hat{\mathbf{x}}(t_{k+1}|t_k)\}$$

where

$$\hat{\mathbf{x}}(t_{k+1}|t_k) = \hat{\mathbf{x}}(t_k|t_k) + \int_{t_k}^{t_{k+1}} \mathbf{f}\left(\hat{\mathbf{x}}(t|t_k),t\right)dt \tag{31}$$

$$\mathbf{P}(t_{k+1}|t_k) = \boldsymbol{\Phi}[t_{k+1},t_k;\hat{\mathbf{x}}(t_k|t_k)]\mathbf{P}(t_k|t_k)\boldsymbol{\Phi}^T[t_{k+1},t_k;\hat{\mathbf{x}}(t_k|t_k)] \tag{32}$$

$$\mathbf{K}\{t_{k+1};\hat{\mathbf{x}}(t_{k+1}|t_k)\} = \mathbf{P}(t_{k+1}|t_k)\mathbf{M}^T\{t_{k+1};\hat{\mathbf{x}}(t_{k+1}|t_k)\}] \times \Big[\mathbf{M}\{t_{k+1};\hat{\mathbf{x}}(t_{k+1}|t_k)\}\cdot$$

$$\mathbf{P}(t_{k+1}|t_k)\mathbf{M}^T\{t_{k+1};\hat{\mathbf{x}}(t_{k+1}|t_k)\} + \mathbf{R}(k+1)\Big]^{-1} \tag{33}$$

and

$$\boldsymbol{\Phi}(t_{k+1},t_k;\hat{\mathbf{x}}(t_k|t_k)) = \boldsymbol{\Phi}_{k+1} \rightarrow \mathbf{I} + \mathbf{F}(t_k)\triangle t \tag{34}$$

$$f_{ij}(t;\hat{\mathbf{x}}(t_k|t_k)) \doteq \left[\frac{\vartheta f_i(x(t),t)}{\vartheta x_j}\right]_{x(t)=\hat{\mathbf{x}}(t_k|t_k)} \tag{35}$$

$$m_{ij}(t_k;\hat{\mathbf{x}}(t_k|t_k)) \doteq \left[\frac{\vartheta h_i(x_{t_k},t_k)}{\vartheta x_j}\right]_{x(t)=\hat{\mathbf{x}}(t_k|t_k)} \tag{36}$$

where matrix $\mathbf{K}$ in Eqs. 29 - 33 represents the Kalman gain, $\mathbf{P}$ is the error covariance matrix, $\boldsymbol{\Phi}$ is an approximation of the state transition matrix and $\mathbf{R}$ denotes the noise covariance matrix. The notation $\mathbf{A}(t_{k+1}|t_{k+1})$ in the above equations indicates the estimated value of $\mathbf{A}$ at time t_{k+1} after $\mathbf{y}_{t_{k+1}}$ has been processed.

The results of the estimation are displayed in Table 2 together with the initial values. Time histories of the response thus estimated are shown in Figs. 2 - 5 for the two windspeeds. The behavior of the numerical convergence of the optimal estimates of the components of the matrices $\mathbf{C}^*$ are shown in Fig. 6 for $W = 25$ ft/sec and Fig. 7 for $W = 75$ ft/sec.

The matrices $\mathbf{C}$, $\mathbf{K}$, $\mathbf{C}_M$ and $\bar{\mathbf{C}}$, which have the obvious physical significance can be retrieved from the optimal $\mathbf{C}^*$, $\mathbf{K}^*$, $\mathbf{M}^*$ and $\mathbf{D}^*$ (Eqs. 4 - 7) thus estimated. For this reason, the matrices $\mathbf{C}$, $\mathbf{K}$, $\mathbf{C}_M$ and $\bar{\mathbf{C}}$ have been and will be referred to as "structural parameters". To accomplish this, it can be shown (Paliou, C. and Shinozuka, M., 1987) that the value of only one effective mass is needed. Finally, the standard deviations of the estimated responses are obtained and the results are given in Table 3.

Numerical Results and Conclusions

From the results shown in Figs. 2 - 5, it can be seen that the specific filter can track the structural responses exceptionally well, particularly in the case of lower windspeed. For the higher wind velocity small discrepancies are observed particularly near the peaks and troughs of the response time histories. This is due to the higher level of nonlinearity that the structure exhibits under higher average windspeeds.

The behavior of numerical convergence of the estimated structural coefficients (Figs. 6-7) exhibits a very interesting trend. At the beginning there are great fluctuations in their values. After the first 20 sec of observation time, the estimated coefficient values become more or less stable as they begin to converge to their final values. Windspeed does not seem to affect the number nor the amplitude of fluctuations; at the lower wind velocity, however, the coefficients show a more stable behavior as they approach their converged values. Table 2 displays the optimal estimations and true values of these coefficients. It is important to note here, that the estimated values for each windspeed are the ensemble average of five sets of optimal estimations arising from five sets of statistically identical but individually different excitation time histories of $\dot{v}(t)$ and $\ddot{v}(t)$ and observation histories $Y(t)$, generated by Eqs. 20-24. Since the Kalman filter was highly sensitive to the initial error covariance matrices, each set participating in the ensemble average represented the best result of a number of independent executions using the same excitation and observation histories but different covariance matrices. Bearing in mind that judgment on the estimations is a trial and error process, each time a set of coefficients is calculated, a time domain analysis is performed and the resulting responses are compared with the response histories of the original nonlinear equation of motion.

To examine the results of the equivalent linearization technique proposed in this study, the components of the estimated coefficient matrices, K^*, M^* and C^* have been compared with the corresponding (henceforth, referred to as "true" or "real") values assigned to the structural parameters of the original nonlinear system used to produce the response histories which served as the observation vector, after adding a noise component, in this analysis. The components of the coefficient matrix $\mathbf{D}^*$ thus estimated have been compared with the values computed by the conventional equivalent linearization procedure according to the method developed by Krylov and Bogliubov and used by Malhotra and Penzien (1970), Paliou et al. (1986) and Shinozuka et al. (1977).

From the comparisons, the following conclusions can be drawn: (1) The estimation performed in this study resulted in stiffness coefficients which on the average deviated by 0.6% (lower, for W=25 ft/sec) and 5.5% (higher, for W=75 ft/sec) from the corresponding "true" values. (2) The components of the inertia coefficient matrix $\mathbf{M}^*$ were in good agreement with those used in the original nonlinear equation of motion, deviating 1.2% and 15% for the two wind velocities, respectively. (3) The components of $\mathbf{D}^*$ showed an average deviation of the order of 30% for W = 25 ft/sec and 15% for W = 75 ft/sec from which it can be concluded that the conventional frequency domain analysis overestimated the damping more in lower windspeeds than in higher ones; and finally, (4) As in most methods of estimation, the structural damping coefficients were poorly estimated, with average deviations ranging from 55-180% for the case of lower windspeed and 35-260% for that of the higher windspeed. These relatively poor results appear to be due to the fact that the actual nonlinear equation of motion is approximated by linearizing the damping factors.

The standard deviations (S.D.) of the structural displacements and velocities for both the original nonlinear system and the proposed linear system have been computed in the time domain using (i) a sample of size one and (ii) a sample of size one hundred, respectively. The results reveal that the standard deviations of the linear system underestimate those of the actual nonlinear system in both cases. These deviations range from 8% to 15% depending on the case considered, the value of the windspeed and the type of structural response.

In order to compute the standard deviation of the structural response by frequency domain analysis, modal analysis has been performed on both linear systems:

one obtained using the procedures in this study and the other using the conventional equivalent linearization method developed by Krylov and Bogliubov and used by Malhotra & Penzien (1970), Paliou et al. (1986) and Shinozuka et al. (1977). Comparison indicates that the proposed linear system results in lower values of the standard deviations than the conventional equivalent linearization method by an average of 7.5% for W = 25 ft/sec and 7% for W = 75 ft/sec, for both structural displacements and velocities.

The values of standard deviations thus computed are displayed in Table 3. Another interesting point here is that, comparing the results of the time domain analysis using the "true" values of coefficients and a sample of size one hundred with those of the frequency domain analysis using Krylov and Bogliubov's method, the frequency domain analysis produces higher standard deviations than the time domain analysis (by 3.5%) for both structural displacements and velocities at the lower windspeed , whereas it produces lower values for the displacements (by 0.5%) and the velocities (by 5.5%) at the higher windspeed. On the other hand, using the proposed linear system, the frequency domain analysis always underestimates the corresponding results of the time domain analysis by 4% for W = 25 ft/sec and 10% for W = 75 ft/ sec, for both displacements and velocities. The better agreement is obtained from the use of the conventional linearization method in this respect. This is due to the fact that in this case all the structural parameters maintain their exact values and only the linearized coefficient matrix $\mathbf{D}^*$ is estimated. In the proposed linear system, however, all the matrices are assumed to be unknown and are estimated by the indentification process. Therefore, a larger deviation is not unexpected. Another interesting conclusion is that if one averages the standard deviations obtained by the frequency domain analyses performed on the two linear systems mentioned above, the result is very close to that of the time domain analysis using the "true" coefficients and a sample of size one hundred. More specifically, these deviations are of the order of 0.1% for W = 25 ft/sec and 3.5% for W = 75 ft/sec, for both structural displacements and velocities.

Summarizing, it can be concluded that the results obtained using the Kalman filtering method are very satisfactory from the engineering point of view. Furthermore, the proposed identification process consists of an easy algorithm and is quite efficient in terms of computer time. These features offer an efficient alternative to the conventional linearization method developed by Krylov and Bogliubov and used by Malhotra & Penzien (1970), Paliou et al. (1986) and Shinozuka et al. (1977). In addition, starting with a structure with unknown structural parameters, the proposed identification procedure produces very good estimates of the response time history and more importantly reliable values of the structural parameters.

References

1. Shinozuka, M., Yun, C.-B. and Vaicaitis, R., 1977, "Dynamic Analysis of Fixed Offshore Structures Subjected to Wind Generated Waves," *Journal of Structural Mechanics*, Vol. 5, No. 2, pp. 135-146.

2. Yun, C.-B. and Shinozuka, M., 1980, "Identification of Nonlinear Structural Dynamic Systems," *Journal of Structural Mechanics*, Vol. 8, No. ST2, pp. 187-203.

3. Shinozuka, M., Yun, C.-B. and Imai, H., 1982, "Identification of Linear Structural Dynamic Systems," *Journal of the Engineering Mechanics Division*, ASCE, Vol. 108, No. EM6, pp. 1371-1390.

4. Shinozuka, M., 1974, "Digital Simulation of Random Processes in Engineering Mechanics with the Aid of FFT Technique," *Stochastic Problems in Mechanics*, University of Waterloo Press, pp. 277-286.

5. Paliou, C., Shinozuka, M. and Chen, Y.-N., 1986, "Reliability and Durability of Marine Structures," accepted for publication in the *Journal of Structural Engineering*, ASCE.

6. Paliou, C. and Shinozuka, M., 1985, "Reliability of Offshore Structures," Technical Report submitted to American Bureau of Shipping under Contract No. ABS CU00190901, Columbia University, New York.

7. Paliou, C. and Shinozuka, M., 1987, "Identification of Equivalent Linear Systems," submitted for publication in the *Journal of Engineering Mechanics*, ASCE.

8. Kalman, R. E., 1960, "A New Approach to Linear Filtering and Prediction Problems," *Journal of Basic Engineering*, ASME, Vol. 82, pp. 35-45.

9. Kalman, R. E. and Bucy, R. S., 1961, "New Results to Linear Filtering and Prediction Theory," *Journal of Basic Engineering*, ASME, Vol. 83, pp. 95-108.

10. Hoshiya, M. and Saito, E., 1984, "Structural Identification by Extended Kalman Filter," *Journal of Engineering Mechanics*, ASCE, Vol. 110, No. 12, pp. 1757-1770

11. Malhotra, A. and Penzien, J., 1970, "Non-deterministic Analysis of Offshore Structures," *Journal of Engineering Mechanics*, ASCE, EM6 (paper 7777), pp. 985-1003.

12. Jazwinski, A. H., 1970, "*Stochastic Processes and Filtering Theory*," Academic Press, New York, N. Y.

13. Gelb, A., 1974, "*Applied Optimal Estimation*," MIT Press, Cambridge, Massachusetts.

14. Lewis, L. L., 1986, "*Optimal Estimation*," John Wiley & Sons, Inc., New York, N.Y.

Table 1a. Structural Parameters

Mass M $[lb]$		Stiffness K $[lb/sec^2]$		Damping C $[lb/sec]$		Inertia C_M $[lb]$		Drag C_D $[lb/ft]$	
800	0	3000	-3000	150	-89	320	0	48	0
0	4000	-3000	-10000	-89	600	0	2000	0	240

Table 1b. Structural Coefficients

$M^{-1} \cdot K$ $[1/sec^2]$		$M^{-1} \cdot C$ $[1/sec]$		$M^{-1} \cdot C_M$ $[]$		$M^{-1} \cdot C_D$ $[1/ft]$	
3.75	-3.75	0.187	-0.111	0.4	0	0.06	0
-0.75	2.50	-0.022	0.150	0	0.5	0	0.06

Table 2 Estimated Values of Parameters

Param.	Exact Values	Init. Values	Estim. Values $W = 25$ ft/sec	Estim. Values $W = 75$ ft/sec
C^*_{11}	0.187	0.400	0.295	0.258
C^*_{21}	-0.022	0.000	-0.063	0.080
C^*_{12}	-0.111	0.000	-0.172	-0.299
C^*_{22}	0.150	0.400	0.249	0.097
K^*_{11}	3.750	5.000	3.730	3.944
K^*_{21}	-0.750	0.000	-0.737	-0.744
K^*_{12}	-3.750	0.000	-3.766	-4.234
K^*_{22}	2.500	4.000	2.488	2.578
M^*_{11}	0.400	0.200	0.405	0.353
M^*_{22}	0.500	0.300	0.506	0.597
D^*_{11}	0.092 *	0.300	0.047	
D^*_{22}	0.051 *	0.300	0.038	
D^*_{11}	0.441 *	0.300		0.357
D^*_{22}	0.330 *	0.300		0.289

* These parameters were computed by the conventional linearization method.

Table 3a. Standard Deviations of Mass Displacements

and Mass Velocities, W = 25 ft/sec

Stand. Deviat.	Time Domain 1 Sample	Time Domain 100 Samples	Freq. Domain	Parameters
Displacements				
Mass 1	0.642 ft	0.618 ft	0.640 ft	exact
Mass 1	0.577 ft	0.569 ft	0.592 ft	estimated
Mass 2	0.416 ft	0.400 ft	0.414 ft	exact
Mass 2	0.376 ft	0.371 ft	0.385 ft	estimated
Velocities				
Mass 1	0.764 ft/sec	0.729 ft/sec	0.755 ft/sec	exact
Mass 1	0.682 ft/sec	0.666 ft/sec	0.692 ft/sec	estimated
Mass 2	0.492 ft/sec	0.470 ft/sec	0.486 ft/sec	exact
Mass 2	0.445 ft/sec	0.435 ft/sec	0.450 ft/sec	estimated

Table 3a. Standard Deviations of Mass Displacements

and Mass Velocities, W = 75 ft/sec

Stand. Deviat.	Time Domain 1 Sample	Time Domain 100 Samples	Freq. Domain	Parameters
Displacements				
Mass 1	2.130 ft	2.050 ft	2.044 ft	exact
Mass 1	1.860 ft	1.810 ft	1.837 ft	estimated
Mass 2	1.360 ft	1.300 ft	1.291 ft	exact
Mass 2	1.240 ft	1.200 ft	1.218 ft	estimated
Velocities				
Mass 1	1.530 ft/sec	1.520 ft/sec	1.445 ft/sec	exact
Mass 1	1.280 ft/sec	1.310 ft/sec	1.329 ft/sec	estimated
Mass 2	0.968 ft/sec	0.952 ft/sec	0.892 ft/sec	exact
Mass 2	0.831 ft/sec	0.843 ft/sec	0.856 ft/sec	estimated

Note : 1 ft = 0.3048 m

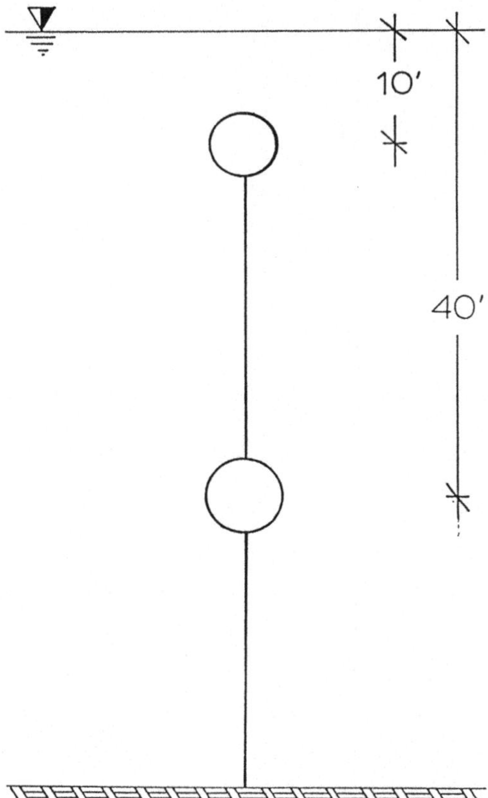

Fig. 1 Offshore Structure Idealized as a 2 D.O.F. system

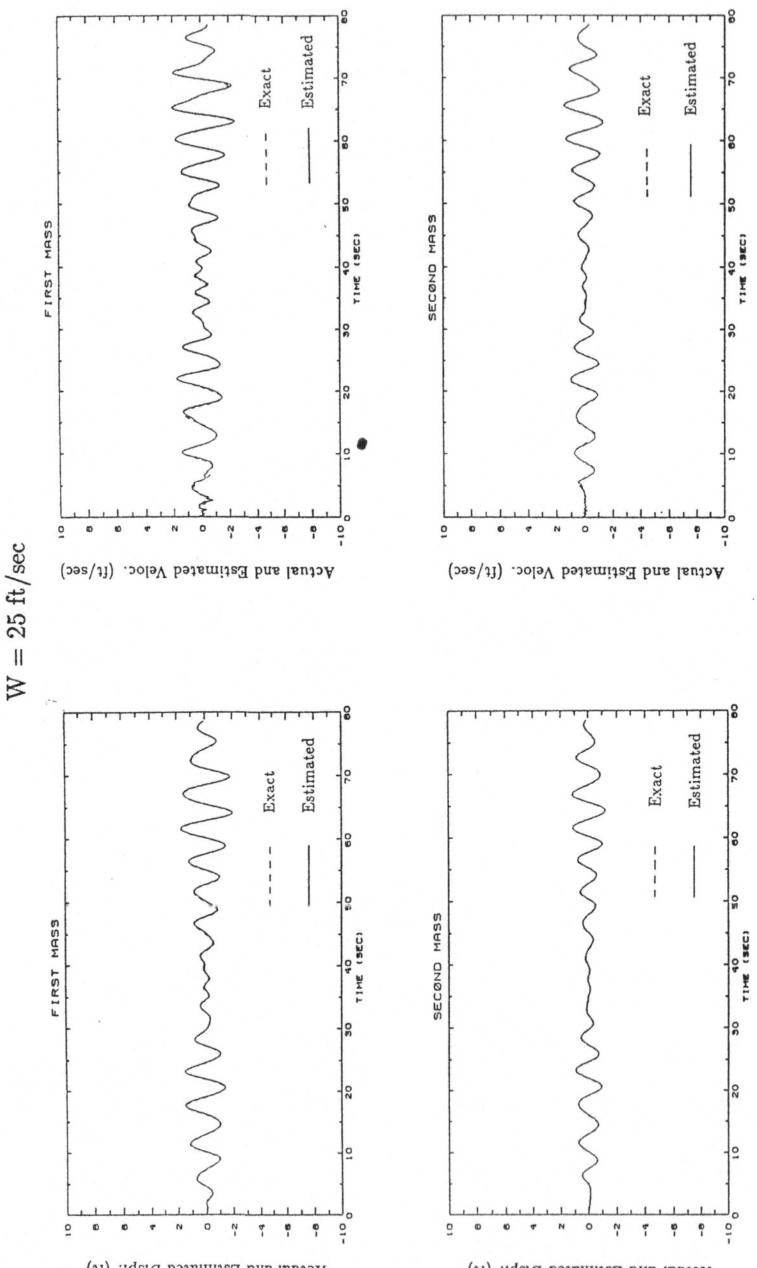

Fig. 2 Actual and Estimated Mass Displacements vs. Time   **Fig. 3** Actual and Estimated Mass Velocities vs. Time

W = 75 ft/sec

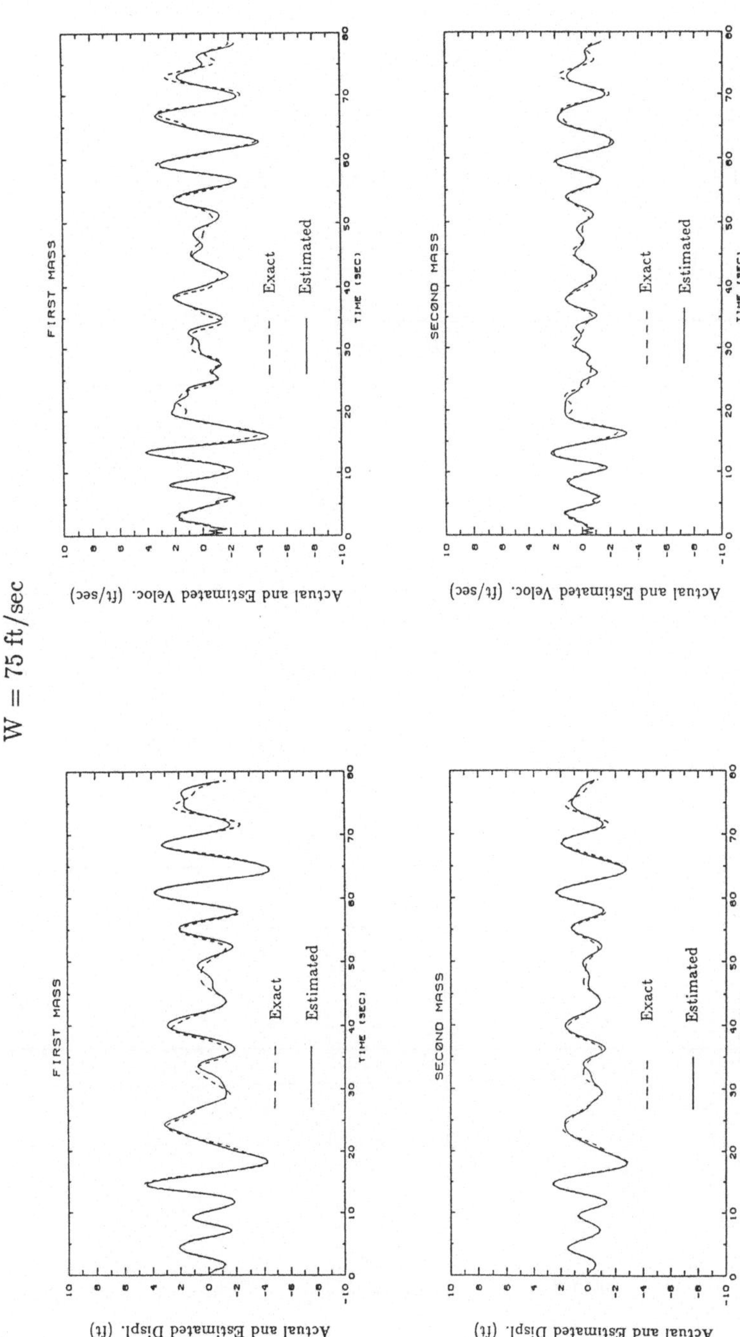

Fig. 4 Actual and Estimated Mass Displacements vs. Time   Fig. 5 Actual and Estimated Mass Velocities vs. Time

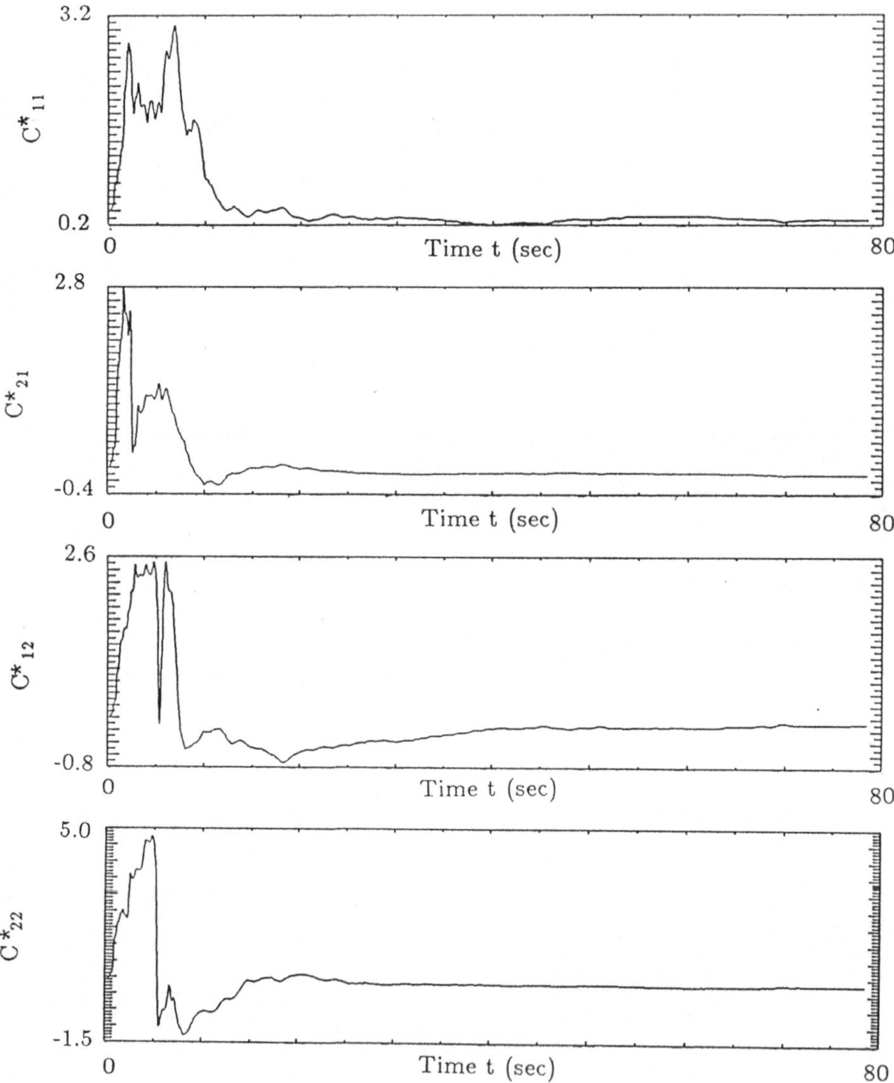

Fig. 6 Estimated Structural Coefficients ($\mathbf{C}^*$) vs. Time, W = 25 ft/sec

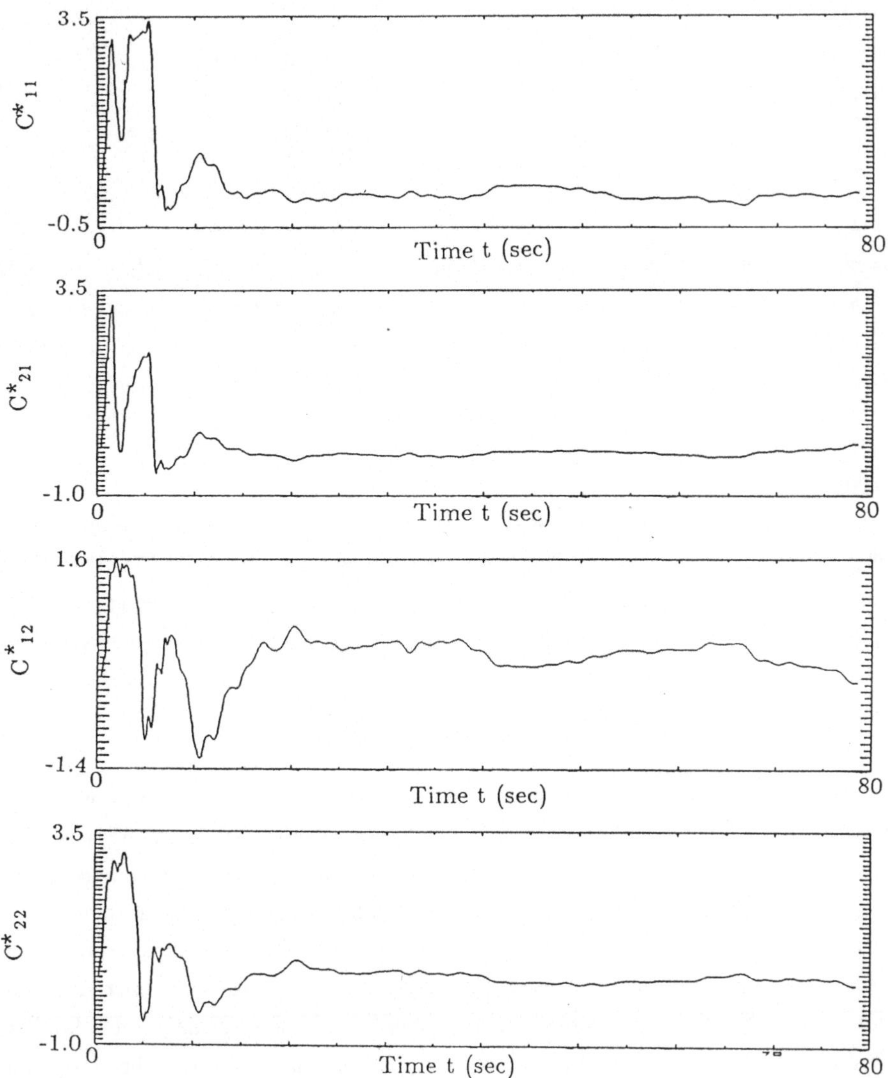

Fig. 7 Estimated Structural Coefficients (**C***) vs. Time, W = 75 ft/sec

Reliability of an Identification System for Predicting Incipient Capsize Due to Chaotic Rolling Motion of a Ship

E. Yarimer and L.N. Virgin
Department of Civil and Municipal Engineering
University College London

Abstract

The nonlinear dynamical roll motion equation for a ship is identified using a standard recursive estimation technique. The quality of the damping parameter estimates is found to be affected by the input noise and by the effect of finite window width; however the estimates are not adversely affected by the approach of escape induced by critical parameter values.

1. Introduction

The prediction of dangerous excursions in dynamical response of floating vessels in real time is a subject of obvious topicality. Both state estimation and system identification are applicable to this problem, although with different objectives: in the former approach state estimation and feedback controls would be used in order to limit the instantaneous response; in the latter approach the aim is to predict whether the vessel's motion is slowly entering a regime in which abnormal behaviour can be expected, and if so to take preventive action.

Our interest in the present study is centred on the latter approach. Recent studies of nonlinear dynamical stability have clarified the role of the control parameters in taking a nonlinear system through regimes of qualitatively different behaviour. Moreover methodologies have been developed to relate the parameter values to stability indices and so to delineate the critical parameter ranges. Of course, these are still subject to the usual uncertainties arising from the modelling simplifications used, e.g. neglect of coupling with other degrees of freedom, poor choice of functional representation of nonlinearities, etc. Similarly, identification results are, by definition, of a statistical nature. Thus, when combining identification techniques with stability theory, special care must be taken over the reliability level of the predictions.

Another approach to the prediction of large roll-angle excursions is that employed in Refs.1,2. From a knowledge of the form, and coefficient values of the nonlinear roll-angle equation, the authors deduce the probability that a certain measure of the roll angle envelope process remains below a threshold level, and, from that, the mean first-passage time of the response across a given high level. In this approach the roll angle is assumed to be a narrow-band random process and no direct reference is made to the dynamical stability of the underlying periodically forced system. A variant of this approach is the parametric excitation model, in which the restoring moment is treated as a random process (Ref. 3).

The identification scheme used in the present study (recursive ordinary least squares) is one of the simplest and most basic. This choice was made possible by the reasonable assumption that in this one-degree-of-freedom system, the state can be observed without error, so only a parameter identification is required. Other possible approaches to identification are the Extended Kalman Filter (Refs. 4,5) and, if the dynamics were linear, the ARMA model estimation by correlation matching (Ref. 6).

2. Dynamics of Ship Roll Motion

Equations of motion describing the coupled 3 degree-of-freedom (sway, roll and yaw) dynamic behaviour of a ship are available (Ref. 7). Nonlinear versions of these equations have been discussed (Ref. 8). These are generally difficult to implement. However it is commonly accepted that uncoupled roll motion can be studied in isolation as an adequate representation of significant dynamic behaviour. The following model,

$$m\ddot{\phi} + g(\dot{\phi}) + h(\phi) = f\cos\omega t + \varepsilon \tag{1}$$

has been used in dynamic stability studies of the rolling of a ship (Refs. 9,10). In this equation the coefficient m is the mass plus added mass, g is the nonlinear damping function, both being frequency dependent in general. The applied moment intensity coefficient f is also related to frequency through the hydrodynamics of wave forces. The nonlinearity of g in ϕ will be neglected here. On the other hand, the nonlinear nature of the righting moment term h is crucial in the mechanism of capsize. This term will be assumed to have the form $C\phi - D\phi^2$ causing h to drop to zero at $\phi = C/D$. The real restoring moment function would be an odd-degree polynomial in ϕ. However a static bias in ϕ must be incorporated to account for a slight asymmetry in the equilibrium position. The effect is to shift the stiffness curve by a fixed amount. With ϕ measured from the new zero-crossing position the stiffness is an asymmetric function and it is sufficient for the present purpose to represent this by the above quoted quadratic expression. Fig. 1 shows this curve, and the corresponding potential energy function.

Roll response can be visualised as the oscillation of a ball on this surface, under the action of the external driving force. This oscillation may be periodic at fundemantal frequency, or it may contain subharmonic components and in certain special circumstances it may follow a cascade of period-doubling bifurcations leading to chaos. A gradual change in the forcing amplitude (or frequency) will take the system from one of these regimes into another, and the passage may be accompanied by transients. Thus, even in the absence of noise, the system may cross over the potential energy hump and "escape" to infinity, corresponding to a capsize event.

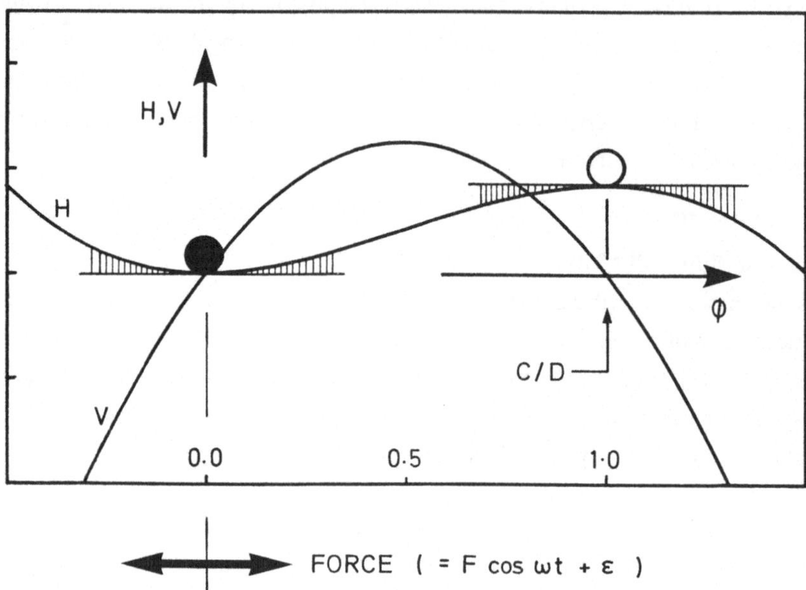

Fig. 1   Schematic diagram of restoring force and potential energy versus roll angle.

Figure 2 shows the generic ways in which the system may experience the onset of dynamic instability. The control variables are the forcing amplitude F and forcing frequency ω. Under increasing ω a cyclic fold instability causes a jump from a to higher amplitude resonant response at b. The system may also undergo a flip bifurcation at point c, which results in a doubling of the period of the response. The motion now exhibits alternating peaks of low and high level (Fig. 3). Further period doublings follow the first one for very small increases of the control ω , each bifurcation being accompanied by an increase of the peak response amplitude. It is possible to obtain similar instabilities by varying F under constant ω, and this is the option used in this study. Furthermore, under the chosen parameter values, this case study is confined to the flip bifurcation. It is important to appreciate that the critical value of F causing instability and eventual ecsape is considerably less than the statically applied force

185

required to push the system beyond the angle of vanishing stability. This is due to the phenomenon of dynamic amplification, and would not be as pronounced away from resonance.

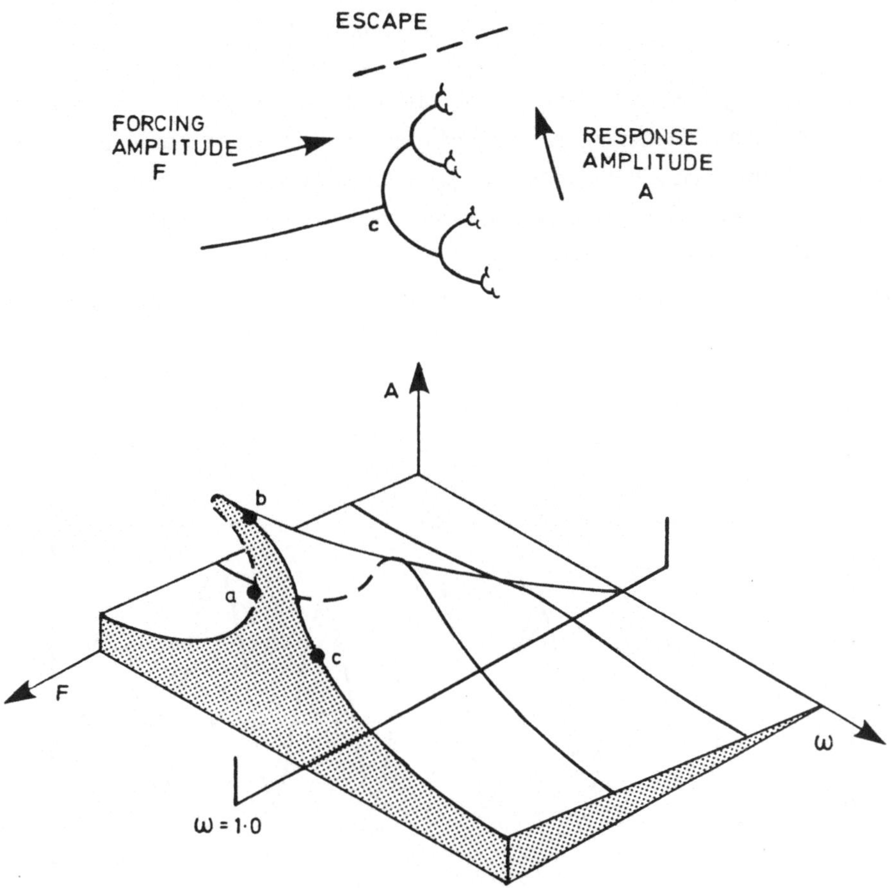

*Fig. 2   Bottom: Amplitude response of the harmonic solution
as a function of the control parameters F and ω.
Top: Cascade of period-doubling bifurcations.*

For a marine structure containing a certain type of bilinear stiffness, it has been shown that the inclusion of input noise has the effect of attenuating the peaks of the subharmonic response (Ref. 11). However, for the present system with its softening spring characteristic, the noise alone may be sufficient to enhance the displacement amplitudes and cause escape, when F is still below the threshold of bifurcation to subharmonics.

3. Generating the Time Series

A simplified version of the equation of motion still allowing the main features of capsizing behaviour to be reproduced is the following:

$$\ddot{\phi} + 0.1\dot{\phi} + \phi - \phi^2 = F \cos t + \varepsilon \tag{2}$$

The corresponding linear oscillator has an effective damping ratio of 0.05, a natural frequency of $1/2\pi$ and the forcing frequency of $1/2\pi$, thus inducing resonance. From previous studies, it is known that with the forcing frequency fixed at this value, the nonlinear system performs small amplitude periodic oscillation for values of F up to approximately 0.18, when it bifurcates into period two. In this regime, the response is of the type shown in Fig.3 , and passes through further period doublings and chaos for only a very small increase in F. In general chaotic motion is bounded and not necessarily a route to instability in the large. However this particular system has been observed to "escape" immediately following entry into chaos (Ref. 12).

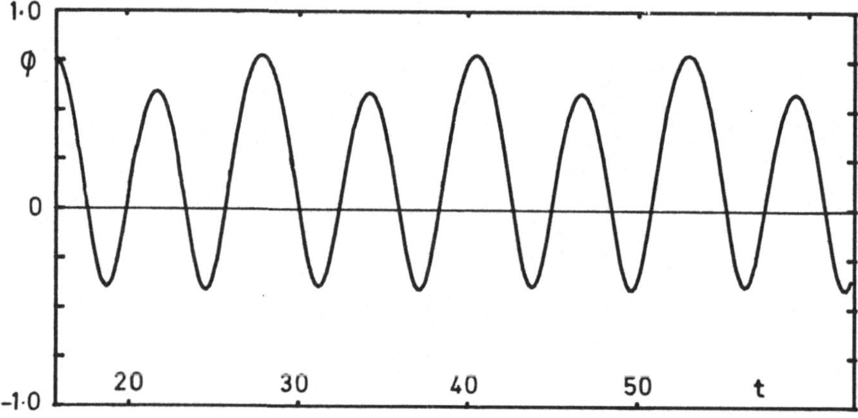

Fig. 3   Time series illustrating a subharmonic motion of order
2. (Case 9 of Table 1.)

The time series were produced by numerically integrating eq. 2 using a 4th order Runge-Kutta scheme. Noise was incorporated into the forcing term at each time step as random pulses from a Uniform distribution. For small values of F the choice of initial conditions for the integration is relatively unimportant, since an initial transient will decay onto the fundamental solution. However as F is increased, there is a likelihood of the displacement escaping over the zero stiffness limit without settling into the periodic solution, if the starting conditions are not sufficiently close to that periodic solution. This sensitivity increases as F approaches its critical value. This is illustrated in Fig. 4a when F=0.19 and the system was started close to, but not on the attractor. The actual initial conditions are recorded in Table 1. The phenomenon can occur in the absence of noise, although noise makes it more probable. Thus Fig. 5a shows the system under the same value of F, and started exactly on the attractor, but subjected to a relatively small level of noise.

4. Stability Prediction

The results of the present study suggest that the parameter estimation process itself is stable even in a dynamically unstable regime. In an actual hazard monitoring system the estimated parameters must be used to predict the impending onset of instability: one possible method would be to compare the estimated parameters with some theoretically determined critical values. At the present time only approximate analytical solutions are available to chart the regions of instability in the space of model parameters (Ref. 13). These critical domains have not yet been fully mapped in the exact numerical sense, and if they were to be explored in the future it seems unlikely that an exhaustive range of dynamical models would be covered by that process, since the available tools are computationally rather expensive.

An alternative method would consist of an on-line analysis of stability running concurrently with the identification routine; thus, given a nonlinear equation of motion, the problem is to arrive at a stability assessment via a quick algorithm. Recent research into the numerical determination of Lyapunov exponents for dynamical systems promises to make quick assessments possible. The writers have not had the opportunity to evaluate this method so far, but include it among their future plans.

5. Identification Procedure and Results

The recursive least squares scheme was used, with the model written as:

$$
\begin{aligned}
(\phi_N - 2\phi_{N-1} &+ \phi_{N-2}) \\
&+ B.dt.(1.5\phi_{N-1} - 2.0\phi_{N-2} + 0.5\phi_{N-3}) \\
&+ C.dt^2.\phi_{N-1} - D.dt^2\phi_{N-1}^2 \\
&= F.\cos\left(\frac{2\pi}{T}(N-1)dt\right).dt^2 \\
&\quad + G.\sin\left(\frac{2\pi}{T}(N-1)dt\right).dt^2 + \varepsilon.dt^2
\end{aligned}
\tag{3}
$$

where dt is the time step, taken as 1/40 of the forcing period T, the same as the spacing of the data points from the simulation program. Note that the first derivative has been approximated by a 3-step backward difference approximator for improved accuracy. Also note that a sine term has been introduced in order to account for the possible phase difference between the model and the time series presented for identification. This device was found to be important to the success of the scheme: the term F in Eq. 2 is estimated as the modulus of (F,G) of Eq. 3. Equation 3 was rewritten in the form:

$$y_N = X_N^{tr}.\theta + c_N + \Xi \tag{4}$$

with the parameter vector θ containing (B, C, D, F, G) and the scalar observation y_N consisting of ϕ_N itself. The vector of regressors X_N was taken as the appropriate multipliers of θ_i in Eq. 3. c_N consisted of the "offset" terms $2.\phi_{N-1} - \phi_{N-2}$.

The classical updating equations were used (Ref. 14) with or without a finite data window. The finite data window allows past data beyond the window to be forgotten, and is intended to be used when the parameters are evolving with time. In all the numerical experiments the unknowns $\{H\}$ were started from the values (0.0 0.0 0.0 0.0 0.0), and the normalised covariance matrix from $(P) = \mathrm{diag}\,(\,10^9\ 10^9\ 10^9\ 10^9\ 10^9\,)$. Double precision (32 bit word length) arithmetic was employed in the calculations. Since the data points were closely spaced at (1/40 T), identification steps were taken at every 2 data points.

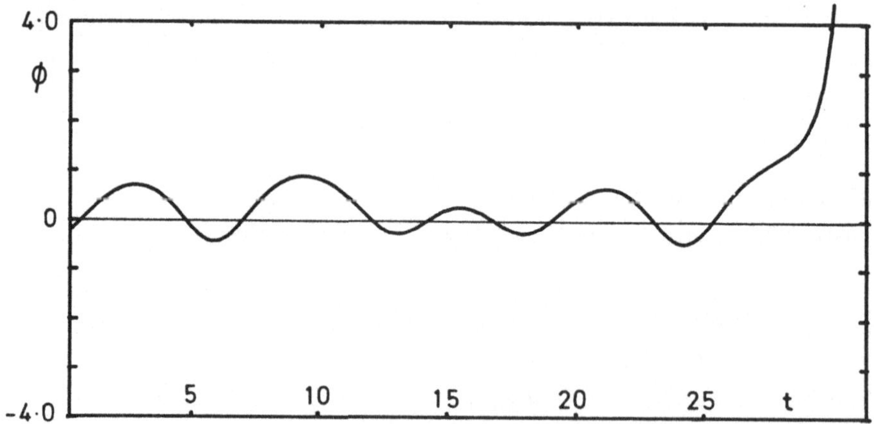

Figure 4a. Time series showing escape due to initial transient (Case 13).

Table 1 shows the selection of cases tried and the outcomes. These initial trials were run without a finite data window. The noise parameter E gives *as a multiple of F* the half range of the Uniform distribution from which the random pulses were drawn in the simulation phase.

The overall trend was of rapid convergence to accurate estimates of B,C,D,F and G in cases with zero noise (first column of Table 1), with convergence becoming less steady and less rapid with the introduction of increasing amounts of noise. However it is noted that the recursive identification process never became numerically unstable because of noise before the system itself did so (i.e. escaped).

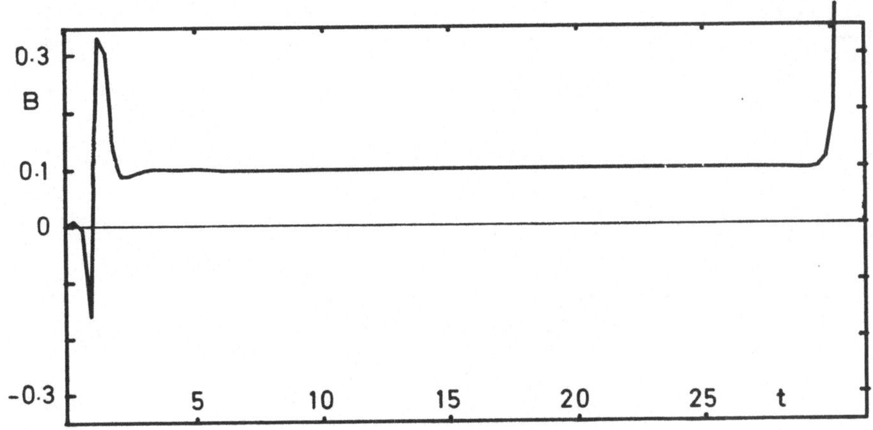

Figure 4b. Estimate of the damping coefficient B for Case 13.

Of special interest is Case 13, where escape took place in the absence of noise, due to poor initial conditions (Fig. 4a). Fig. 4b shows the progress of the iterations for the damping coefficient B for this case (the true value of B was 0.1 in all cases). Another interesting case is No. 9P5, where escape occurred almost at the end of the available record (Figs. 5a, 5b). From these Figures it can be seen that the identification of parameter B was converging well even at the very onset of escape. In other cases it was observed that parameter estimates held their accuracy until within 10.dt (or 1/4 T) of the capsize event.

6. Reliability Considerations

In order for the proposed parameter estimation technique to serve its purpose, it must be implemented with a finite data window, and it must yield the estimates with acceptable error margins. In ordinary least squares estimation the error is quantified by the covariance matrix (**P**) computed by the algorithm.

E F	0.0	0.5	2.5	5.0	10.0	50.0
0.05 (-0.1835, 0.2748)	Case 1 A	Case 1P2 A	Case 1P5 A	Case 2 N	Case 3 N	Case 4 NE
0.12 (-0.3263, 0.2564)	Case 5 A	Case 5P2 A	Case 5P5 A	Case 6 N	Case 7 NE	Case 8 XE
0.19 (-0.3193397, 0.34607)	Case 9 A	Case 9P2 N	Case 9P5 NE	Case 10 NE	Case 11 XE	Case 12 XE
0.19 (-0.2500, 0.34607)	Case 13 NE	Case 13P2 NE	Case 13P5 NE	Case 14 XE	Case 15 XE	Case 16 XE

Table 1. Non-windowed runs

Legend:
A   -   Accurate parameter estimates
N   -   Estimates noise-affected but acceptable
NE   -   Escape after estimates reached acceptable values
XE   -   Escape before estimates reached acceptable values

The initial values of displacement and velocity shown in the first column (except those in the last row) ensure that the motion starts on the steady state solution i.e. without initial transients.

Window width	=Forcing cycles	μ_B	var(B)	P_{11}	var(cdt^2).P_{11}
240	6	0.111	6.47 E -3	0.58 E 4	4.5 E -3
480	12	0.108	2.10 E -3	0.27 E 4	2.1 E -3
720	18	0.103	0.88 E -3	0.20 E 4	1.4 E -3
960	24	0.103	0.53 E -3	0.15 E 4	1.2 E -3

Table 2. Windowed runs for Case 2.

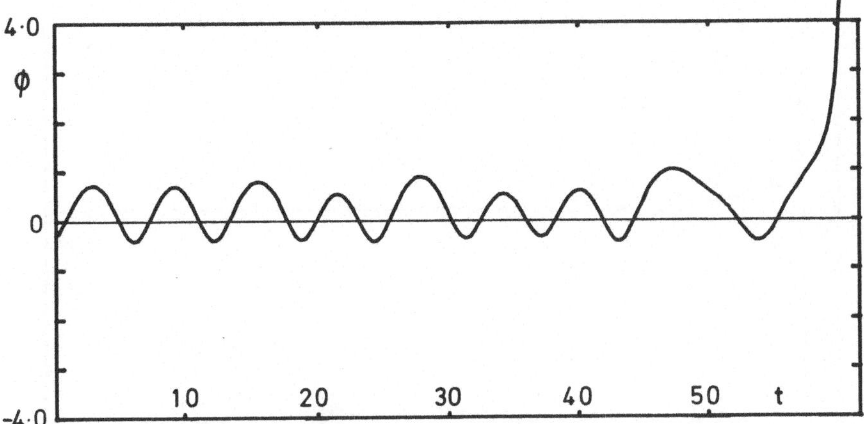

Figure 5a. Time series showing escape due to noise (Case 9P5).

To obtain the covariance matrix of the estimates, (**P**) must be scaled by the variance σ_Ξ^2 of the noise term Ξ in eq. 4. However the standard derivation of (**P**) assumes that the regressors vector **X** is non-random, whereas here the regressors contain the past values of the random roll angle process ϕ. In general, the values of P_{ij} converge slowly in a recursive algorithm, and it is difficult to pick out their correct asymptotic value; different starting values could lead to different results in otherwise identical runs. Moreover it is found that the variance of the noise Ξ in equation (4), as estimated from the residuals in the identification routine, does not match $(dt)^4$ times the variance of ε in equation (2). This is thought to be due to the fact that the response of the 1 DOF oscillator only reflects the input noise present in the vicinity of its resonant frequency. Therefore, in this application the theoretical variance of the estimates could not be trusted as an error measure; instead, it was decided to evaluate the statistics of the estimated parameter from a long computer run, using the time-averaging method. Since a constant data window width was maintained, the θ process was stationary during such a run. This experiment was conducted with different data window widths, but only the problem labelled as Case 2 in Table 1 was considered. The results are shown in Table 2.

The results in Table 2 pertain to the damping coefficient B, which turns out to be the most sensitive of all the parameters to noise and window width. In this table the colums headed μ_B and var(B) contain the statistics of B as obtained from time-averaging. The quantities P_{11} are the corresponding element of matrix (**P**) as calculated by the recursive algorithm. It is seen that the variability of the estimated B improves with window length. For the longest window used, the coefficient of variation of B was 0.22, a value which suggests that an even wider window should be chosen in practice, for this level of noise in the forcing term. From the last

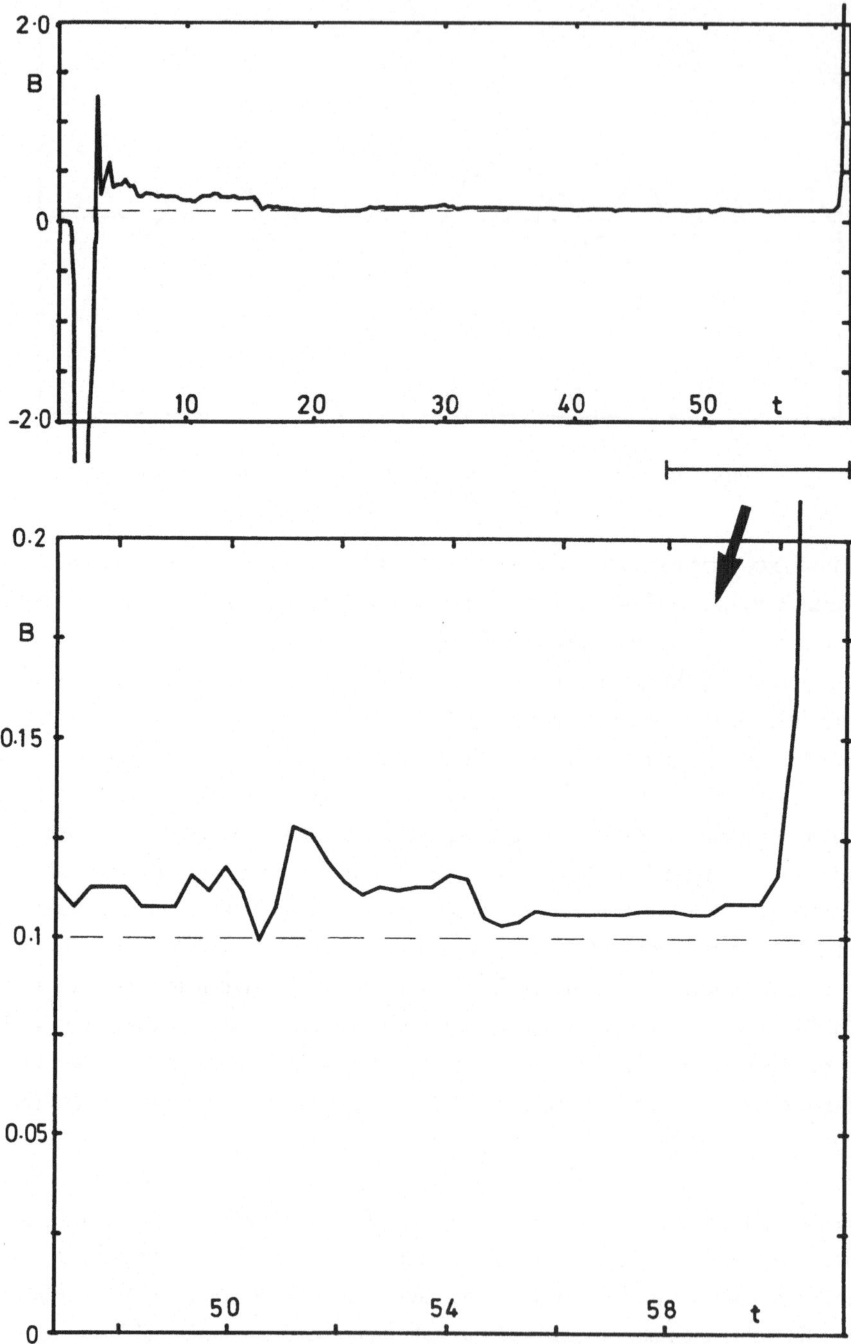

Figure 5b. Estimate of the damping coefficient for Case 9P5.

two columns, it can be seen that the predicted variances are correct in terms of order-of-magnitude but not exactly identical with the observed variances. In view of the difficulty in judging convergence of **P**, this agreement could be fortuitous. The noise variance used in the last column was itself estimated by time averaging the square of the residual term in equation (3); this value (7.7 E -7) differed considerably from the source noise variance introduced in the numerical integration of Eq.(2), which was $var(\varepsilon.dt^2)$ =1.266 E -5. The probable reason for this discrepancy has been discussed above.

7. Conclusions

The recursive ordinary least squares algorithm was successful in identifying the parameters of an equation modelling the nonlinear roll response of a vessel under sinusoidal excitation and stationary white noise disturbances. The damping coefficient B was found to be the most sensitive parameter. In some cases of high noise level the estimates of B had not converged sufficiently when the capsize event took place. By contrast, the errors in all other parameters were minimal. The finite data-window version of the algorithm was also implemented with evolving conditions in mind. However the error in B increased both with the addition of noise and reduction in window length. Future extensions of this work should include a) a more faithful mathematical model of ship dynamics and wave loading, b) a cross-link with suitable dynamic stability prediction methodology.

8. References

1. J.B. Roberts and R.G. Standing, "A Probabilistic Model of Ship Roll Motions for Stability Assessment", Int. Conf. on The Safeship Project: Ship Stability and Safety, RINA, 1986, Vol.1, Paper No. 6.

2. J.J.H. Brouwers, "Stability of a Non-linearly Damped Second-Order System with Randomly Fluctuating Restoring Coefficient", Int. J. Non-Linear Mechanics, Vol. 21, No.1, pp.1-13, 1986

3. J.F. Dunne and J.H. Wright, "Predicting the Frequency of Occurrence of Large Roll Angles in Irregular Seas", Trans. RINA, 1986, Vol. 126.

4. M. Hoshiya and E. Saito, "Structural Identification by Extended Kalman Filter", J. of Engng. Mech., 1984, Vol. 110, No.12, pp.1757-1770.

5. C-B. Yun and M. Shinozuka, "Identification of Nonlinear Structural Dynamic Systems", J. Struct. Mech., 1980, Vol.8(2), pp.187-203.

6. N.K. Lin, "Real Time Estimation of Ship Motion using ARMA Filtering Techniques", Proc. Sixth Int. Symp. Offshore Mechanics and Arctic Engineering (OMAE), Houston, 1987, ASME, Vol.2, pp.325-329.

7. R.E.D. Bishop, W.G. Price and P. Temarel, "General Linear Antisymmetric Motions of a Rigid Ship", Int. Conf. on The Safeship Project: Ship Stability and Safety, RINA, 1986, Vol.1, Paper No. 5.

8. A.K. Brook , "The role of Simulation in Determining the Roll Response of a Vessel in an Irregular Seaway", Int. Conf. on The Safeship Project: Ship Stability and Safety, RINA, 1986, Vol.1, Paper No. 7.

9. J.G.H. Wright and W.B. Marshfield, "Ship Roll Response and Capsize Behaviour in Beam Seas", Trans. RINA, 1980, Vol. 122, p.129.

10. L.N. Virgin, "The Nonlinear Rolling Response of a Vessel Including Chaotic Motions Leading to Capsize in Regular Seas", Applied Ocean Research, 1987, Vol.9, No.2, pp.89-95.

11. E.R. Jefferys, "Non-linear Marine Structures with Random Inputs", Proc. Sixth Int. Symp. Offshore Mechanics and Arctic Engineering (OMAE), Houston, 1987, ASME, Vol.2, pp.311-318.

12. J.M.T. Thompson, S.R. Bishop and L.M. Leung, "Fractal Basins and Chaotic Bifurcations from a Potential Well", Physics Letters A, 1987, Vol. 121, No.3, pp.116-120.

13. G. Schmidt and A. Tondl, "Non-Linear Vibrations", 1986, Cambridge University Press.

14. G.C. Goodwin and R.L. Payne, "Dynamic System Identification: Experiment Design and Data Analysis", 1977, Academic Press, N.Y.

Correlation of Analysis and Test in Modeling of Structures: Assessment and Review

Samir R. Ibrahim

Professor of Mechanical Engineering and Mechanics
Old Dominion University
Norfolk, Virginia 23508, U.S.A.

ABSTRACT

For any structure, mechanical system or component, an accurate mathematical model is a necessity for the prediction of loads and responses, establishing stability margins, design of control system and integrity monitoring, among other uses. Analytical modeling techniques, even through fairly advanced, possess enough uncertainties to justify testing of the full scale actual structure, or a prototype, in order to verify, improve or optimize an existing analytical model. Test data, or identified parameters, are usually used to identify a dynamic model-mass, stiffness and damping distribution or optimize an existing analytical model.

In this paper, the problem of dynamic modeling of structures and model updating is discussed. The existing approaches are reviewed in order to evaluate the existing state of the art. Present limitations are summarized and recommendations for future work are presented.

INTRODUCTION

Analytical dynamic modeling of structures is usually carried out during the design stage. The analyst constructs a mathematical model in accordance with assumptions concerning the distribution of mass and stiffness and, usually, neglecting damping. The resulting model is a set of differential equations that may or may not be reasonably representative of the actual structure. Whatever inaccuracies that may be present in analytical models are mainly due to:

1. The approximation of boundary conditions.

2. The classical lumping of distributed parameter systems. (The solution is never unique.)

3. The estimation of the different physical properties of different structural materials.

4. The lack of damping representation.

5. Inadequate modeling of joints and couplings.

In spite of such familiar and common shortcomings of analytical modeling, the advantages over the experimental approach are numerous as well as evident. In addition to lower cost, modifications to optimize the dynamic behavior of the structure can be made during the design stage without the need for a prototype. Unfortunately, the uncertainties inherent in the existing analytical modeling techniques make experimental modeling, or model verification, a necessity. This is especially true for structures whose missions, integrity, safety or control critically depend on the structure's dynamic characteristics.

Direct experimental modeling (physical parameters identification) of structures would have seemed to be the natural way to augment analytical modeling. In this area, efforts to experimentally determine or derive the equations of motion of the structure, directly from its responses to input forces or initial excitation, have not been, so far, very successful. The techniques developed for such a purpose have the same levels, if not more, of uncertainties found in analytical modeling.

For reasons of uncertainties in both analytical modeling and direct identification of structures' mass, stiffness and damping distributions, the modal approach has naturally emerged as a classical approach in vibration testing. This is mainly due to its mathematical simplicity and physical observability. To this effect, most activities in structural dynamics are generally classified under the term modal analysis.

The experimentally identified, or measured, modal parameters - natural frequencies, mode shapes and damping factors - beside being directly related to the structures' physical parameters are as well useful, by themselves, in numerous applications.

Recently, the importance of these parameters as the basis for analytical model updating is becoming increasingly apparent.

Observing the trends in the literature, accurate identification of modal parameters as important as it is, is no longer the object and goal.

Rather, it is, as it should be, a stepping stone towards the ultimate goal of structural dynamics; an accurate mathematical model.

Analytical Model

Structures, as distributed parameter systems, should be described by partial differential equations. Because of the complexity of modern structures, the traditional beams, plates and shells governing equations become inadequate, if not incapable, of describing such structures. Few exceptions may hold to be true in cases such as truss structures. For such reasons, lumping of structures is presently an acceptable approach. Finite Element Modeling (FEM) is the most commonly used in analytical modeling. The FEM is a coupled set of second order ordinary differential equations in the form:

$$[M]\ \{\ddot{x}\} + [K]\{x\} = \{f(t)\}$$

It is quite common for such models to possess thousands of degrees of freedom to accurately represent a fairly complex structure. It is as common for these models to offer no information on damping.

In spite of the tremendously large size of FEM's, they are known to predict a fairly small number of modes that could be close to the structure's actual, or measured, modes. Usually in the order of one hundred modes, or less.

A better modeling of structures should contain information on damping. Damping itself is a problem of considerable complexity. In practice, different types of damping are usually approximated by viscous type damping. A more representative model for a structure with damping takes the form:

$$[M]\{\ddot{x}\} + [C]\{\dot{x}\} + [K]\{x\} = \{f(t)\} \tag{2}$$

An analytical model, thus far, challenges the dynamicist with the following major problems:

1. Possible inaccuracies.

2. Large model size.

3. Lack of damping representation.

Test Data

A test performed on the actual structure or a prototype is usually conducted to verify, improve, or merely evaluate an analytical model. The test data for such a purpose are either:

1. Input/output time or frequency functions.

2. Modal parameters λ's and $\{\psi\}$'s.

In either case, for practical limitations, the simultaneous force inputs to a structure are limited in number as well as frequency range. Also the number of measurements, even though usually larger than the number of force inputs, is limited and in most cases is much smaller than the number of degrees of freedom in the analytical model. The number of identified modes is restricted by the identification algorithms. A typical large test may comprise, for example, the following:

$$
\begin{array}{ll}
\text{Number of input forces} & 5 \\
\text{Number of measurements} & 200 \\
\text{Number of identified modes} & 50
\end{array}
$$

As in analytical modeling, test data or parameters present the following problems:

1. The number of measurement degrees of freedom is limited and may be different from those of the analytical model.

2. Some types of degrees of freedom, such as rotational ones, cannot be readily measured with present technology.

3. The number of identified mode shapes is limited.

4. The damped mode shapes, complex, are identified while the analysis is based on the normal mode theory.

5. Measured data or parameters contain levels of errors.

6. Some modes of the structure under test may not be excited in a certain test or, if excited, may not be identified.

Mathematical Model Identification/Updating

The problem here is classified into two main categories depending on the type of data or information used. The first category is test based only and no analytical model information exist or used. In such a case a test mathematical model is identified or derived from experimental data or experimentally identified modal parameters.

If the identified parameters are used, the approach is termed as modal approach. On the other hand, if input and output information are directly used to identify a mathematical model, the approach is then labeled as non-modal approach.

The second category makes use of the available analytical model together with test information to update or improve the existing analytical model. This category can also be based on modal or non-modal approaches. Figure 1 shows the schematic diagrams for the four approaches, namely:

I.1　Test based modal approach

I.2　Test based non-modal approach

II.1　Test-Analysis based modal approach

II.2　Test-Analysis based non-modal approach.

The distinct difference between the test based and the test-analysis categories is that the first one is limited to the test degrees of freedom, while in the second category, the updated model can have more degrees of freedom than the test.

DEVELOPMENT OF THE STATE-OF-THE-ART

The early seventies mark a noticable change in the overall subject of dynamic modeling. Both analysis and test techniques started at this time to formulate well defined directions and trends. In the area of analysis finite element techniques matured to the level of usefulness in dynamic analysis as a fairly dependable tool. Dynamic testing as well, started to advance from the classical resonance testing techniques to the science, and art, of modal identification.

During this era, as dictated by the needs of the growing space and aeronautical programs, the demand for an accurate mathematical representation of complex structures was too critical to ignore. The seventies as well witnessed a clear change in direction of dynamic modeling from test based approach to test-analysis correlation.

To contain the literature prior to that time, the survey paper by Young and On [1] concluded that there existed no technique which has really been proven or generally accepted as a means of deriving from test data

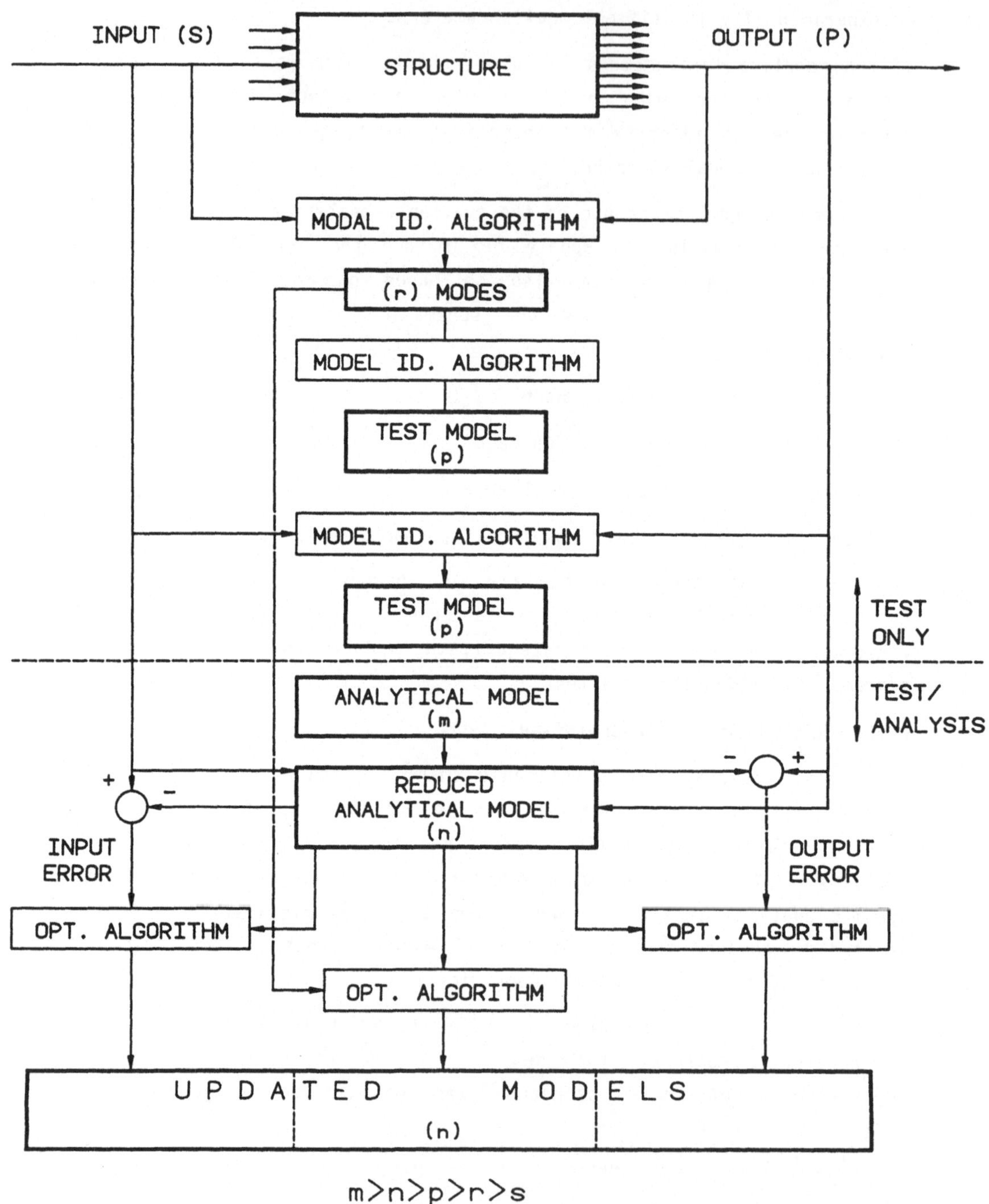

Figure 1: Schematic Diagram of Different Approaches
for Math Model Derivation/Optimization

a useful analytical model for a structure under test. A similar conclusion was also reached in a symposium on system identification of vibrating structures [2] in 1972.

Berman and Flannelly [3] marked the very beginning of the development of the state-of-the-art of dynamic modeling. They addressed the key and most important issue in the subject area; the completeness of the dynamic model. Even though they concluded that: It is felt that the surface has just been scratched and there is much to be learned about how the incomplete model behaves and how to use it to best advantage, they pointed out to the main obstacle in both test based and test-analysis based modeling and model updating techniques; the incompleteness of test data set by itself as compared to the analytical model.

If the number of measurements in the test is p, the number of modes excited or identified from the test is r and the number of degrees of freedom of the analytical model is n, the key issue in the completeness of the model may be simply stated as:

$$r < p < n \qquad\qquad (3)$$

For test based category of dynamic modeling completeness is achieved when r=p and in fact, if test data is relatively accurate, an exact and unique lumped parameter model or r, or p, degrees of freedom is derived. Such a model is far more superior to any finite element model of the same degrees of freedom.

Since a complete test based modal approach model is highly unlikely, work in this area ceased to continue in that direction.

Test based non-modal approaches did not have much success either due to the same problem of model completeness. Leuridan [4] and Coppolino [5], for example, offered two sensible approaches in time and frequency domains respectively. However, their work, even though they identify some mass, stiffness and damping matrices, is more of modal identification algorithms than modeling techniques.

Most of the subsequent work can be classified as test-analysis based, modal or non-modal. As a matter of fact, the majority of work in this area can be described as efforts to complete the incomplete test data set. Some valuable survey and review papers, to complete this study, have been offered by Wada and Chen [6], Caesar [7], Pal and Schidtberg

[8], Heylen and Sas [9] and Link [10].

GOVERNING EQUATIONS

The following equations are satisfied by both analytical and updated models:

$$[M]\{\bar{x}\} + [K]\{x\} = \{f(t)\} \tag{4}$$

$$[M][\phi][\Omega^2] = [K][\phi] \tag{5}$$

$$[\phi]^T[M][\phi] = [I] \tag{6}$$

$$[\phi]^T[K][\phi] = [\Omega^2] \tag{7}$$

If complex modes are implemented, then the governing equations become:

$$[M]\{\ddot{x}\} + [C]\{\dot{x}\} + [K]\{x\} = \{f(t)\} \tag{8}$$

$$\begin{bmatrix} 0 & I \\ -M^{-1}K & -M^{-1}C \end{bmatrix} \begin{Bmatrix} \phi \\ \lambda\phi \end{Bmatrix} = \lambda \begin{Bmatrix} \phi \\ \lambda\phi \end{Bmatrix} \tag{9}$$

$$[M_U] = [M_A] + [\Delta M] \tag{10}$$

$$[K_U] = [K_A] + [\Delta K] \tag{11}$$

$$[C_U] = [\Delta C] \tag{12}$$

Where the subscript A denotes analytical, the subscript U denotes updated and Δ designates perturbation. In the above equations, the order is usually and arbitrarily selected by the analyst. It can be the same order n of the finite element model, n_C for a condensed model where $n_C < n$ or p for a model with the same degrees of freedom as the test measurements.

The problem of condensing or reducing the order of the finite element model is critical to the subject of model updating.

MODEL REDUCTION

The test coordinates are, in most cases, limited in number and are insufficient to describe the dynamic behavior of the entire structure. Internal components are usually inaccessible for measurements and rotational degrees of freedom are not measured in the test. On the other hand, the analytical model is too large and over describe the structure with detailed coordinates that are of minor importance to the study of the dynamic behavior of structures. For these reasons several approaches have been proposed to eliminate the undesired degrees of form from the FEM. Such a process is termed model condensation or reduction.

Guyan [11] was the first to propose an approach based on the static influence coefficient as:

$$\{f\} = [K]\,\{x\} \qquad\qquad (13)$$

and by setting the forces corresponding to the degrees of freedom to be eliminated to zero, the reduced stiffness matrix is computed from the partitioned original stiffness matrix. Kaufman and Hall [12], Kidder [13], Sowers [14], Downs [15], Miller [16], Anderson and Hallauer [17], Kim and Anderson [18] and Wolf and Molnar [19] offered alternate techniques, or basis, for system matrices reduction.

COMPLETING MODAL VECTORS

The measured modal vectors pose three problems in using them for model updating. They contain errors, they represent complex modes while normal modes are used in orthogonality relations (equations (5) to (7)) and they do not contain all the degrees of freedom of the analytical model.

Baruch and Bar Itshack [20], Ibrahim [21], Zhang and Lallement [22], Niedbal [23] and Natke and Roterts [24] suggested approaches to compute normal modes from measured complex modes. These approaches must be used with care since the complexity of measured mode shapes could be the result of experimental or identification errors as shown by Deblauwe and Allemang [25]. If well formulated, these approaches can be envisioned to have smoothing effects, if not to compute normal modes.

To complete and to smooth the measured modal vectors, the approach presented by Baruch and Meirovitch [26] can be used for simpler structures. More sophisticated approaches are presented by Allemang [27], Berman and Nagy [28], O'Callahan et al [29] and Vold [30].

Again, in using such smoothing algorithms or completing techniques the limitations and assumptions inherent in them should be always and carefully considered. The assumptions such as that the measured modes are linear combinations of the analytical ones or that they satisfy the analytical model, may further adversely affect the model updating process.

COMPLETING THE MODEL

At this stage, it is assumed that the FEM has been reduced to a desired size and degree of freedom and that the experimentally determined modal parameters are fairly accurate and compatible with the analytical model.

Most model optimization techniques make use of equations (5) - (7), but these equations are not sufficient to solve for the perturbations $[\Delta M]$, $[\Delta K]$ and $[\Delta C]$ in equations (10) to (12). Additional criteria, function optimization or conditions are needed. Without efforts to complete the model, pseudo inverse and statistical approaches were used but proved to be impractical.

$$\varepsilon = \| M_A^{-1/2} \Delta M M_A^{-1/2} \| \tag{14}$$

from which $[\Delta M]$ is computed from the equations:

$$[\Delta M] = [M_A][\phi_E][m_A]^{-1} [I - m_A][\phi_E]^T[M_A] \tag{15}$$

where

$$[m_A] = [\phi_E]^T[M_A][\phi_E] \tag{16}$$

and the subscript E denotes experimentally determined quantities.

To correct the stiffness matrix, equations (5) and (7) are used together with minimizing the norm

$$\varepsilon = \| M_A^{-1/2} \Delta K M_A^{-1/2} \| \tag{17}$$

from which ΔK is computed as:

$$[\Delta K] = -[K_A\phi_E\phi_E^T M_U + M_U\phi_E\phi_E^T K_A] + [M_U\phi_E][\phi_E^T K_A\phi_E + \Omega^2][\phi_E^T M_U] \tag{18}$$

Berman's method has the advantages of not requiring prior analytical eigensolution and is non-iterative. However, it may produce extraneous modes in the frequency range of interest and it changes all elements of the matrices. Caeser [31,32] added some additional constraints, to Berman's direct matrix updating method, to preserve the total mass of the structure and to control interface forces and effective masses.

Using the higher modes of the analytical model to complete the set of measured modes is another approach, in dealing with incomplete dynamic model, that was suggested by Link [10,33] and Ibrahim [34]. Link refers to this approach as matrix mixing. This procedure is also direct, non-iterative and guarantees no extraneous modes within the frequency range of the measured set of modes.

SENSITIVITY BASED MODEL OPTIMIZATION TECHNIQUES

The use of sensitivity analyses has gained quite a popularity in model updating during recent years. Most techniques use the first order approximation of Taylor series. Assuming two sets of dependent parameters Q and p, these parameters can be related through the relation:

$$Q_{k+1} = Q_k + \sum_{j=1}^{q} \left[\frac{\partial Q_k}{\partial p_j} (p_{j,k+1} - p_{j,k}) \right] \tag{19}$$

where k indicates an iteration level. With the availability of an analytical model, such a model is usually used as a first iteration. The process is continued until a convergence criteria is met; usually the minimum difference between the measured parameters and those computed from the updated model.

These techniques require extensive computation. The advantages here are the ability to select the elements to be modified and, as well, the possible use of weighting functions to emphasize areas of major, minor or no change. Several techniques of this type are currently in use [35-41].

NON-MODAL MODEL OPTIMIZATION

In this approach the experimental inputs and outputs, rather than the modal parameters, are used to optimize an existing analytical model or possibly to derive a totally experimental mathematical model.

Natke [42] through his comparison between the two different experimental methods, modal and non-modal, used in modal improvement, he discussed his work using the measured input and output data to improve, by estimation, the parameters of an analytical model with approximately known uncertain positions.

Starting with known correction matrices $[\Delta M]_j$, $[\Delta C]_j$ and $[\Delta K]_j$ which designate the terms to be updated, at any iteration step i, the updated matrices take the form:

$$[M_U]_i = [M_U]_{i-1} + \sum_{j=1}^{N_m} \Delta a_j^i [\Delta M]_j \tag{20}$$

$$[C_U]_i = [C_U]_{i-1} + \sum_{m=1}^{N_c} \Delta a_m^{\ i} \ [\Delta C]_m \qquad (21)$$

$$[K_U]_i = [K_U]_{i-1} + \sum_{n=1}^{N_k} \Delta a_n^{\ i} \ [\Delta K]_n \qquad (22)$$

where

$$\Delta a_j^{\ i} = a_j^{\ i} - a_j^{\ i-1}, \cdots \qquad (23)$$

and

$$N_m + N_c + N_k \leqslant 3n^2 \qquad (24)$$

The coefficients a's are computed from minimizing the difference between the model and the test inputs or outputs. Iteration is continued until a convergence criteria is met.

Fritzen [43] paralleled this previous procedure and extended it for use with transfer function measurement data.

Neither Natke nor Fritzen discussed or imposed any conditions on the input forces, their spectra or number as related to those of the analytical model.

Badenhausen [44] presented a time domain approach for modal identification considering the incompleteness of his model.

DISCUSSION

This study has discussed some of the pioneering and plausible efforts that have tackled and offered some solutions to a very complex problem. It was not the intention to offer a complete survey or review of the subject of model updating but rather to emphasize the dimensionality and the size of the problem. The purpose of this paper was to further define the problem, present, in a general form, the existing approaches and most important to question the suitability of the present state of the art in order to stress the need for further work until appropriate answers are reached. The past advances in modal identification must be followed by some breakthroughs to correctly, efficiently and rigorously solve the existing problems in model updating.

It may be fairly accurate to state that working techniques to directly identify mass, stiffness and damping distributions are very scarce if not altogether nonexisting. This is an area of considerable interest, especially for control system design. For several cases an accurate model of a limited number of degrees of freedom, for example to include sensors locations and control inputs, may be sufficient for such specific applications.

On the other hand, using modal parameters identified from testing to update analytical models has witnessed more activities during the past ten years. Again it seems as if we are just scratching the surface as Berman stated fifteen years ago in his discussions on the 'incomplete model'.

The common problems, issues or difficulties that face the dynamicist, in the area of model derivation, updating or optimization, may be summarized as:

1. The size of the analytical Finite Element Model is very large.

2. Model reduction or condensation techniques are not fully developed, understood or justified.

3. Modal identification algorithms identify a limited number of modes at a limited number of coordinates.

4. Test coordinates and degrees of freedom are incomplete and are different from those of the analytical model.

5. The normal mode approach is frequently used even for structures that have highly complex modes (non-proportional damping)

6. There is no established scientific criteria for designating or selecting the elements of the analytical model to be changed.

7. The algorithms to use modal test data to improve analytical models are not sufficiently rigorous and lack physical justification or meaning.

8. The updated model, in some cases, is no more than a group of numbers and lacks physical significance.

9. There is no established and acceptable criteria to judge the soundness or accuracy of the updated model as far as input/output or stresses and strains are concerned. The ability of the improved model to reproduce the same set of modes, that were used for updating, is insufficient and redundant.

These very points may, as well, help in directing future research efforts in the badly needed subject of model updating.

ACKNOWLEDGEMENTS

This study was conducted while the author was on his sabbatical leave year from Old Dominion University. The author is very grateful to the host institutions during that year. Namely, the Kotholic University of Leuven, Leuven, Belgium; University of Besancon, France; and the European Space Technology and Research Center (ESTEC/ESA), Noordwijk, the Netherlands.

REFERENCES

[1] Young, J.P. and On, F.J., "Mathematical Modeling Via Direct Use of Vibration Data," National Aeronautic and Space Engineering and Manufacturing Meeting, Society of Automotive Engineers, Los Angeles, California, October 1969, Reprint No. 690615.

[2] Pilkey, W. D. and Cohen, R. (eds), "System Identification of Vibrating Structures, Mathematical Models from Test Data," American Society of Mechanical Engineers, New York, 1972.

[3] Berman, A. and Flannelly, W. G., "Theory of Incomplete Models of Dynamic Structures," AIAA Journal, Vol. 9, Aug. 1971, pp. 1481 - 1487.

[4] Leuridan, J., "Some Direct Parameter Model Identification Methods Applicable for Multiple Input Modal Analysis," Ph.D. Dissertation, University of Cincinnati, Ohio, U.S.A., 1984, 384 pp.

[5] Coppolino R.N., "A Simultaneous Frequency Domain Technique for Estimation of Modal Parameters from Measured Data," Society of Automotive Engineers, Paper 811046, Presented at Aerospace Congress Exposition, Anaheim, California, October 1981.

[6] Wada, B.K. and Chen, J.C., "Test and Analysis for Structural Dynamic Systems," Proceedings of the 4th International Modal Analysis Conference, February 1986, pp. 632-647.

[7] Caesar, B., "Update and Identification of Dynamic Mathematical Models," Proceedings of the 4th International Modal Analysis Conference, February 1986, pp. 394-401.

[8] Pal, T.G. and Schidtberg, R.A., "Combining Analytical and Experimental Modal Analysis for Effective Structural Dynamic Modeling," Proceedings of the 1st International Modal Analysis Conference, 1982, pp. 265-271.

[9] Heylen, W. and Sas, P., "Review of Modal Optimization Techniques," Proceedings of the 11th International Seminar on Modal Analysis, Katholieke Universiteit Leuven, Belgium, September 1986, paper W2-1.

[10] Link, M., "Survey on Selected Update Procedures," Proceedings of the 11th International Seminar on Modal Analysis, Katholieke Universiteit, Leuven, Belgium, September 1986, paper W2-2.

[11] Guyan, R.J., "Reduction of Stiffness and Mass Matrices," Journal of AIAA, Vol. 3, No. 2, Feb. 1965, pp. 380.

[12] Kaufman, S. and Hall, D.B., "Reduction of Mass and Load Matrices," Journal of AIAA, Vol. 6, No. 3, March 1068, pp. 550-551.

[13] Kidder, R.L., "Reduction of Structural Frequency Equations," Journal of AIAA, Vol. 11, No. 6, June 1973, pp. 892.

[14] Sowers, J.D., "Condensation of Free Body Mass Matrices Using Flexibility Coefficients," AIAA Journal, Vol. 16, No. 3, March 1978, pp. 272-273.

[15] Downs B., "Accurate Reduction of Stiffness and Mass Matrices for Vibration Analysis and a Rationale for Selecting Master Degrees of Freedom," ASME Journal of Mechanical Design, Vol., September 1979, pp. 1-5.

[16] Miller, A.C., "Dynamic Reduction of Structural Models," ASCE Journal, Structural Division, Vol. 106, 1980, pp. 2097-2108.

[17] Anderson, L.R. and Hallauer, W.L., Jr., "A Method of Order Reduction for Structural Dynamics Based on Riccati Iteration," AIAA Journal, Vol. 19, No. 6, June 1981, pp. 796-800.

[18] Kim, K.O. and Anderson, W.J., "Generalized Dynamic Reduction in Finite Element Dynamic Optimization," AIAA Journal, Vol. 22, No. 11, November 1984, pp. 1616-1617.

[19] Wolf, F.W. and Molnar, A.J., "Reduced System Models Using Modal Oscillators for Subsystems Rationally Normalized Modes," Shock and Vibration Bulletin, Bulletin 44, 1974, pp. 111-118.

[20] Baruch, M. and Bar Itshack, I.Y., "Optimal Weighted Orthogonalization of Measured Modes," AIAA Journal, Vol. 16, No. 4, April 1978, pp. 346-351.

[21] Ibrahim, S.R., "Computation of Normal Modes from Identified Complex Modes," AIAA Journal, Vol. 21, No. 3, March 1983, pp. 446-451.

[22] Zhang, Q. and Lallement G., "Comparison of Normal Eigenmodes Calculation Methods Based on Identified Complex Eigenmodes," to appear in the Journal of Spacecraft and Rockets, Nov-Dec. 1986.

[23] Niedbal, N., "Analytical Determination of Real Normal Modes from Measured Complex Responses," SAE paper 84-0095, Palm Springs, California, Dec. 1984.

[24] Natke, H.G. and Rotert, D., "Determination of Normal Modes from Identified Complex Modes," Z. Flugwiss, Weltrantorsch, 9, 1985, Helt Z.

[25] Deblauwe, F. and Allemang, R.J., "A Possible Origin of Complex Modal Vectors," Proceedings of the 11th International Seminar on Modal Analysis, Paper No. A2-3, Katholieke Universiteit Leuven, Leuven, Belgium, Sept. 1986.

[26] Baruch, H. and Mirovitch, L., "Identification of Eigensolutions of Distributed Parameter Systems" Proceedings of the 23rd AIAA SDM Conference, Paper No. 82-0771, April 1982, pp. 574-581.

[27] Allemang, R.J., "Investigation of Some Multiple Input/Output Frequency Response Experimental Modal Analysis Techniques, Ph.D. Dissertation, University of Cincinnati, Cincinnati, Ohio, 1980.

[28] Berman, A. and Nagy, E., "Improvement of a Large Dynamic Analytical Model Using Ground Vibration Test Data," Proceedings of the AIAA SDM Conference, Paper No. 82-0743, May 1982, pp. 301-306.

[29] O'Callahan, J., Avitable, P., Lieu, I.W. and Madden, R., "An Efficient Method of Determination of Rotational Degrees of Freedom from Analytical and Experimental Modal Data," Proceedings of the 4th International Modal Analysis Conference, Union College Schenectady, N.Y., February 1986, pp. 50-58.

[30] Vold, H., "Spatial Filtering of Experimental Mode Shapes," Proceedings of the 11th International Seminar on Modal Analysis, Katholieke Universiteit Leuven, Leuven, Belgium, Paper No. W2-8, September 1986.

[31] Caesar, B., "Analysis Test Correlation of Dynamic Mathematical Models," Proceedings of the DFVLR 2nd International Symposium on Aeroelasticity and Structural Dynamics, Aachen, West Germany, April 1985, pp. 617-642.

[32] Caesar, B., "Update of Mass and Stiffness Matrices Using Modal Test Data," Proceedings of the 11th International Seminar on Modal Analysis, Katholieke Universteit Leuven, Leuven, Belgium, Paper No. W2-4, September 1986.

[33] Link, M., "Identification of Physical System Matrices Using Incomplete Vibration Test Data," Proceedings of the 4th International Modal Analysis Conference, Union College, Schenectady, N.Y., February 1986.

[34] Ibrahim, S.R., "Dynamic Modeling of Structures from Measured Complex Modes," AIAA Journal, Vol. 21, No. 3, March 1983, pp. 898-901.

[35] Heylen, W., "Model Optimization with Measured Modal Data by Mass and Stiffness Changes," Proceedings of the 4th International Modal Analysis Conference, Union College, Schenectady, N.Y., February 1986, pp. 94-100.

[36] Filled R., Lallement G., Piranda, J.P. and Zhang, C., "Parametric Correction of Regular Non-Dissipative Finite Element Models," Proceedings of the 11th International Seminar on Modal Analysis, Katholieke Universiteit Leuven, Leuven, Belgium, September 1986, Paper No. W2-6.

[37] Liping, H., Kecheng, C. and Zhandi Xu, "Method of Modifying Structural Dynamic Model by Means of Incomplete Complex Modes Identified from Experimental Data," The 4th International Modal Analysis Conference Proceedings, pp. 434-438, 1986.

[38] Beliveon, J.G., Massoud, M., Baurassa, P., Lauzier, C., Vignerson, F. and Soucy Y., "Statistical Identification of the Dynamic Parameters of an Astromast from Finite Element and Test Results Using Bayesian Sensitivity Analysis," The 1st International Modal Analysis Conference Proceedings, 1982, pp. 89-95.

[39] Blakely, K.D. and Wallon, W.B., "Selection of Measurement and Parameter Uncertainties for Finite Element Model Resivison," The 1st International Modal Analysis Conference Proceedings, 1982, pp. 82-88.

[40] Fuh, J.S., Gustavson, B. and Berman, A., "Error Estimation and Compensation in Reduced Dynamic Models of Large Space Structures," AIAA Paper No. 86-08371.

[41] Gysin, H.P., "A Critical Application of the Error Matrix Method for Localization of Finite Element Modeling Inaccuracies," The 4th International Modal Analysis Conference Proceedings 1986.

[42] Natke, H.G., "Improvement of Analytical Models with Input/Output Measurements Contra Experimental Modal Analysis," Proceedings of the 4th International Modal Analysis Conference, February 1986, pp. 409-413.

[43] Fritzen, C.P., "Identification of Mass, Damping and Stiffness Matrices of Mechanical Systems," ASME Journal of Vibrations, Acoustics, Stress and Reliability in Design, Vol. 108, January 1986, pp. 9-16.

[44] Badenhausen, K., "Time Domain Identification of Incomplete Physical System Matrices Using a Condensed Estimation Equation," The Fourth International Modal Analysis Conference Proceedings, 1986, pp. 402-408.

Localization Techniques

G. LALLEMENT, Laboratoire de Mécanique Appliquée
Université de Franche-Comté, Besançon, 25030, France

ABSTRACT : Three techniques for the qualitative selection of sub-domains containing dominant modeling errors or structural damages are described. These techniques are based on the use of the observed and calculated eigensolutions.

NOMENCLATURE

$U \in \mathbb{R}^{n,m}$	matrix U, belonging to the set of real matrices of order nxm
$V \in \mathbb{C}^{n,m}$	matrix V, belonging to the set of complex matrices of order nxm
$\triangleq$	by definition
$\simeq$	approximately equal to
U^T	transpose of the matrix U
U^+	Moore-Penrose pseudo-inverse of the matrix A
DOF	abreviation for Degree of Freedom
c	number of DOF observed on the structure (c = number of sensors)
k_i	localization indicator with respect to the stiffness of the ith sub-structure
l	number of sub-structures introduced for the localization
m	number of eigensolutions identified
m_i	localization indicator with respect to the mass of the ith sub-structure
n	number of eigensolutions calculated from the model $M^{(c)}$; $K^{(c)}$
N	number of DOF of the model $M^{(c)}$; $K^{(c)}$
I_N	unit matrix of order N

<u>Quantities related to the initial estimation</u>

$M^{(c)}; K^{(c)}$	$\in \mathbb{R}^{N,N}$ mass and stiffness matrices of the initial finite element estimation, both are symmetric, positive definite (extension to the case $K^{(c)}$ semi-definite is immediate)
$M_i^{(c)}; K_i^{(c)}$	$\in \mathbb{R}^{N,N}$ mass and stiffness matrices of the ith sub-structure defined for the localization; both are symmetric
$Y^{(c)}; \Lambda^{(c)}$	$\in \mathbb{R}^{N,N}$ complete modal and spectral matrices of the initial estimation; these matrices satisfy the orthonormality relations : $Y^{(c)T} M^{(c)} Y^{(c)} = I_N$; $Y^{(c)T} K^{(c)} Y^{(c)} = \Lambda^{(c)}$
$y_\nu^{(c)}$	$\in \mathbb{R}^{N,1}$ ν^{th} column of $Y^{(c)}$
$_1 y_\nu^{(c)}$	$\in \mathbb{R}^{c,1}$ sub-vector of $y_\nu^{(c)}$ corresponding to the c observed DOF

$$Y^{(c)} = \begin{bmatrix} Y'^{(c)} & \\ & Y''^{(c)} \end{bmatrix} = \begin{bmatrix} {}_1Y^{(c)} & {}_3Y^{(c)} \\ {}_2Y^{(c)} & {}_4Y^{(c)} \end{bmatrix}$$

$Y'^{(c)} \in \mathbb{R}^{N,n}$ calculated sub-matrix

$Y''^{(c)}$ unknown; ${}_1Y^{(c)} \in \mathbb{R}^{c,n}$

$$\Lambda^{(c)} = \begin{bmatrix} \Lambda'^{(c)} & \\ & \Lambda''^{(c)} \end{bmatrix} = \begin{bmatrix} \Lambda_1^{(c)} & \\ & \Lambda_2^{(c)} \\ & & \Lambda''^{(c)} \end{bmatrix}$$

$\Lambda_1^{(c)} \in \mathbb{R}^{m,m}$; $\Lambda_2^{(c)} \in \mathbb{R}^{n-m,n-m}$

$\Lambda'^{(c)} = \Lambda_1^{(c)} \oplus \Lambda_2^{(c)}, \in \mathbb{R}^{n,n}$

$\lambda_\nu^{(c)}$ ν^{th} diagonal element of $\Lambda^{(c)}$; $\Lambda^{(c)} = \mathrm{diag}\,\{\lambda_\nu^{(c)}\}$

Quantities related to the structure

M ; K $\in \mathbb{R}^{N,N}$ mass and stiffness matrices of the conservative structure associated with the physical structure ; symmetric ; definite positive

ΔM ; ΔK $\in \mathbb{R}^{N,N}$ mass and stiffness matrices of correction, both symmetric ; $M \triangleq M^{(c)} + \Delta M$; $K \triangleq K^{(c)} + \Delta K$

Y $\in \mathbb{R}^{N,N}$ modal matrix of the conservative structure

Λ $\in \mathbb{R}^{N,N}$ spectral matrix of the conservative structure

$Y = \begin{bmatrix} Y_1 & Y_2 \end{bmatrix} = \begin{bmatrix} {}_1Y & {}_3Y \\ {}_2Y & {}_4Y \end{bmatrix}$; $Y_1 \in \mathbb{R}^{N,m}$ sub-matrix of the modal matrix partially identified

${}_1Y \in \mathbb{R}^{c,m}$ identified sub-matrix of the modal matrix; ${}_2Y \in \mathbb{R}^{N-c,m}$

$\Lambda = \begin{bmatrix} \Lambda_1 & \\ & \Lambda_2 \end{bmatrix}$; $\Lambda = \mathrm{diag}\,\{\lambda_\nu\}$, $\in \mathbb{R}^{N,N}$; $\Lambda_1 \in \mathbb{R}^{m,m}$

$\hat{C} = \begin{bmatrix} C & \\ C & C'' \end{bmatrix}$ $\hat{C} \in \mathbb{R}^{N,N}$ matrix of linear combination

$C \in \mathbb{R}^{n,m}$; $C'' \in \mathbb{R}^{N-n,m}$.

1. INTRODUCTION

This paper deals with two related problems in the domain of Parametric Identification :

- The qualitative localization of dominant modeling errors in Conservative Finite Element Models.

- The localization of structural modifications dues to damages.

Three kinds of methods developed leading to the spatial localization of regions presenting dominant modeling errors are set out. The last one can be immediately applied to both kinds problems. The two first one's are more specific to parametric correction problems of Finite Element Models.

Why should we present the three methods instead of confining our attention on the last one which, at present, is giving the best results ? For four reasons : a) to get the "workshop feeling"; b) to outline the development of the basic ideas ; c) to avoid other users eventual time losses ; d) to underline the weak points for which improvements could be foreseen.

The three methods make use of the two series of eigensolutions :
- Calculated one's from conservative, regular, discrete and symmetric mathematical models $M^{(c)}$; $K^{(c)} \in R^{N,N}$ representing the initial estimation.
- Deducted one's from observations made on the structure itself by applying
a Modal Identification Technique. Its eigensolutions are those of the associated conservative structure to the dissipative one. Therefore, the transformation of identified complex eigensolutions to real eigensolutions of the
associated conservative structure is necessary. Several methods allow this
transformation $|1-8|$.
A method directly using the complex eigensolutions is also equated and it's
possibilities are shown in the Annex .

A method using particular solutions observed on the structure (ex : particular solutions due to a deterministic excitation) can also be contemplated
$|9|$. But up to now, we've prefered the use of normal eigenmodes because it
has the following advantages :
- it splits into two groups conservative and dissipative unknowns,
- it uncouples interactions between generalized masses and generalised dampings,
- it leads to a better convergence radius when a minimization of the output
errors formulation is employed.

In the other hand, we will be confronted with difficulties dues to uncertainties on the identified generalized masses (problems of correctly scaling
identified eigenvectors) to which users don't give-except in aeronautics-
enough importance.

To condense the discussion, notations are stated only once in the general
nomenclature.

2. DIRECT LOCALIZATION OF DOMINANT ERRORS OF THE INITIAL ESTIMA-
TION FROM EQUILIBRIUM AND ORTHONORMALITY EQUATIONS.

This procedure was developped in 1976 under a contract of the CNRS $|10|$.

2.1 - Simultaneous localization of Dominant Errors in the Correction Matrices
ΔM ; ΔK .

The m observed eigensolutions satisfy the dynamic equilibrium equations :
$$[K^{(c)} + \Delta K] \, Y_1 - [M^{(c)} + \Delta M] \, Y_1 \, \Lambda_1 = 0 \tag{1}$$
As the Modal Submatrix $Y_1 \in R^{N,m}$ is only partially observed, the following

approximation is made :

$$Y_1 \overset{\sim}{=} Y'^{(c)} \, C \, , \text{ where : } Y'^{(c)} \in \mathbb{R}^{N,n} \, ; \, n \geqslant m \, ; \, C \in \mathbb{R}^{n,m} \tag{2}$$

Therefore we obtain :

$$\Delta M \, Y_1 \, \Lambda_1 \, - \, \Delta K \, Y_1 \overset{\sim}{=} L_1$$

$$\text{where : } L_1 \in \mathbb{R}^{N,m} \, ; \, L_1 \overset{\Delta}{=} K^{(c)} \, Y'^{(c)} \, C \, - \, M^{(c)} \, Y'^{(c)} \, C \, \Lambda_1 \tag{3}$$

The "Localization Matrix" L_1 can be calculated if an estimation of the linear combination matrix C can be constructed. This estimation is made in Least Squares sense from the observed modal submatrix $_1Y \in \mathbb{R}^{c,m}$; $N>c\geqslant n\geqslant m$:

$$_1Y \cong \, _1Y'^{(c)}C \quad \text{where } \, _1Y'^{(c)} \in \mathbb{R}^{c,n} \text{ is a submatrix of } Y'^{(c)} \text{ corresponding to}$$

c lines from the observed DOF and is supposed to be of maximal rank n.

It follows : $C = \, _1Y'^t \, _1Y$

Thus, the localization equates the errors on the inputs. The way it is equated is shown on Fig.1 followed by the next remarks :

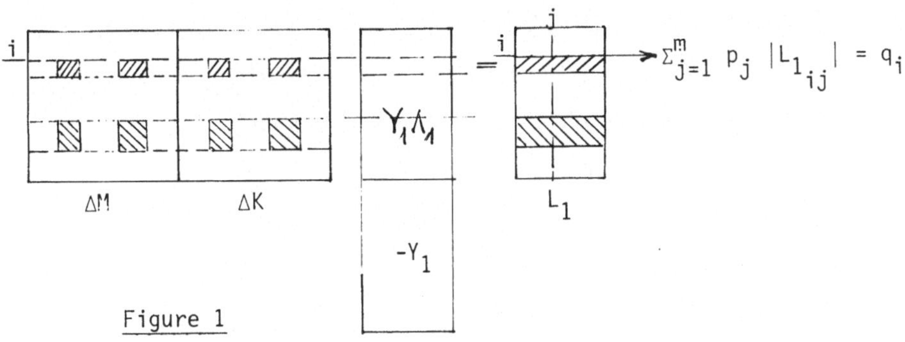

Figure 1

- If the uth line on the $[\Delta M | \Delta K]$ matrix is all zeros (which means there are no modeling errors on the uth DOF), the uth line of matrix L_1 is also empty.

- On the contrary :

a) If a line of L_1 has small elements, it means that the elements of the corresponding line of $[\Delta M | \Delta K]$ are also small. If a line vector from $[\Delta M | \Delta K]$ is orthogonal to the m columns of matrix $\begin{bmatrix} Y_1\Lambda_1 \\ -Y_1 \end{bmatrix}$ it means that corresponding modeling error cannot be observed, in other words, it has no effect on the m identified modes.

b) If a line of L_1 has large elements, the corresponding line of $[\Delta M | \Delta K]$ is also dominant.

--) In practice, the DOF presenting the dominant modeling errors are selected from the examination of a localization vector q(N,1) which components q_i can be, for exemple :

$$q_i = \sum_{j=1}^{m} p_j \, |L_{1ij}| \quad , \text{ or : } \quad q_i = \sum_{j=1}^{m} p_j \, L_{1ij}^2$$

Weighting scalars p_j ($p_j \geqslant 0$) are statistically founded weighting scalars and are chosen in function of : the degree of confidence accorded to each identified eigenmode, to it's technical relevance, etc...

--) Lines are grouped by blocks (or by structural elements) when N is large to reduce the dimension of q and to average the errors on the observed data.

2.2 - Relative influence of Correction Matrices ΔM ; ΔK.

The principle of this calculation can be established, first of all in the case of complete modal bases, from orthonormality relationships :

$$Y^{(c)T} \, M^{(c)} \, Y^{(c)} = I_N \qquad Y^T \left[M^{(c)} + \Delta M \right] Y = I_N \tag{5}$$

$$Y^{(c)T} \, K^{(c)} \, Y^{(c)} = \Lambda^{(c)} \qquad Y^T \left[K^{(c)} + \Delta K \right] Y = \Lambda \tag{6}$$

and from the linear transformation : $Y = Y^{(c)} \, \hat{C}$.

By hypothesis, let us admit that the two following conditions are satisfied :

a) Matrices ΔM ; ΔK can be considered as perturbation matrices :

$$\| \Delta M \| \ll \| M^{(c)} \| \qquad \| \Delta K \| \ll \| K^{(c)} \| \ .$$

b) Matching between the two sets of eigenvectors has been done :

    - Homologous vectors are ordered in the same way in matrices $Y^{(c)}$ and Y .
    - Homologous vectors have the same sign.

Then, it can be written : $\hat{C} \triangleq I_N + A$ (7) where : $\| A \| \ll \| I_N \|$, and the following relations are obtained from orthonormality relationships :

$$\hat{C}^T \, \hat{C} + \hat{C}^T \, \Delta\tilde{M} \, \hat{C} = I_N \ , \tag{8}$$

$$\hat{C}^T \, \Lambda^{(c)} \, \hat{C} + \hat{C}^T \, \Delta\tilde{K} \, \hat{C} = \Lambda, \tag{9}$$

Or : $A + A^T + \Delta\tilde{M} + (AA^T + \Delta\tilde{M}A + A^T\Delta\tilde{M} + A^T\Delta\tilde{M}A) = 0$ (10)

$$\Lambda^{(c)} + \Lambda^{(c)}A + A^T\Lambda^{(c)} + \Delta\tilde{K} + (A^T \, \Lambda^{(c)}A + \Delta\tilde{K}A + A^T\Delta\tilde{K} + A^T \, \Delta\tilde{K}A) = \Lambda \tag{11}$$

where : $\Delta\tilde{M} \triangleq Y^{(c)T} \, \Delta M \, Y^{(c)}$; $\Delta\tilde{K} \triangleq Y^{(c)T} \, \Delta K \, Y^{(c)}$.

Linearizing, these two relations become :

$$A + A^T + \Delta\tilde{M} \cong 0 \tag{10'}$$

$$\Lambda^{(c)}A + A^T\Lambda^{(c)} + \Delta\tilde{K} \cong \Delta\Lambda \quad \text{where : } \Delta\Lambda \triangleq \Lambda - \Lambda^{(c)} \tag{11'}$$

Hence we can state the two following remarks :

- If $\Delta M = 0$, A is an antisymmetric matrix. More generally, a tendency to anti-symmetry on matrix A means : ΔM "small" or not observable.

- If $\Delta K = 0$; $\Lambda^{(c)}A + (\Lambda^{(c)}A)^T = \Delta\Lambda$. More generally, if extradiagonal elements of matrix $\Lambda^{(c)}A$ have an antisymmetric tendency, ΔK is "small" or not observable.

If the matching of modes has not been satisfed or is impossible to satisfy, relationships (8) and (9) are used :

$$\text{if} \quad \begin{vmatrix} \Delta M = 0 & : & \hat{C}^T\,\hat{C} = I_N \ , \\[2ex] \Delta K = 0 & : & \hat{C}^T\,\hat{\Lambda}^{(c)}\hat{C} = \Lambda \ . \end{vmatrix}$$

And in a more general way, the tendency of one of the left-hand side matrices to be diagonal gives some information on the relative importance of ΔM with respect to ΔK (and vise versa).

2.2.1 - Practical considerations related to the observation of a Incomplete Modal Base.

Matrices which take part in the preceeding relationships are partitioned in submatrices as follows :

$$Y = \begin{bmatrix} Y_1 & Y_2 \end{bmatrix} \ ; \ Y^{(c)} = \begin{bmatrix} Y'^{(c)} & Y''^{(c)} \end{bmatrix} \ ; \ \hat{C} = \begin{bmatrix} C \\ C' \end{bmatrix} C'' \ ; \ \Lambda = \begin{bmatrix} \Lambda_1 \\ & \Lambda_2 \end{bmatrix} \ ; \ \Lambda^{(c)} = \begin{bmatrix} \Lambda'^{(c)} \\ & \Lambda''^{(c)} \end{bmatrix}$$

where :
$$C \in \mathbb{R}^{n,m}$$
$$\Lambda_1 \in \mathbb{R}^{m,m}$$
$$\Lambda'^{(c)} \in \mathbb{R}^{n,n}$$

In Y, only the sub-matrix $Y_1 \in \mathbb{R}^{N,m}$ is partially identified. It is expressed as : $Y_1 = Y'^{(c)}C + Y''^{(c)}C'$ where in $Y'^{(c)} \in \mathbb{R}^{N,n}$, n has to be chosen in such a way that the linear combination matrix C' is negligible with respect to C, ($Y''^{(c)}$ is unknown, therefore C' can't be calculated) hence : $Y_1 \overset{\sim}{=} Y'^{(c)}C$, where C is estimated by Least Squares from the observed modal submatrix $_1Y \in \mathbb{R}^{C,m}$.

We obtain from (12) and (13) the only useful equations :

$$\begin{vmatrix} \text{for } \Delta M = 0 & : & C^T C + C'^T C' = I_m \\ \text{for } \Delta K = 0 & : & C^T \Lambda'^{(c)} C + C'^T \Lambda''^{(c)} C' = \Lambda_1 \end{vmatrix}$$
$$(15)$$
$$(16)$$

As matrix C' can't be calculated, to make use of (15), in other words, to control the diagonal tendency of $C^T C$ it is necessary to satisfy : $\|C'^T C'\| \ll \|C^T C\|$. In fact, this condition can be easily satisfied in practice.

The use of equation (15) is much more delicate. Even though the elements of C' are small indeed, large eigenvalues contained in $\Lambda''^{(c)}$ can lead to $\|C'^T \Lambda''^{(c)} C'\|$ which is not small with respect to $\|C^T \Lambda'^{(c)} C\|$. The necessary conditions to use (16) are set out in $|11|$.

2.2.2 - Use of relations (10') and (11')

If the m eigenvectors can be matched, the partitioning of matrices A and $\Lambda'^{(c)}$ coherent with that of $\hat{C}$ leads to :

$$\hat{C} = \begin{array}{|c|} \hline C \\ \hline C' \end{array}\ C'' \quad \triangleq\quad \begin{array}{|c|} \hline C_1 \\ \hline C_2 \\ \hline C' \end{array}\ C'' \quad \triangleq\quad \begin{array}{|c|} \hline I_m \\ \hline I_{N-m} \end{array} + \begin{array}{|c|} \hline A_1 \\ \hline A_2 \\ \hline A_3 \end{array}\ A''$$

where :
$$C_1 ;\ A_1 \in \mathbb{R}^{m,m} ;$$
$$C_2 = A_2 \in \mathbb{R}^{n-m,m} ;$$
$$C' = A_3 \in \mathbb{R}^{N-n,m} ;$$

$$\Lambda'^{(c)} = \Lambda_1^{(c)} \oplus \Lambda_2^{(c)}, \quad \text{where} :\ \Lambda_1^{(c)} \in \mathbb{R}^{m,m} ;\ \Lambda_2^{(c)} \in \mathbb{R}^{n-m,n-m}$$

Taking into count only the useful part of (15) and (16), in other words : when $\Delta M=0$: $C^T C \overset{\sim}{=} I_m$, or when $\Delta K=0$: $C^T \Lambda'^{(c)} C \overset{\sim}{=} \Lambda_1$, we immediately obtain :

$$\text{for } \Delta M=0 \quad : \quad A_1 + A_1^T + (A_1^T A_1 + A_2^T A_2) \overset{\sim}{=} 0 \tag{17}$$

$$\text{for } \Delta K=0 \quad : \quad \Lambda_1^{(c)} A_1 + A_1^T \Lambda_1^{(c)} + (A_1^T \Lambda_1^{(c)} A_1 + A_2^T \Lambda_2^{(c)} A_2) \overset{\sim}{=} \Lambda_1 - \Lambda_1^{(c)} \triangleq \Delta\Lambda_1 \tag{18}$$

Neglecting the influence of second order terms in these expressions, we get the same conclusions as before :

-an antisymmetric A_1 matrix means : $\Delta M=0$. Similarely, Matrix A_1 has an anti-symmetric tendency as ΔM gets "smaller",
- a $\Lambda_1^{(c)} A_1$ antisymmetric matrix means : $\Delta K=0$. In the same way, matrix $\Lambda_1^{(c)} A_1$ has an antisymmetric tendency as ΔK gets "smaller".

2.2.3 - Quantitative Contributions of Error Matrices ΔM and ΔK in Matrices A_1 and $\Delta\Lambda_1$.

When the approximation (14) is satisfied : $Y_1 \overset{\sim}{=} Y'^{(c)} C$, it is shown that the useful part of (8 ; 9) is :

$$C^T C + C^T \Delta\tilde{M}_1 C = I_m \qquad\qquad \Delta\tilde{M}_1 \triangleq Y_1^{(c)T} \Delta M\ Y_1^{(c)} \tag{19}$$

$$C^T \Lambda'^{(c)} C + C^T \Delta\tilde{K}_1 C = \Lambda_1 \quad \text{where:} \quad \Delta\tilde{K}_1 \triangleq Y_1^{(c)T} \Delta K\ Y_1^{(c)} \tag{20}$$

Restraining our attention only to the principal part (linearization) and when matching of eigenvectors has been possible, the above equations are written :

$$A_1 + A_1^T + \Delta\tilde{M}_1 \overset{\sim}{=} 0 \tag{21}$$

$$\Lambda_1^{(c)} A_1 + (\Lambda_1^{(c)} A_1)^T + \Delta\tilde{K}_1 \overset{\sim}{=} \Delta\Lambda_1 \tag{22}$$

Obviously, these expressions are the same as those found by the Classical Perturbation Method expanded to first order terms. But as a consequence of this linearization, the principle of superposition can be applied :

$$\Delta\Lambda_1 \triangleq \Delta\Lambda_1^M + \Delta\Lambda_1^K ;\ A_1 \triangleq A_1^M + A_1^K ,$$

where : $\Delta\Lambda_1^M ;\ A_1^M$ respectively represent the contribution of the error matrix ΔM in $\Delta\Lambda_1$ and A_1,

$\Delta\Lambda_1^K$; A_1^K respectively represent the contribution of the error matrix ΔK in $\Delta\Lambda_1$ and A_1.

The following relationships are then easily stablished :

in $\Delta\Lambda_1$: $\Delta\Lambda_1^M = \text{diag}\{2\lambda_\nu A_{\nu\nu}\}$; $\Delta\Lambda_1^K = \text{diag}\{\Delta\lambda_\nu - 2\lambda_\nu A_{\nu\nu}\}$, $\nu = 1,2,\ldots,m$

in A_1^M : $A_{\nu\nu}^M = A_{\nu\nu}$; $A_{\nu\sigma}^M = \lambda_\sigma(A_{\nu\sigma} + A_{\sigma\nu})/(\lambda_\sigma - \lambda_\nu)$; $\nu \neq \sigma$; $\nu,\sigma = 1,2,\ldots,m$

in A_1^K : $A_{\nu\nu}^K = 0$; $A_{\nu\sigma}^K = (\lambda_\nu A_{\nu\sigma} + \lambda_\sigma A_{\sigma\nu})/(\lambda_\nu - \lambda_\sigma)$; $\nu \neq \sigma$; $\nu,\sigma = 1,2,\ldots,m$.

Knowing matrices A_1 and $\Delta\Lambda_1$, calculation of matrices $\Delta\Lambda_1^M$; $\Delta\Lambda_1^K$; A_1^M ; A_1^K is immediate. The importance of each of this four matrices specifies the contributions of ΔM and ΔK respectively.

2.3 - Remarks on this Localization Techniques of Modeling Errors.

A discussion on the possibilities of practical application of these Techniques has been presented in |11| and can't be studied in detail in this article. The essential conclusions are :

a) These procedures are only applicable to models $M^{(c)}$; $K^{(c)}$ of moderate size ($N \ll 10^2$) and to structures which have quite uniform characteristics (not having strong variations of the order of magnitude of the elements $M_{ij}^{(c)}$ and $K_{ij}^{(c)}$ from matrices $M^{(c)}$ and $K^{(c)}$).

b) This techniques can be used to detect structural damages only if a reference model of the form $M^{(c)}$; $K^{(c)}$ is known as well as two states of the structure : before damages (Y_1 ; Λ_1) and after damages ($\tilde{Y}_1$; $\tilde{\Lambda}_1$).

c) The information contained in observed modal deformations is condensed (cf.4) in the linear combination matrix $C \in R^{n,m}$ only. Roughly speaking, this information is "smashed" and reduced to the projection components on some adjacent eigenvectors $y_\nu^{(c)}$. Reconstruction of the information on the DOF that have not been observed is intended only from the c pickup measurements available. This fact constitutes the strongest disadvantage of this methods. In our opinion, chosing representation bases other than submatrix $Y'^{(c)}$ (interpolation polynomials, finite differences methods, ...) doesn't help to overcome the problem.

d) Systematic Errors on Identified Generalised Masses (or there equivalent like, the scaling errors on the identified eigenvectors) are usually of a higher order of magnitude than that of the orientation of eigenvectors :

Y_1 is not really identified but : $Y_1 D$, where $D \triangleq I_m + \varepsilon$.

Unknown matrix $\varepsilon = \text{diag}\{\varepsilon_\nu\}$, usually: $\varepsilon_\nu \sim .2$ to $.4$, characterises the Generalized Masses Errors. These systematic errors lead to important ones in the localization equations which are based on norm relationships. This problem, its consequences and the partial remedies proposed (ex. : renormalisation of identified eigenvectors with respect to the estimated mass matrix $M^{(c)}$) are developed in $|11|$ and $|16|$.

All these negative aspects don't mean that these localization techniques are completely useless, but that they present certain practical limitations and present some delicate points needing special attention.

3. Localization of Dominant Errors in the Stiffness Matrix from its Spectral Decomposition $|13|$

The basic ideas relie on an extension of a method proposed by Sidhu and Ewins $|14|$. The basic idea is : from the definition of Stiffness Error Matrix ΔK : $\Delta K \triangleq K - K^{(c)}$, which can be written as :

$$\Delta K = K^{(c)} \left[K^{(c)-1} - K^{-1} \right] K = K \left[K^{(c)-1} - K^{-1} \right] K^{(c)},$$

we can infer an iteratif calculation proceeding from matrix ΔK :

$$\Delta K^{(1)} = K^{(c)} \left[K^{(c)-1} - K^{-1} \right] K^{(0)} \quad , \text{ where : } K^{(0)} \triangleq K^{(c)}$$

$$\Delta K^{(i)} = K^{(c)} \left[K^{(c)-1} - K^{-1} \right] K^{(i-1)} \quad , \text{ where : } K^{(i-1)} \triangleq K^{(c)} + \Delta K^{(i-1)}$$

and where the static flexibility matrices between parentheses are calculated from the eigensolutions of the finite element model and of the structure itself.

The two stage iterative proceeding that follows is based on a limited developement of the static flexibility matrix :

$$K^{-1} = \left[K^{(c)} + \Delta K \right]^{-1} = K^{(c)-1} - K^{(c)-1} \Delta K K^{(c)-1} + K^{(c)-1} \Delta K K^{(c)-1} + 0(\Delta K^3) \qquad (23)$$

The first stage, taking into account only first-order terms ΔK, can be written

$$\Delta K \cong K^{(c)} \left[K^{(c)-1} - K^{-1} \right] K^{(c)} \qquad (24)$$

By inverting the second orthonormal relationship (6), we obtain

$$K^{-1} = Y\Lambda^{-1} Y^T = \begin{bmatrix} {}_1Y & {}_3Y \\ {}_2Y & {}_4Y \end{bmatrix} \begin{bmatrix} \Lambda_1^{-1} & \\ & \Lambda_2^{-1} \end{bmatrix} \begin{bmatrix} {}_1Y^T & {}_2Y^T \\ {}_3Y^T & {}_4Y^T \end{bmatrix} = \sum_{\nu=1}^{N} \frac{y_\nu y_\nu^T}{\lambda_\nu} \ .$$

The submatrices $\Lambda_1 \in \mathbb{R}^{m,m}$; ${}_1Y \in \mathbb{R}^{c,m}$ are known (identified) through dynamic testing of the structure. The submatrices Λ_2 ; ${}_2Y$; ${}_3Y$; ${}_4Y$ are unknown and assumed non-identified.

Given a sufficiently rapid convergence due to the presence of the $1/\lambda_\nu$ term, we can use an approximate expression for K^{-1} given by the contribution of the only m eigenmodes identified

$$K^{-1} \simeq \begin{bmatrix} {}_1Y \\ {}_2Y \end{bmatrix} [\Lambda_1^{-1}] [{}_1Y^T \quad {}_2Y^T].$$

By a similar process, the approximate expressions for matrices ΔM and M^{-1} are obtained

$$\Delta M \simeq M^{(c)} [M^{(c)-1} - M^{-1}] M^{(c)} \tag{25}$$

$$M^{-1} = YY^T \simeq \begin{bmatrix} {}_1Y \\ {}_2Y \end{bmatrix} [{}_1Y^T \quad {}_2Y^T]. \tag{26}$$

In equation (26) no convergence with eigenmodes of increasing order is obtainable. Equation (26) does not imply the relationship, generally not satisfied

$$\left\| \begin{bmatrix} {}_1Y \\ {}_2Y \end{bmatrix} [{}_1Y^T \quad {}_2Y^T] \right\| \gg \left\| \begin{bmatrix} {}_3Y \\ {}_4Y \end{bmatrix} [{}_3Y^T \quad {}_4Y^T] \right\|.$$

The approximation of matrix M^{-1} is hence only justified by :

a) The practical impossibility of identifying the N-m eigenmodes of matrix $\begin{bmatrix} {}_3Y \\ {}_4Y \end{bmatrix}$

b) The partial compensation in equation (25) of the N-m "forgotten" modes due to the matrix $[M^{-1} - M^{(c)-1}]$.

This initial formulation is similar to the one suggested previously by Sidhu |13|. In ref. |13| the influence of the unknown submatrix ${}_2Y$ on matrices ΔK and ΔM is disregarded. Such a procedure may entail serious mistakes. In the following, we suggest :

a) Taking into account the submatrix ${}_2Y$ through an iterative process

b) An estimate of matrices ΔM and ΔK including the higher order terms in their series expansion

c) The introduction of indicators by nodes (or group of nodes) of the finite element grid improving the qualitative localisation by spatial area.

In the cases where a preliminary condensation on the dof of the initial model |15| is carried out, the suggested iterative process allows a less stringent condensation and a more accurate restitution of the topography of error matrices.

3.1 - Locating the dominant errors (first approximation)

The dominant modeling errors are localized through an iterative computation of matrices ΔK ; ΔM defined by equations (24) and (25). This iterative process is imposed by the necessity of evaluating the modal submatrix ${}_2Y$.

For the initial estimates, the following matrices are considered

$$_2Y^{(0)} = {_2}Y^{(c)}$$

$$\Delta K^{(0)} = K^{(c)} \left[\begin{array}{c|c} {_1}Y^{(c)}\Lambda_1^{(c)-1}{_1}Y^{(c)T}{_{-1}}Y\Lambda_1^{-1}{_1}Y^T & {_1}Y^{(c)}\Lambda_1^{(c)-1}{_2}Y^{(c)T}{_{-1}}Y\Lambda_1^{-1}{_2}Y^{(0)T} \\ \hline {_2}Y^{(c)}\Lambda_1^{(c)-1}{_1}Y^{(c)T}{_{-2}}Y^{(0)}\Lambda_1^{-1}{_1}Y^T & {_2}Y^{(c)}\Lambda_1^{(c)-1}{_2}Y^{(c)T}{_{-2}}Y^{(0)}\Lambda_1^{-1}{_2}Y^{(0)T} \end{array} \right] K^{(c)} \tag{27}$$

$$\Delta M^{(0)} = M^{(c)} \left[\begin{array}{c|c} {_1}Y^{(c)}{_1}Y^{(c)T}{_{-1}}Y_1Y^T & {_1}Y^{(c)}{_2}Y^{(c)T}{_{-1}}Y_2Y^{(c)T} \\ \hline {_2}Y^{(c)}{_1}Y^{(c)T}{_{-2}}Y^{(0)}{_1}Y^T & {_2}Y^{(c)}{_2}Y^{(c)T}{_{-2}}Y^{(0)}{_2}Y^{(0)T} \end{array} \right] M^{(c)}. \tag{28}$$

Having computed matrices $\Delta K^{(c)}; \Delta M^{(c)}$, matrix $Z^{(0)}(\lambda_\nu) \in R^{N,N}$ is formed :

$$Z^{(0)}(\lambda_\nu) = [K^{(c)} + \Delta K^{(0)}] - \lambda_\nu [M^{(c)} + \Delta M^{(0)}].$$

This matrix satisfies : $Z^{(0)}(\lambda_\nu) \, y_\nu \cong 0$, $\nu = 1,2,\ldots,m$.

The preceding equation allows the evaluation of the first estimate $_2Y^{(1)} = [\cdots {_2}y_\nu^{(1)} \cdots]$ from the division into submatrices

$$Z^{(0)}(\lambda_\nu) = [Z_1^{(0)}(\lambda_\nu) \quad Z_2^{(0)}(\lambda_\nu)] \; ; \quad y = [{_1}y_\nu^T \quad {_2}y_\nu^{T(1)}]^T$$

such that :

$$Z_1^{(0)}(\lambda_\nu) \, {_1}y_\nu + Z_2^0(\lambda_\nu) \, {_2}y_\nu^{(1)} = 0 \; , \quad \nu = 1,2,\ldots,m$$

which implies, in terms of the least squares :

$$_2y_\nu^{(1)} = - Z_2^{(0)\dagger}(\lambda_\nu) \, Z_1^{(0)}(\lambda_\nu) \, {_1}y_\nu, \quad \nu = 1,2,\ldots,m$$

If the condition : $\| {_2}Y^{(1)} - {_2}Y^{(0)} \| / \| {_2}Y^{(0)} \| \leqslant \alpha$ (α being a given positive scalar) is met, the procedure is stopped after computing estimates $\Delta K^{(1)}$ and $\Delta M^{(1)}$, obtained by replacing $_2Y^{(0)}$ by $_2Y^{(1)}$ in equations (27) and (28). In the contrary case, the convergence process is reiterated.

3.1.1 - Localization indicators

The matrices ΔK ; ΔM, established in this way, are very approximate and mainly full matrices that do not have the general form of sparse or band matrices imposed upon $M^{(c)}; K^{(c)}$ through the finite element modelization. Hence, they must be used for the sole purpose of searching for the dominant blocks or submatrices that they contain. In the example of a nodal localization of the finite element grid, the indicators α_{mi} and α_{ki} are established for every node i with six dof :

$$\alpha_{mi} = \sum_{j=1}^{6} \frac{|\Delta mi|j}{|mi|j} \; ; \quad \alpha_{ki} = \sum_{j=1}^{6} \frac{|\Delta ki|j}{|ki|j}$$

where : $|\Delta mi|j$ and $|mi|j$ are the sums of the absolute values of the elements of the j_{th} row of component block (6x6) corresponding to the ith node in ΔM and $M^{(c)}$ respectively and where $|\Delta ki|j$ and $|ki|j$ are defined in the same fashion in ΔK and $K^{(c)}$. After ordering the α_{mi} (α_{ki}), $i=1,2,\ldots$ by decreasing values, the numbers of the nodes presenting the dominant errors of the mass (of stiffness) modelization are selected.

In practice, it is useful to first group the α_{mi} and α_{ki} corresponding to neighbouring nodes in the finite element grid in order to produce superelements (or substructures), and then to identify the dominant superelements.

3.1.2 - Convergence of the expansion of the static flexibility matrix

Before considering the possibility of taking into account second order terms in the series expansion of equation (23) it is useful to specify its convergence. A convergence criterion may be established from the eigenvalue problem

$$\left[\Delta K - \phi_\nu K^{(c)}\right] q_\nu = 0 \tag{29}$$

in which the spectral matrix $\phi = \text{diag}\,\{\phi_\nu\}$ and modal matrix $Q = \left[\ldots q_\nu \ldots\right]$ satisfy the orthogonality relationships

$$Q^T K^{(c)} Q = I_N \;;\quad Q^T \Delta K Q = \phi$$

By inverting these last two relations, we obtain

$$K^{(c)-1} = QQ^T \;;\; \Delta K = Q^{T-1} \phi Q^{-1}$$

and by introducing these expressions into equation (23), we obtain

$$K^{(c)-1} = Q \left[I_N - \phi + \phi^2 - \phi^3 + \ldots\right] Q^T$$

which obviously proves that if one of the eigenvalues of equation (29) has a magnitude greater than one, the convergence of the expansion of K^{-1} is no longer ensured.

3.1.3 - Taking into account second order terms

The accuracy of the localization of dominant modelization errors may be improved without excessive increase in computation cost by the inclusion of second order terms in ΔM and ΔK in the expressions of matrices K^{-1} and M^{-1}.

$$K^{-1} \simeq K^{(c)-1} - K^{(c)-1} \Delta K K^{(c)-1} + K^{(c)-1} \Delta K K^{(c)-1} \Delta K K^{(c)-1}$$

$$M^{-1} \simeq M^{(c)-1} M^{(c)-1} \Delta M M^{(c)-1} + M^{(c)-1} \Delta M M^{(c)-1} \Delta M M^{(c)-1}.$$

These two equations are nonlinear with respect to ΔK and ΔM, thus an iterative process is again applied in order to determine them.

First stage : Matrices $\Delta K^{(o)}$; $\Delta M^{(c)}$ are calculated from equations (27) and
(28)

Second stage : matrices $E_k^{(o)}$ and $E_m^{(o)}$ are calculated :

$$E_k^{(0)} \triangleq K^{(c)-1} \, \Delta K^{(0)} \, K^{(c)-1} \, \Delta K^{(0)} \, K^{(c)-1} \ ,$$

$$E_m^{(0)} \triangleq M^{(c)-1} \, \Delta M^{(0)} \, M^{(c)-1} \, \Delta M^{(0)} \, M^{(c)-1} \ ,$$

where :

$$K^{(c)-1} = \begin{bmatrix} {}_1Y^{(c)} \\ {}_2Y^{(c)} \end{bmatrix} [\Lambda_1^{(c)}] [{}_1Y^{(c)T} \ {}_2Y^{(c)T}] \ ; \quad M^{(c)-1} = \begin{bmatrix} {}_1Y^{(c)} \\ {}_2Y^{(c)} \end{bmatrix} [{}_1Y^{(c)T} \ {}_2Y^{(c)T}]$$

Third stage. Matrices $\Delta M^{(1)}$ and $\Delta K^{(2)}$ are then given by :

$$\Delta K^{(1)} = \Delta K^{(0)} + \theta_k \, K^{(c)} \, E_k^{(C)} \, K^{(c)}$$

$$\Delta M^{(1)} = \Delta M^{(0)} + \dot{\theta}_m \, M^{(c)} \, E_m^{(0)} \, M^{(c)}$$

where : θ_k $(\theta_k < 1)$; $\theta_m < 1)$ are positive scalars, allowing the convergence to be controlled. This process is reiterated until the conditions

$$\| \Delta K^{(i)} - \Delta K^{(i-1)} \| \leqslant a_k \ ; \quad \| \Delta M^{(i)} - \Delta M^{(i-1)} \| \leqslant a_m,$$

is obtained, where a_k and a_m are given positive quantities.

3.1.4 - Influence of errors contained in the identified eigenmodes

A simple qualitative estimate of the influence of modal identification errors is performed here by introducing the following hypotheses :

a) The errors on the identified eigenfrequencies are completely negligible
b) The errors on the identified eigenvectors may be represented by replacing the exact modal submatrices ${}_1Y$ and ${}_2Y$ by matrices ${}_1\hat{Y}$ and ${}_2\hat{Y}$ such that:

$${}_1\hat{Y} = {}_1Y[I_m + \varepsilon_1] \ ; \quad {}_2\hat{Y} = {}_2Y[I_m + \varepsilon_2] \ ; \quad \varepsilon_1 \ ; \quad \varepsilon_2 \in \mathbb{R}^{m,m} \tag{30}$$

By introducing (30) into (27) and (28), we obtain :

$$\Delta \hat{K}^{(i)} = \Delta K^{(i)} + \underline{\Delta K}^{(i)}$$

$$\Delta \hat{M}^{(i)} = \Delta M^{(i)} + \underline{\Delta M}^{(i)}$$

where :

$$\underline{\Delta K}^{(i)} = -K^{(c)} \begin{bmatrix} {}_1Y(\varepsilon_1\Lambda_1^{-1}+\Lambda_1^{-1}\varepsilon_1^T)_1Y^T & {}_1Y(\varepsilon_1\Lambda_1^{-1}+\Lambda_1^{-1}\varepsilon_2^T)_2Y^{(i)T} \\ {}_2Y^{(i)}(\varepsilon_2\Lambda_1^{-1}+\Lambda_1^{-1}\varepsilon_1^T)_1Y^T & {}_2Y^{(i)}(\varepsilon_2\Lambda_1^{-1}+\Lambda_1^{-1}\varepsilon_2^T)_2Y^{(i)T} \end{bmatrix} K^{(c)} + 0(\varepsilon^2)$$

$$\underline{\Delta M}^{(i)} = -M^{(c)} \begin{bmatrix} {}_1Y(\varepsilon_1+\varepsilon_1^T)_1Y^T & {}_1Y(\varepsilon_1+\varepsilon_2^T)_2Y^{(i)T} \\ {}_2Y(\varepsilon_2+\varepsilon_1^T)_1Y^T & {}_2Y^{(i)}(\varepsilon_2+\varepsilon_2^T)_2Y^{(i)T} \end{bmatrix} M^{(c)} + 0(\varepsilon^2)$$

The first equation shows that if :

$$\| \varepsilon_1 \| \cong \| I - \Lambda_1^{(c)-1} \Lambda_1 \| \quad \text{then}$$

$$\| \underline{\Delta K}^{(i)} \| \cong \| \Delta K^{(i)} \| \ .$$

A correct localization of stiffness errors thus implies that the average values of errors on the identified eigenvectors (including the systematic norm errors (i.e. generalized mass   errors) are less than the average of relative difference $|(\lambda_\nu^{(c)} - \lambda_\nu) / \lambda_\nu^{(c)}|$.

This condition is very strict. It implies a systematic control of the accuracy of the generalized identified masses, either through the initial computation model $M^{(c)}$, or through additional experimental techniques $|16|$.

3.2 - Application possibilities

This localization technique has been tested on simple structures such as truss and simple welded beamsor tube assemblies (in forms of an L, an H and an A) observed in plane bending vibrations. Correct results have been obtained in the following cases :

- known additional masses voluntarilyadded and not taken in account in the initial estimation,
- locali ation and modeli zation of the stiffness of the welded assemblies between adjacent beams,
- locali zation of cracks in the welded assemblies.

The proposed method brings important improvements to a preceding formulation and extends its range of applications. It has however, the following drawbacks :

a) Its application is limited to finite element models of the moderate order N (N $\overset{\sim}{=}$ few 10^2 DOF)
b) It mainly uses the generalized identified masses and requires that these be identified accurately.
c) It doesn't need to explicitly match calculated and identified eigensolutions -in theory- but its application is quite difficult when the eigensolutions corresponding to the largest eigenfrequencies contained in $(\Lambda_1 ; Y_1)$ have a high spectral density $\left[\text{problems of definition of the upper boundaries of the matrices } (\Lambda_1 ; Y_1) \text{ and } (\Lambda_1^{(c)} ; Y_1^{(c)})\right]$.

4. Localization of Dominant Errors of Stiffness and Mass Matrices Based on The Sensitivity   Equations $|17|$

This technique makes use of the sensivity relationships between identified eigensolutions and the correction global mass and stiffness parameters of substructures (or "macro finite-elements"). These parameters act as indicators to localize dominant errors or structure damages.

This procedure belongs to the general identification class : "Minimization of the errors on the output" and is considered as a more robust and better adapted one for large and complicated structures with a large number of DOF ($N \sim 10^3$ to 10^4 DOF).

4.1 - Localization indicators

The localization is effected by "macro-elements". It is assumed that the matrices $M^{(c)}$; $K^{(c)}$ can be represented by :

$$M^{(c)} = \sum_{i=1}^{1} M_i^{(c)} \quad ; \quad K^{(c)} = \sum_{i=1}^{1} K_i^{(c)} \, ,$$

where $M_i^{(c)}$; $K_i^{(c)} \in R^{N,N}$ are the initial estimation of the elementary mass and stiffness matrices of the "macro-element" or sub-ensemble i of the structure. Define the parameters m_i ; k_i, $i = 1,2,\ldots,1$ intervening linearly in the mass and stiffness matrices of the macro-elements :

$$M = \sum_{i=1}^{1} m_i M_i^{(c)} = \sum_{i=1}^{1} (1+\Delta m_i) M_i^{(c)} \; ; \; K = \sum_{i=1}^{1} k_i K_i^{(c)} = \sum_{i=1}^{1} (1+\Delta k_i) K_i^{(c)}$$

The parameters m_i ; k_i are the qualitative localization indicators. In the results of the localization process :

- if m_i and $k_i \simeq 1$, the sub-domain characterized by $M_i^{(c)}$; $K_i^{(c)}$ does not present modeling errors

- if m_j and (or) $k_j << 1$ (or $>> 1$), this sub-domain presents the dominant modeling errors.

The indicators m_i ; k_i can be considered as correctors to the local kinetic and potential energies. The linearity hypothesis is justified since it is always possible to separate the terms in a macro-element which originate from energies of different natures (bending ; shear ; longitudinal, torsion...).

4.2 - Qualitative estimation of localization indicators.

The identified sub-eigenvector $_1 y_\nu \in R^{c,1}$, $\nu = 1,2,\ldots,m$ is linearized with respect to the indicators m_i and k_i by a first order Taylor series expansion ;

$$_1 y_\nu \overset{\sim}{=} {}_1 y_\nu^{(c)} + \sum_{i=1}^{1} \frac{\partial\, _1 y_\nu^{(c)}}{\partial m_i} \Delta m_i + \sum_{i=1}^{1} \frac{\partial\, _1 y_\nu^{(c)}}{\partial k_i} \Delta k_i$$

Hence : $\Delta_1 y_\nu \overset{\Delta}{=} {}_1 y_\nu - {}_1 y_\nu^{(c)} = S_{y_\nu} \Delta p$, $\nu = 1,2,\ldots,m$ $\qquad\qquad$ (31)

where $\left| \, S_{y_\nu} \overset{\Delta}{=} \left[\frac{\partial\, _1 y_\nu^{(c)}}{\partial k_1} ; \ldots ; \frac{\partial\, _1 y_\nu^{(c)}}{\partial k_1} \; \middle| \; \frac{\partial\, _1 y_\nu^{(c)}}{\partial m_1} ; \ldots ; \frac{\partial\, _1 y_\nu^{(c)}}{\partial m_1} \right] , \in R^{c,21} \right.$

$\qquad\qquad \left| \, \Delta p \overset{\Delta}{=} [\Delta k_1 ; \ldots \Delta k_1 \; | \; \Delta m_1 ; \ldots ; \Delta m_1] \; \in R^{21,1} \right.$

Similarly for the eigenvalue λ_ν, $\nu = 1,2,\ldots,m$, the Taylor series expansion leads to :

$$\Delta\lambda_\nu = \lambda_\nu - \lambda_\nu^{(c)} = S_{\lambda_\nu} \Delta p \; , \tag{32}$$

$$\text{where : } S_{\lambda_\nu} = \left[\frac{\partial\lambda_\nu^{(c)}}{\partial k_1} \cdots \frac{\partial\lambda_\nu^{(c)}}{\partial k_1} \; \middle| \; \frac{\partial\lambda_\nu^{(c)}}{\partial k_1} \cdots \frac{\partial\lambda_\nu^{(c)}}{\partial m_1} \right] \in \mathbb{R}^{1,21}$$

Regrouping equations (31) and (32) yields :

$$\Delta z = \tilde{S} \Delta p \; , \text{ where : } \Delta z \in \mathbb{R}^{(c+1)m,1} \; ; \; \tilde{S} \in \mathbb{R}^{(c+1)m,21} \tag{33}$$

This relation has the general form :

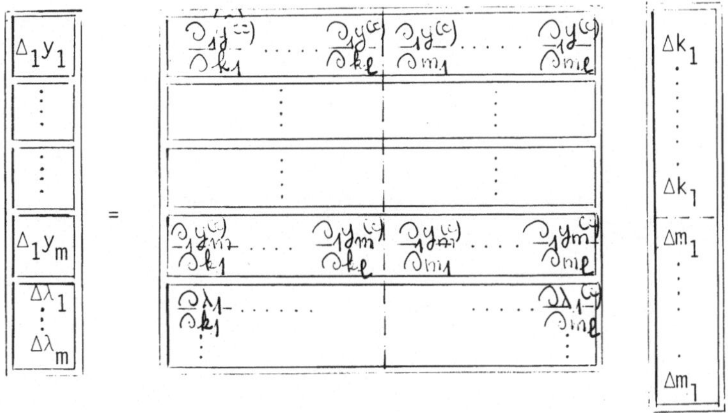

It is solved with respect to $\Delta p \in \mathbb{R}^{21,1}$ and is restricted to the over-determined case (1 of the order : (c+1) m/4). In practice, the solution of (33) is effected after the introduction of dimensionless quantities, ie after the introduction of a diagonal weighting matrix

$$W^{-1} = \text{diag} \{ \ldots ; \; a_\nu \, \|_1 y_\nu\|^{-1} \; ; \; \ldots \quad \ldots, \; b_\nu \, \lambda_\nu^{-1} \; ; \; \ldots \}$$

where a_ν ; b_ν, $\nu = 1,2,\ldots,m$ are statistically founded weighting coefficients of the νth eigensolution. Equation (33) becomes :

$$b = S \, q, \text{ where : } b \overset{\Delta}{=} W^{-1} \Delta z \; ; \; S \overset{\Delta}{=} W^{-1} \tilde{S} \; ; \; q \overset{\Delta}{=} \Delta p \tag{34}$$

4.3 - Solution of equation (34) with respect to vector q

The solution of (34) is the most delicate step and is effected in the "Least Squares sense" |18|, but after :

- examination of the modal kinetic and strain energies repartitions in each sub-domain
- elimination or regrouping of sub-domains associated with low energies
- examination of the contribution of each sub-domain to the "distance" b

- elimination or regrouping of sub-domains weakly linearly independent of the others. We use at the present time a "combination" of the following ideas :

a) Evaluation of each sub-domain i contribution c_i to the "distance" b :

$$c_i = \frac{\|P_i b\|}{\|b\|} \quad , \quad P_i \triangleq s_i \left[s_i^T s_i \right]^{-1} s_i^T \, , \text{ where } s_i \text{ is the } i^{th} \text{ column of the matrix S}$$

b) Use of a modified QR factorization of S for a detection of the quasi-linearly dependent contributions

c) Based on the global direct solution approach : $q = S^+ b$ (and after taking into account of ideas a) and b). Let : $e_j \triangleq \frac{\|s_j \, q_j\|}{\|b\|}$ where q_j is the jth component of the vector q. If q_j and e_j are large, the sub-domain j presents dominant modeling errors.

d) Best sub-space approach : searching for the best sub-space of S of a given dimension d corresponding to q_s, s = 1,2,...,d, which best reduce the

error $\|\varepsilon_d\|$ with : $\varepsilon_d \triangleq b - S_d \, q_d$ where : S_d is a sub-matrix of S, q_d is the corresponding sub-vector of b. An analysis of the errors $\|\varepsilon_d\|$ obtained with sub-spaces of increasing dimension, d = 2,3,..., permits the selection of the most probable dominant sub-domains.

4.4 - Remarks

a) The first derivatives of the eigensolutions of the initial estimation $M^{(c)}$; $K^{(c)}$ with respect to the indicators m_i ; k_i are obtained by taking the derivatives of the equations : $\left[K^{(c)} - \lambda_\nu^{(c)} M^{(c)} \right] y_\nu^{(c)} = 0$; $y_\nu^{(c)T} M^{(c)} y_\nu^{(c)} = 1$, $\nu = 1,2,...,m$.

b) Equations (31,32) implies the pairing of identified and calculated eigensolutions. In many cases, only the m ($m \stackrel{\wedge}{=} 4$ to 6) first eigensolutions can be paired in this way. For these m eigensolutions, two cases can be distinguished.

- The frequency spectrum is sparse in the band of interest : pairing could then be directly effected by the usual procedures |19| or by the Modal Assurance Criteria |20 ; 21|

- The frequency spectrum includes some zones of high density : the pairing is effected after frequency separation by numerical modifications introduced simultaneously in the mathematical model $M^{(c)}$; $K^{(c)}$ and in the structure by the intermediary of the identified eigensolutions |22|.

c) Norm systematic errors on identified eigenvectors (or, its equivalent, identified generalized mass errors) affect equation (31). We propose to consider them as supplementary unknowns introduce in (33).

Let : $_1y_\nu^{ex} \triangleq (1 + \alpha_\nu)\ _1y_\nu$

where: $\begin{vmatrix} _1y_\nu^{ex} \in R^{c,1} & , \text{exact eigensubvector, unknown} \\ _1y_\nu \in R^{c,1} & , \text{identified eigensubvector containing a norm error} \\ \alpha_\nu & , \text{real scalar which characterizes the norm error.} \end{vmatrix}$

Introducing this expression in (31), we obtain :

$$_1y_\nu^{ex} - {_1y_\nu^{(c)}} = (1 + \alpha_\nu)\ _1y_\nu - {_1y_\nu^{(c)}} = S_{y_\nu} \Delta p$$

$$_1y_\nu - {_1y_\nu^{(c)}} \triangleq \Delta_1 y_\nu = \left[S_{y_\nu} \ \vdots \ _1y_\nu \right] \begin{bmatrix} \Delta p \\ \alpha_\nu \end{bmatrix} \tag{35}$$

For structures made of a single material, a practical consideration is made concerning the overall error made on Young's Modulus. To represent this distributed error, a new p_E parameter of global affinity is introduced : $E^{ex} = p_E E.$ The derivatives of the eigenvectors and the eigenvalues with respect to p_E are :

$$\frac{\partial y_\nu^{(c)}}{\partial p_E} = 0 \quad ; \quad \frac{\partial \lambda_\nu^{(c)}}{\partial p_E} = \lambda_\nu^{(c)} \tag{36}$$

Introducing (35;36) in (33) this last expression becomes :

$$\begin{bmatrix} \Delta_1 y_1 \\ \vdots \\ \Delta_1 y_m \\ \hline \Delta\lambda_1 \\ \vdots \\ \Delta\lambda_m \\ \hline 0 \end{bmatrix} = \begin{bmatrix} S_y & \begin{matrix} _1y_1 \\ 0 \end{matrix} & \begin{matrix} & \\ & _1y_m \end{matrix} \\ \hline S_\lambda & \begin{matrix} \lambda_1^{(c)} \\ \vdots \\ \lambda_m^{(c)} \end{matrix} & 0 \\ \hline 0 & \theta_E \quad 0 & \\ & 0 & \theta_\nu \end{bmatrix} \begin{bmatrix} \Delta p \\ \hline \Delta p_E \\ \hline \alpha_1 \\ \vdots \\ \alpha_m \end{bmatrix}$$

where : θ_E and θ_ν are positive scalars controling the values of Δp_E and α_ν, $\nu = 1,2,\ldots,m$; submatrices $S_y \in R^{c\,m,21}$ and $S_\lambda \in \mathbb{R}^{m,21}$ respectively contain the derivatives of eigenvectors and eigenvalues with respect to the localization indicators.

4.4 - Applications Possibilities

In present, this localization technique is being used as a preliminary proce-cure for parametric correction of Finite Element Models for industrial applications ($N \sim 10^3$ to 10^4 DOF). Its basic inconvenients are :

- the high adaptation cost to every specific finite element code serving to build the substructure matrices $M_i^{(c)}$; $K_i^{(c)}$, i = 1,2,...,l
- its limitations dues to the linearization of the "distances" between calculated eigensolutions and observed ones
- the necessity to match calculated and identified eigensolutions.

Its essential advantages are :

- that it takes in account the information contained in the "topology" of matrices $M^{(c)}$; $K^{(c)}$ and that it respects it,
- it doesn't need the reconstruction of modal information which hasn't been observed,
- it has a strong robustness with respect to non-systematic errors contained in the identified eigensolutions,
- it has a great practical flexibility:the possibility to introduce weighting coefficients based on physical considerations ; the possibility to simultaneously introduce, by the same general process, data concerning the observed static behaviour.
- it can be immediately adapted to localization problems of structure damages. In these cases, the initial estimation $M^{(c)}$; $K^{(c)}$ should have a dynamic sensibility close to that of the real structure, and vector b should be formed from the difference between two successive states.

5 - CONCLUSION

Our present experience reveals that modeling error localization or structure damages localization based on the sensitivities of eigensolutions is more convenient for technical applications. In spite of the additional cost introduced by the use of identified eigensolutions, we prefer this method to the others using information on free or forced observed responses, because of its ability to represent model-structure much larger distances in the case of spectrums with separated frequencies. Otherwise, a method to separate eigenmode which have close frequencies can be applied

An adaptation to finite element codes ANSYS (USA), MEF (FRANCE) and SAP IV (USA) has been done. Only further experience can tell its real possibilities in industrial applications.

ANNEX : Dominant Error Localization of Mass and Stiffness Matrices from a Spectral Decomposition Based on Complex Eigensolutions.

The principle is identical to the one developped in paragraph 3, but it is based on a state space formulation : $x(t) = [y(t)^T ; \dot{y}(t)^T]^T$. The following discussion only outlines the main ideas.

Assuming that energy dissipation is of the viscous kind, the autonomus structure has the following model : $UX = AXS$, where :

$$A \triangleq \begin{bmatrix} -K & \\ & M \end{bmatrix} \; ; \; U \triangleq \begin{bmatrix} B & M \\ M & \end{bmatrix} ; A;U \in \mathbb{R}^{2N,2N} ; X = \begin{bmatrix} Y \\ YS \end{bmatrix} \in \mathbb{C}^{2N,2N} ; S = \mathrm{diag}\{s_\nu\} \in \mathbb{C}^{2N,2N}$$

and it also verifies the orthonormality relationships :

$$X^T UX = N = \mathrm{diag}\,\{n_\nu\} \in \mathbb{C}^{2N,2N} \; ; \; X^T AX = NS \in \mathbb{C}^{2N,2N}$$

Model $A^{(c)}$, $U^{(c)} \in \mathbb{R}^{2N,2N}$ representing the initial estimation has the same general form :

$$A^{(c)} \triangleq \begin{bmatrix} -K^{(c)} & \\ & M^{(c)} \end{bmatrix} \qquad U^{(c)} \triangleq \begin{bmatrix} B^{(c)} & M^{(c)} \\ M^{(c)} & \end{bmatrix}$$

This model also satisfies the orthonormality relationships :

$$X^{(c)T} U^{(c)} X^{(c)} = N^{(c)} = \mathrm{diag}\,\{n_\nu^{(c)}\} \in \mathbb{C}^{2N,2N} \; ; \; X^{(c)T} A^{(c)} X^{(c)} = N^{(c)} S^{(c)} \in \mathbb{C}^{2N,2N}$$

By introducing the error matrices :

$$\Delta U \triangleq U - U^{(c)} \; ; \; \Delta A \triangleq A - A^{(c)} \; ,$$

we can express them as a first order development as :

$$\Delta U \simeq U^{(c)} [U^{(c)-1} - U^{-1}] U^{(c)} = U^{(c)} [X^{(c)} N^{(c)-1} X^{(c)T} - XN^{-1} X^T] U^{(c)}$$

$$\Delta A \simeq A^{(c)} [A^{(c)-1} - A^{-1}] A^{(c)} = A^{(c)} [X^{(c)} N^{(c)-1} S^{(c)-1} X^{(c)T} - XN^{-1} S^{-1} X^T] A^{(c)}$$

By introducing in these relations the change of base : $X = X^{(c)} Q$, we finally obtain expressions for the error matrices ΔU and ΔA :

$$\Delta U \triangleq \begin{bmatrix} \Delta B & \Delta M \\ \Delta M & \end{bmatrix} \simeq Z\,[N^{(c)-1} - QN^{-1}Q^T]\,Z^T, \text{ where : } Z \triangleq U^{(c)} X^{(c)} \qquad (I)$$

$$\Delta A \triangleq \begin{bmatrix} -\Delta K & \\ & \Delta M \end{bmatrix} \simeq Z\,[S^{(c)} N^{(c)-1} - S^{(c)} Q N^{-1} S^{-1} Q^T S^{(c)}]\,Z^T \qquad (II)$$

The right-hand side member of this equation can be evaluated from calculated eigensolutions (using the initial estimation) and the identified ones.

Partitioning into submatrices the right-hand side member of these equations as defined by the left-hand side member, and applying procedures like those developped in paragraphs 2 and 3, we obtain :

- Two groups of localization equations for Error matrix ΔM,
- One group of equations for the localization of Error Matrix ΔK
- Two groups of equality constraints equations corresponding to null sub-matrices.

We should underline that the chosen definition for matrices A and U is not unique. If we take into account the following definition of the autonomous model : AX = UXS

where :
$$\hat{A} \triangleq \begin{bmatrix} & -K \\ -K & -B \end{bmatrix} \quad ; \quad \hat{U} \triangleq \begin{bmatrix} -K & \\ & M \end{bmatrix}$$

We obtain the orthonormality relations :

$$X^T \hat{U} X = NS \quad ; \quad X^T \hat{A} X = NS^2$$

and the two following new expressions for Error Matrices.

$$\hat{\Delta U} \triangleq \begin{bmatrix} -\Delta K & \\ & \Delta M \end{bmatrix} \cong \hat{Z}[N^{(c)-1}S^{(c)-1} - QN^{-1}S^{-1}Q^T] \hat{Z}^T, \text{ where : } \hat{Z} \triangleq \hat{U}^{(c)}X^{(c)} \qquad (III)$$

$$\hat{\Delta A} \triangleq \begin{bmatrix} & \Delta K \\ \Delta K & \Delta B \end{bmatrix} \cong \hat{Z} [N^{(c)-1} - S^{(c)}QN^{-1}S^{-2}Q^TS^{(c)}]\hat{Z}^T \qquad (IV)$$

Non systematic errors contained in the identified eigensolutions act differently in (I ; II) and (III ; IV) and it is useful to simultaneously use them for localization purpose.

REFERENCES

|1| S.R. IBRAHIM, Determination of normal modes from measured complex modes, Shock Vib. Bull., n°5, 1982, 13-17

|2| Q. ZHANG, G. LALLEMENT, R. FILLOD, Passage des solutions propres complexes identifiées aux solutions propres de la structure conservative associée, Meca-Maté-Elec., n°404, Avril 1984, 18-23

|3| Q. ZHANG, G. LALLEMENT, New method of determining the associated conservative structure from the identified eigensolutions, Proc. 3rd IMAC, Orlando, FI, Jan.1985, 322-328

|4| Q. ZHANG, G. LALLEMENT, Simultaneous determination of normal modes and generalized damping matrix from complex modes, Proc. 2nd ISASD, Aachen, April 1-3, 1985, 529-535

|5| N. NIEDBAL, Advance in ground vibration testing using a combinaison of phase resonance and phase separation methods, Proc. 2nd ISASD, Aachen, April 1-3, 1985, 523-528

|6| H.G. NATKE, D. ROTERT, Determination of normal modes from identified complex modes, Z. Flugwiss. Weltraumforsch, 9, 1985, Heft 2

|7| Q. ZHANG, G. LALLEMENT, Three normal modes calculation methods based on identified complex eigenmodes, Proc. 10th Modal Analysis Seminar Leuven, Belgium, oct.1985

|8| Q. ZHANG, G. LALLEMENT, Comparison of normal eigenmodes calculation methods based on identified complex eigenmodes, J.Spacecraft and Rockets, vol.24, n°1, Ja-Feb.1987, 69-71

|9| H.G. NATKE (Ed.), Identification of Vibrating structures, Springer Verlag, New York, 1982, 510 pages

|10| L.P. BUGEAT, R. FILLOD, G. LALLEMENT, J. PIRANDA, Ajustement du comportement dynamique d'un modèle discret conservatif, Rapport final contrat CNES-LMA n°76 - CNES - 3002

|11| G. LALLEMENT, Recalage de modèles discrets structuraux : localisation des sous-structures présentant les défauts dominants de modélisation, Rapport Technique pour AMD-BA, Août 1983, 32 p.

|12| L.P. BUGEAT, R. FILLOD, G. LALLEMENT, J. PIRANDA, Ajustment of a conservative non gyroscopic model from measurement, Shock and Vib. Bull., 48, Part 3, sept.1978, 71-81

|13| Q. ZHANG, G. LALLEMENT, Dominant error localization in a finite element model of a mechanical structure, Mechanical Systems and Signal Processing (1987), 1(2), 141-149

|14| J. SIDHU,   , Reconciliation of predicted and measured modal properties of structures, Ph.D.Thesis, Imperial College, Dept. of Mech.Eng., London, 1983

|15| J. SIDHU, D.J. EWINS, Correlation of finite element and modal test studies of a practical structure, Pr. 2nd IMAC, 1984, Orlando, Fl, 756-762

|16| Q. ZHANG, G. LALLEMENT, R. FILLOD, Modal identification of self-adjoint and non self-adjoint structures by additional masses techniques, ASME Paper 85-DET-109, 1985

|17| Q. ZHANG, G. LALLEMENT, R. FILLOD, J. PIRANDA, A complete procedure for the adjustment of a mathematical model from the identified complex modes, Proc. IMAC V, April 6-9 1987, London, 1183-1190

|18| C.L. LAWSON, R.J. HANSON, Solving least squares problems, Prentice-Hall, Inc., Englewood Cliffs, 1974

|19| L. P. BUGEAT, G. LALLEMENT, Methods of matching of calculated and identified eigensolutions, Strovnicky Casopis, vol.32, n°5, 1981

|20| R.E. HULL, B.I. BEJMUK, Use of general mass contributions in correlation of test and analytical vibration modes, Shock and Vib. Bull. n°43, part.3, 1973, 79-86

|21| R.J. ALLEMANG, D.L. BROWN, A correlation coefficient for modal vector analysis, Proc. of 2nd IMAC, Orlando, U.S.A., 1984, 110-116

|22| Q. ZHANG, Identification modale et paramétrique de structures mécaniques auto-adjointes et non auto-adjointes, Th. Doct. ès Sc., Université de Franche-Comté, Laboratoire de Mécanique Appliquée, Besançon, Janv.1987.

A Two Stage Identification Approach in Updating the Analytical Model of Buildings

MORTEZA A. M. TORKAMANI[1] and AHMAD K. AHMADI[2]

SUMMARY

Ambient response measurements were made on an eighteen-story building at three different stages of construction to detect any changes in the frequencies, mode shapes, and stiffness with construction. The first nine frequencies and corresponding mode shapes, for each stage of construction, are found. A comparison is made among these mode shapes and frequencies and with the mode shapes and frequencies of an analytical model incorporating beams, columns, shear walls, panels and diagonal elements. The added effects, on frequencies and mode shapes, of non-structural elements such as stairs, elevators, claddings, and partition walls are studied. Using Improved Statistical Structural Identification, an attempt is made to study the stiffening effect of non-structural elements by updating the stiffness matrix of the building.

1. INTRODUCTION

During the last two decades advancing computer technology has made modeling and analysis of the high-rise buildings an easy task. In spite of modern computational resources neglecting non-structural elements, such as stairwells, elevator shafts, partition walls, claddings and, etc., in the mathematical modeling of buildings, is common practice. Although, the effect of a single non-structural element on the static or dynamic response of a building may be negligible, the cumulative effect of several could be significant. Full-scale measurements of buildings show that analytical models do not give mode shapes and frequencies which

---

[1] Associate Professor, [2] Former Graduate Student, Department of Civil Engineering, University of Pittsburgh, Pittsburgh, Pennsylvania 15261

concur with test results [9]. This is the direct consequence of improper modeling of some of the structural elements, neglecting the non-structural elements and joint rotations. The results indicate that there is a need to improve the mathematical modeling of buildings such that the gap between the modal responses of analytical and experimental analyses is minimized.

In order to close the gap between analytical and experimental mode shapes and frequencies a building's stiffness and mass matrices may be updated. Computation of the mass matrix of a building is generally based on the mass of dead weight of the elements, components and systems and is more accurate than the stiffness matrix. Therefore, the stiffness matrix of the building should be revised. The Improved Statistical Structural Identification method, which is described in the paper, is used in this research to revise the stiffness matrix of a building.

This paper presents the results of a study to determine the effects of non-structural elements on the frequencies, mode shapes, and stiffness coefficients of a high-rise building. Ambient vibration measurements were conducted on a building at three different stages of construction and after the construction was completed. A set of mode shapes and frequencies were found from data processing of the ambient vibration tests. A second set of mode shape and frequencies were found considering the analytical model. Then, the Improved Statistical Structural Identification method was used to revise the stiffness matrix of the building. A third set of mode shapes and frequencies were calculated using the revised stiffness matrix. A comparison is made among these three sets of modal responses. Descriptions of the building, experimental setup, model of the building, modal responses resulting from the analytical model, vibration tests, revised model, and comparison of these responses are given in the following sections.

Full-scale forced vibration tests have been used extensively to measure the dynamic response of buildings. The most widely used methods of full-scale dynamic testing include ambient vibration and forced vibration from harmonic vibration generators. Foutch[5] and Petrovski et al.[10] compared the results of these two testing methods when applied to buildings. They found close agreement in the mode shapes and frequencies of the buildings from both methods (using non-distructive forces in the case of vibration generator tests). However, the duration of testing and amount of labor needed in vibration generator tests are much greater than for ambient tests. When using the vibration generator test additional problems and difficulties may arise if the building is under construction as was the building tested in this study. For these reasons only ambient vibration testing is considered in this research.

2. METHODS OF PARAMETER ESTIMATION

The equation of motion for undamped free vibration of an n-story building is

$$[M]\,\{\ddot{X}\} + [K]\,\{X\} = \{0\} \tag{1}$$

where,

$\{\ddot{X}\}, \{X\}$ n x 1 vector of accelerations and displacements, respectively,

$[M]$ n x n structural mass matrix,

$[K]$ n x n structural stiffness matrix,

n number of dynamic degrees of freedom.

The mass and stiffness matrices are functions of the structural parameters of the system, denoted as $P_1{}', P_2{}'...,P_m{}'$. Thus, the eigenvalues and eigenvectors of the structure are also be functions of $P_1{}'\ P_2{}'...,P_m{}'$. Therefore Eq. 1 may be expressed as:

$$[[K] - \lambda_j(p)[M]]\{\phi(p)\}_j = \{0\} \tag{2}$$

where,

$\lambda_j(p)$ j^{th} eigenvalue of structure

$\{\phi(p)\}_j$ n x 1 eigenvector (mode shape) for the jth mode

The eigenvalues and eigenvectors (mode shapes) of structures are mathematically well-behaved, and smooth functions, which may be expanded by a Taylor series,

$$\left\{ \begin{array}{c} \{\lambda(p)\} \\ \{\phi(p)\} \end{array} \right\}_{\substack{\{P\}=\{P\} \\ i+1}} = \left\{ \begin{array}{c} \{\lambda(p)\} \\ \{\phi(p)\} \end{array} \right\}_{\substack{\{P\}=\{P\} \\ i}} + [S] \; [\{P\}_{i+1} - \{P\}_i] + \{0^+\} \tag{3}$$

where,

$\{\lambda\}$ l x 1 structural eigenvalue vector $1 < l < n$

$\{\phi(p)\}$ n_1 x 1 structural eigenvector of n_1 elements $1 < n_1 < n^2$

$\{P\}$ m x 1 structural parameters vector

$$[S] = \left[\begin{array}{c} [\partial\lambda/\partial p] \\ \hline [\partial\phi/\partial p] \end{array} \right]$$

rearranging Eq. 3, we obtain Eq. 4.

$$\left\{ \begin{array}{c} \{\Delta\lambda(p)\} \\ \{\Delta\phi(p)\} \end{array} \right\} = [S]\{\Delta P\} \tag{4}$$

In Eq. 4 the unknowns are the elements of $\{\Delta P\}$. If $[S]^{-1}$ exists, the solution to the problem is trivial. In other words, if the number of measured eigenvalues and independent elements of eigenvectors entered in the left-hand side of Eq. 3 is equal to the number of parameters, $[S]$ is a square matrix, i.e., $n_2 = m$. From Eq. 4, it immediately follows that

$$\{\Delta P\} = [S]^{-1} \left\{ \begin{array}{c} \{\Delta\lambda(p)\} \\ \{\Delta\phi(p)\} \end{array} \right\} \tag{5}$$

However, this is seldom the case. For the situation where the number of parameters exceeds the number of measured eigenvalues and eigenvectors entered in the left hand side of Eq. 3 (under-determined systems), special mathematical techniques may be employed to calculate an estimation of the parameters. On the other hand, there are cases where the number of measured eigenvalues and eigenvectors exceed the number of parameters (over-determined systems). Since these measured quantities estimate the eigenvalues and eigenvectors, all of them should be considered in the estimation of the parameters. Thus, the resulting model has characteristics which simulate the system dynamic response over the frequency range of the measured normal mode data. Therefore, a mapping (estimator) matrix [W] is sought such that the transformation, or mapping, equation may be expressed as:

$$\{\Delta P\} = [W]\left\{\frac{\{\Delta\lambda(p)\}}{\{\Delta\phi(p)\}}\right\} \qquad (6)$$

Since there are in general more unknowns than measurements or vise-versa, the choice of [W] is not unique and one must therefore establish criteria which will be used in the selection of an appropriate [W] .

The use of any statistical method for parameter estimation depends upon the randomness of variable parameters, measured frequencies and mode shapes. Statistical concepts and estimation theory form the basis of these methods. The most commonly used methods are: 1. <u>The Least Squares Structural Identification Method (LSM)</u>; 2. <u>The Weighted Least Squares Structural Identification Method (WLSM)</u>; 3. <u>The Statistical Structural Identification Method (SSI)</u>; and 4. <u>The Statistical Structural Identification Using an Altered Error Function Method (SIASTRO)</u>, [7].

The estimator matrix, [W], resulting from the first three methods is shown in Table 1.

Table 1. : Estimator Matrix, [W], for Different Methods.

Method	$\{\Delta P\} = [W]\{\Delta R\}$	
Least	Under determined	$[S]^{I}([S][S]^{T})^{-1}\{\Delta R\}$
Squares	+++++++++++++++++	
Method	Over determined	$([S]^{T}[S])^{-1}[S]^{T}\{\Delta R\}$
Weighted	Under determined	$[S]^{T}([S][S]^{T})^{-1}\{\Delta R\}$
Least Squares	+++++++++++++++++	
Method	Over determined	$([S]^{T}[D_r][S])^{-1} [S])^{T}[D_r]\{\Delta R\}$
Statistical	$([C_p]^{-1} + [S]^{T}[C_r]^{-1}[S])^{-1}[S]^{T}[C_r]^{-1}\{\Delta R\}$	
Structural	or	
Identification	$[C_p][S]^{T}([S][C_p][S]^{T} + [C_r])^{-1}\{\Delta R\}$	

Where, $[C_p]$ and $[C_r]$ are the diagonal covariance matrices of errors on prior parameters and measured responses, respectively. $[D_p]$ and $[D_r]$ are the weighting (confidence) matrices for errors on prior parameters and measured responses such that $[D_p] = [C_p]^{-1}$ and $[D_r] = [C_r]^{-1}$. Note that in the Weighted Least Squares Method, under-determined case, the weighting factor is canceled, indicating that the statistical properties of the measured data do not affect the final result.

The writers have used the Statistical Structural Identification Formula extensively [1,8]. Experience indicates that when the number of parameters is large, the rate of convergence is very slow. However, in some instances, the estimation parameters diverge [1]. In order to resolve this problem, Ref. [7] suggests a new error function in the Bayesian estimation theory for the derivation of the structural identification formula. This is

$$E(P) = (\{R^*\}-\{R(P)\})^T[D_r](\{R^*\}-\{R(P)\})+(\{P\}-\{P\}_0)^T[D_p](\{P\}-\{P\}_0) \quad (7)$$

where,

$\{P\}$	predicated parameters,
$\{P\}_0$	initial parameters,
$\{R^*\}$	experimentally measured response
$\{R(P)\}$	predicted response,
$[D_r]$	confidence matrix of measured responses,
$[D_p]$	confidence matrix of analytical parameters.

Since predicted parameters and responses are not known, an iterative parameter revision may be the logical procedure for solution. In this method, predicted response $\{R(P)\}$ is approximated as an iterative response at step $n+1$, i.e.,

$$\{R(P = P_{n+1})\} = \{R(P)\} \quad (8)$$

Considering Taylor series expansion of Eq. 8 and substituting Eq. 3 into Eq. 7, differentiating the resulting error function with respect to the parameters, and setting the results equal to zero, the following equation is given in Ref. [7].

$${P^*}=\lim_{n\to\infty} {P}_{n+1}={P}_n + ([D_p] + [S]_n^T[D_r][S]_n)^{-1}([D_p]({P}_0 - {P}_n) \quad (9)$$

$$+ [S]_n^T[D_r]({R^*} - {R}_n))$$

where

${P^*}$	final parameters,
${P}_{n+1}$	predicted parameters at iteration step n+1
${R}_{n+1}$	predicted response at iteration step n+1.

2.1 A NEW FORMULA

We suggest that the error function, Eq. 7, be modified to the following form

$$E(P)=({R^*}-{R(P)})^T[D_r]({R^*}-{R(P)})+({P}-{\hat{P}})^T[D_p]({P}-{\hat{P}}) \qquad (10)$$

where

$${\hat{P}}$$ an estimated parameters.

The second term on the right-hand side of Eq. 10 will force the predicted parameters {P}, to converge to the desired parameters which are not known but are estimated.

Considering the method of derivation previously mentioned, replacing {R(P)} by ${R}_{n+1}$', {P} by ${P}_{n+1}$', using Taylor series expansion, substituting Eq. 3 into Eq. 10 and setting the derivative of E(P) with respect to the parameters equal to zero, we obtain the following estimation formula:

$${P}_{n+1}={P}_n+([D_p]+[S]_n^T[D_r][S]_n)^{-1}([D_p]({\hat{P}}-{P}_n)+[S]_n^T[D_r] \qquad (11)$$
$$({R^*} - {R}_n))$$

Equation 11 is expressed in terms of the confidence matrices $[D_r]$ and $[D_p]$. Equation 11 may be expressed in terms of covariance matrices $[C_r]$ and $[C_p]$ in the following form

$${P}_{n+1} = [W]_n({\Delta R^*} - [S]_n({\hat{P}} - {P}_n)) + {\hat{P}} \qquad (12)$$

where

$$[W]_n = [C_p][S]_n^T([S]_n[C_p][S]_n^T + [C_r])^{-1} \qquad (13)$$

and

$${\Delta R^*} = {R^*} - {R}_n \qquad (14)$$

The covariance matrix of vector ${\Delta P}$ may be expressed in the following form

$$Cov({\Delta P}_{n+1}) = [C_p]_n - [C_p]_n[S]_n^T([S]_n[C_p]_n[S]_n^T + [C_r])^{-1}[S]_n[C_p]_n \qquad (15)$$

The New Statistical Structural Identification formula (NSSI) is recommended strongly for cases where the elements of the analytical parameters and measured response have different confidence coefficients, or covariances. In the data processing of the experimental observations we would not have known that one value was estimated more accurately than the other. But in this case the standard deviations obtained from $Cov({\Delta P}_{n+1})$ come to the rescue and show the uncertainty of the estimated parameters. We are naturally interested in the highest accuracy possible in estimating all parameters of the system, but, with frequently meager or incomplete data, it is definitely an asset to have an estimate of the accuracy of each of the final estimated values of the parameters.

It is noteworthy that: 1. if an estimate of ${\hat{P}}$ is calculated by the least squares method, i.e., ${\hat{P}} = {P_n} + [S]_n^T([S]_n([S]_n^T)^{-1} {\Delta R^*}$ for the under-determined cases or ${\hat{P}} = {P}_n + ([S]_n^T[S])^{-1}[S]_n^T{\Delta R^*}$ for the over-determined cases, and then substituting this ${\hat{P}}$ in Eq. 11 or 12, the final result will be the same as the least squares method, except the statistical properties, covariance, of these revised parameters are known; 2. if for ${\hat{P}} - {P}_n$, ${0}$ is substituted into Eq. 11 or 12 the final result will be the statistical structural identification formula given in Ref. [4]; 3. if for ${\hat{P}} - {P}_n$, ${P}_0 - {P}_n$ is substituted into Eq. 11, the final result will be Eq. 11, which is derived in Ref. [7].

2.2 IMPROVED STATISTICAL STRUCTURAL IDENTIFICATION

As stated earlier, for the case of multi-variable parameters when the number of measured frequencies and mode shapes are small, the SSI method has a slow rate of convergence, non-convergence, or divergence. By increasing the number of measured frequencies and mode shapes, the rate of convergence increases; however, this is not possible in practice because the number of measured mode shapes and frequencies are limited.

The error function which is used for the SSI method is presented below:

$$E(p_j) = \{\{\Delta R\} - [S]\{\Delta P\}\}^T [C_r]\{\{\Delta R\} - [S]\{\Delta P\}\} + \{\Delta P\}^T [C_p]\{\Delta P\} \qquad (16)$$

There are two parts in the error function of the SSI method, with one of the two parts outweighting the other. A careful study of this error function reveals that the second term in this function is the dominant term. Therefore this term outweighs the first term causing slow convergence and, in some cases, divergence. This dominant part is the error between $\{P\}_{i+1}$ and $\{P\}_i$. For example, in the first iteration, i.e., i=0, the value of $\{P\}_0$ is known, prior parameters, therefore this prior information will outweigh the other observations and causing the estimated parameters to converge to the initial parameters.

This problem can be corrected by considering more weight in the first part of the error function. Based on this discussion it is suggested that the error function be modified as follows:

$$E(p_j) = \{\{\Delta R\} - [S]\{\Delta P\}\}^T [C_r]^{-1}\{\{\Delta R\} - [S]\{\Delta P\}\} \qquad (17)$$

$$+ \beta\{\Delta P\}^T [C_p]^{-1}\{\Delta P\}$$

Where β is a weighting factor and its value is dependent on the judgment of the analyst.

Using Eq. 17 as an error function and the method of derivation previously mentioned, results in the following estimation formula:

$${\Delta P}_{n+1} = \beta^{-1}[C_p]_n[S]_n^T(\beta^{-1}[S]_n[C_p]_n[S]_n^T + [C_r])^{-1}{\Delta R} \qquad (18)$$

This equation is called <u>The Improved Statistical Structural Identification Formula (ISSI)</u>.

The covariance matrix of the error in revised parameters may be computed from the following formula.

$$Cov(\Delta P)_{n+1} = [C_p]_n - \beta^{-1}[C_p]_n[S]_n^T(\beta^{-1} [S]_n[C_p]_n[S]_n^T \qquad (19)$$

$$+ [C_r])^{-1}[S]_n[C_p]_n$$

3. APPLICATION OF SYSTEM IDENTIFICATION IN THREE-DIMENSIONAL MODEL OF BUILDINGS

In the application of the system identification methods computation of the sensitivity matrix, [S], is required. For computation of [S], an intermediate matrix consisting of the derivatives of the stiffness of the beams and columns with respect to the parameters is needed. In reference [12], in which shear type buildings were considered for analysis, it was easy to trace the location of the stiffness of each beam or column in the assembled stiffness matrix. Therefore, the computation of the sensitivity matrix, [S], was straightforward. However, in the assembled stiffness matrix of a three-dimensional model of a building, it is not possible to trace the location of the stiffness of each column or beam. Moreover, there are many beams and columns in a real building; considering the stiffness of these beams and columns as structural parameters will result in a large number of parameters. On the other hand, if parameters are the entries of the assembled stiffness matrix, for a n-story building with three degrees of freedom at each floor, the number of parameters to be identified is $[3n)^2 + 3n]/2$, which is also very large.

Knowing that, a limited number of the mode shapes and frequencies can be extracted from data processing of the dynamic test data. System identification methods are not able to accurately revise these many parameters. Therefore, there is a need to model the stiffness matrix of the three-dimensional buildings to contain the following properties: 1. it accurately represents the stiffness matrix of the building; 2. it realistically contains the variable parameters of the system ; and 3. it makes the computation of the sensitivity matrix, [S], rather an easy task.

3.1 THREE-DIMENSIONAL MODEL OF BUILDINGS FOR SYSTEM IDENTIFICATION

It is known that the main reason for the differences between the analytical and experimental frequencies and mode shapes of a building is mostly due to inaccuracy in the mathematical modeling of the stiffness matrix. This is because the contribution of the non-structural elements, some of the structural elements, and joint rotations are disregarded. System identification may be used to consider the contribution of these elements and components in the mathematical modeling of the buildings.

In order to apply the system identification to the three-dimensional model of a building, it is assumed that the total stiffness matrix is formed from superposition of two stiffness matrices as stated below:

$$[K] = [k_1] + [K_2] \tag{20}$$

where,

$[K]$ stiffness matrix of building considering structural and non-structural elements

$[K_1]$ stiffness matrix of structural elements

$[K_2]$ stiffness matrix of non-structural and variable structural elements

Since $[K_1]$ is constructed from a mathematical model of the building, which considers known structural elements, it is assumed that the elements of this matrix are independent of the parameters of the system. On the other hand, the matrix $[K_2]$ is constructed from a fictitious model, which considers non-structural elements, variable stiffness structural elements and the contribution of the joint rotations. It is assumed that the elements of this matrix are dependent on the parameters of the system.

At this point, we may go one step further and make the assumption that the elements of the matrix $[K_2]$ are the parameters of the system. The matrix $[K_2]$ should also have the properties which were discussed in the previous section. Based on these considerations, it is assumed that all structural and non-structural elements which are not considered in the mathematical model of the building may be represented by a fictitious model. This model passes through the mass center of each floor and has two translational and one torsional degrees of freedom at each floor. The assembled stiffness matrix of this model is $[K_2]$, which is a banded matrix. The general form of this matrix is given next. Since the parameters are the non-zero elements of matrix $[K_2]$, the sensitivity matrix can be computed without difficulty.

$$\begin{bmatrix}
k_1 & 0 & 0 & -k_1 \\
k_2 & 0 & 0 & -k_2 \\
k_3 & 0 & 0 & -k_3 \\
k_1+k_4 & 0 & 0 & -k_4 \\
k_2+k_5 & 0 & 0 & -k_5 \\
k_3+k_6 & 0 & 0 & -k_6 \\
\cdot & \cdot & \cdot & \cdot \\
\cdot & \cdot & \cdot & \cdot \\
\cdot & \cdot & \cdot & \cdot \\
k_{3n-5}+k_{3n-2} & 0 & 0 & \\
k_{3n-4}+k_{3n-1} & 0 & 0 & \\
k_{3n-3}+k_{3n} & & &
\end{bmatrix}$$

3.2 EXAMPLE

In order to investigate the application of system identification methods to the three-dimensional models and compare the accuracy of the different methods, a two-story building is considered as an example. See Fig. 1. In this example only LSM, SSI, and ISSI are applied to revise the stiffness matrix of the building.

The type of the model which is used for this building is a pseudo three-dimensional model which has two translational and one torsional global degrees of freedom at mass center.

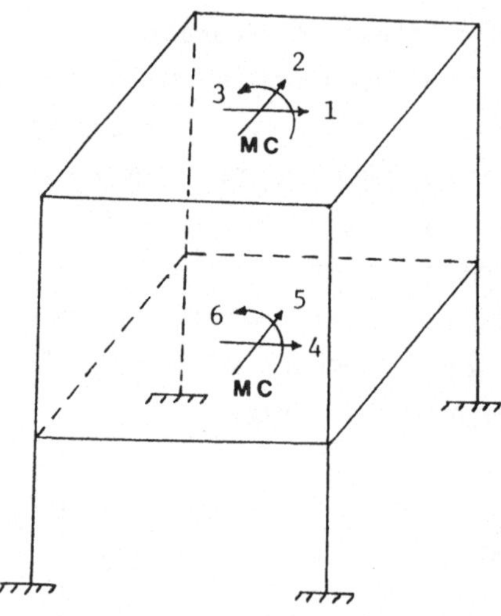

Figure 1: Three-Dimensional Building

The stiffness matrix of the building is obtained by adding the transformed stiffness matrix of each frame at the mass center. The mass matrix of the building is computed by lumping the mass of each floor and half of the columns at the mass center of that floor. The mass matrix and the stiffness matrix, which is divided into two matrices $[K_1]$ and $[K_2]$, are given as:

$$[M] = \begin{bmatrix} 3.3 & & & & & \\ & 3.3 & & & & \\ & & 6810. & & & \\ & & & 3. & & \\ & & & & 3. & \\ & & & & & 4290 \end{bmatrix}$$

245

$$[K_1] = \begin{bmatrix} 900. & 0. & 3250. & -1150. & 0. & 12000. \\ 0. & 170. & -0500. & 0. & -220. & -150. \\ 3250. & -500. & 673000. & -5100. & 540. & -824600. \\ -1150. & 0. & -5100. & 4100. & 0. & -41300. \\ 0. & -220. & 540. & 0. & 1260. & 1500. \\ 12000. & -150. & -825600. & -41300. & 1500. & 9265000. \end{bmatrix}$$

$$[K_2] = \begin{bmatrix} 100. & 0. & 0. & -100. & 0. & 0. \\ 0. & 30. & 0. & 0. & -30. & 0. \\ 0. & 0. & 2000. & 0. & 0. & -2000. \\ -100. & 0. & 0. & 400. & 0. & 0. \\ 0. & -30. & 0. & 0. & 140. & 0. \\ 0. & 0. & -2000. & 0. & 0. & 5000. \end{bmatrix}$$

Table 7 shows the first three measured frequencies and mode shapes of the building. These frequencies and mode shapes are obtained by assuming that the real stiffness matrix of the building is the summation of the structural and non-structural stiffness matrices $[K]$ ($[K] = [K_1] + [K_2]$).

Table 7: First Three Measured Frequencies and Mode Shapes

	1	-.00283	.53990	.00077	.00074	.10640	.00007
Eigs.	2	-.05842	-.03331	.01198	.00920	-.01450	.00123
	3	.52355	-.00071	.00116	.16935	-.00130	.00020

	1	45.45325
Eigenvalues	2	83.81732
	3	184.07291

Table 8 shows the first three frequencies and mode shapes of an analytical model of the building. These frequencies and mode shapes are obtained by assuming that the stiffness matrix of the analytical model is $[K_1]$. Therefore, the differences between these two sets of frequencies and mode shapes are due to disregarding the stiffness matrix $[K_2]$ in the analytical model.

Table 8: First Three Frequencies and Mode Shapes of Analytical Model

	1	$-.00272$	$.54068$	$.00067$	$.00069$	$.10355$	$.00006$
Eigs.	2	$-.07156$	$-.02868$	$.01195$	$.00777$	$-.01447$	$.00124$
	3	$.52164$	$-.00101$	$.00145$	$.17082$	$-.00175$	$.00023$

	1	38.55382
Eigenvalues	2	82.99332
	3	162.95968

Least Squares Method: The first case study is the application of LSM in identification of $[K_2]$ using one and two measured eigenvalues and eigenvectors. An initial value of zero is assumed for all unknown parameters in the first iteration. For the case of one measured eigenvalue and eigenvector LSM has an ill-conditioning problem and parameters do not converge. For the case of two eigenvalues and eigenvectors, after ten iterations all the parameters converge to the real values. Table 9 shows the results after five iterations.

Statistical Structural Identification: The second case study is the application of SSI in identification of $[K_2]$. It is assumed that one and two measured eigenvalues and eigenvectors are available. Covariance matrix of errors in measured data is assumed less than .2% of measured data and is given as:

$$\begin{bmatrix} .2 & & & & & & \\ & 3\times10^{-9} & & & & & \\ & & 12\times10^{-5} & & & & \\ & & & 2\times10^{-10} & & & \\ & & & & 2\times10^{-10} & & \\ & & & & & 45\times10^{-7} & \\ & & & & & & 2\times10^{-12} \end{bmatrix}$$

The initial covariance matrix of error in prior parameters is computed considering the method given in reference [1]. Results of this method are shown in Table 9. Although the error in measured data is very low, parameters 3 and 6 did not change, and parameter 4 has a slow rate of convergence, however, the singularity problem was not observed in this case.

Improved Statistic Structural Identification: The third case study is the application of ISSI in identification of $[K_2]$ using one and two measured eigenvalues and eigenvectors. In this method it is assumed that $[C_r]$, and $[C_p]$ are the same as SSI and $\beta(p)$ is .001. The results are shown in Table 9. As it is shown, the rate of convergence is better and faster than SSI.

Table 9 : Results of Different System Identification Methods after Five Iterations (Three-Dimensional Model)

		k_1	k_2	k_3	k_4	k_5	k_6
Exact Parameters		100	30	2000	300	110	3000
Initial Parameters		1	1	1	1	1	1
LSM	d^2	101	31	2082	304	111	-5641
SSI	c^1	80	31	4	195	91	1
	d	98	23	52	315	147	2
ISSI	e^3	102	31	721	307	111	26
	f^4	101	31	2017	301	111	1460
	g^5	103	29	930	326	115	12
	h^6	101	30	2166	304	113	2228

1 One measured eigenvalue and eigenvector.
2 Two measured eigenvalues and eigenvectors.
3 One measured eigenvalue and eigenvector with constant β.
4 One measured eigenvalue and eigenvector with variable β.
5 Two measured eigenvalue and eigenvector with constant β.
6 Two measured eigenvalue and eigenvector with variable β.

In order to improve the rate of convergence, it is assumed that $\beta(p)$ is changing for each iteration, i.e. $\beta_1(p)=.001$, $\beta_2(p)=.0001$, . . ., $\beta_5(p)=.0000001$. Results of this case are shown in Table 9 and Fig. 2. As it is shown for this case, by using one eigenvalue and eigenvector all parameters converged to the "real" values, except k_6. This parameter will converge to "real" value after two more iterations, or by decreasing $\beta(p)$.

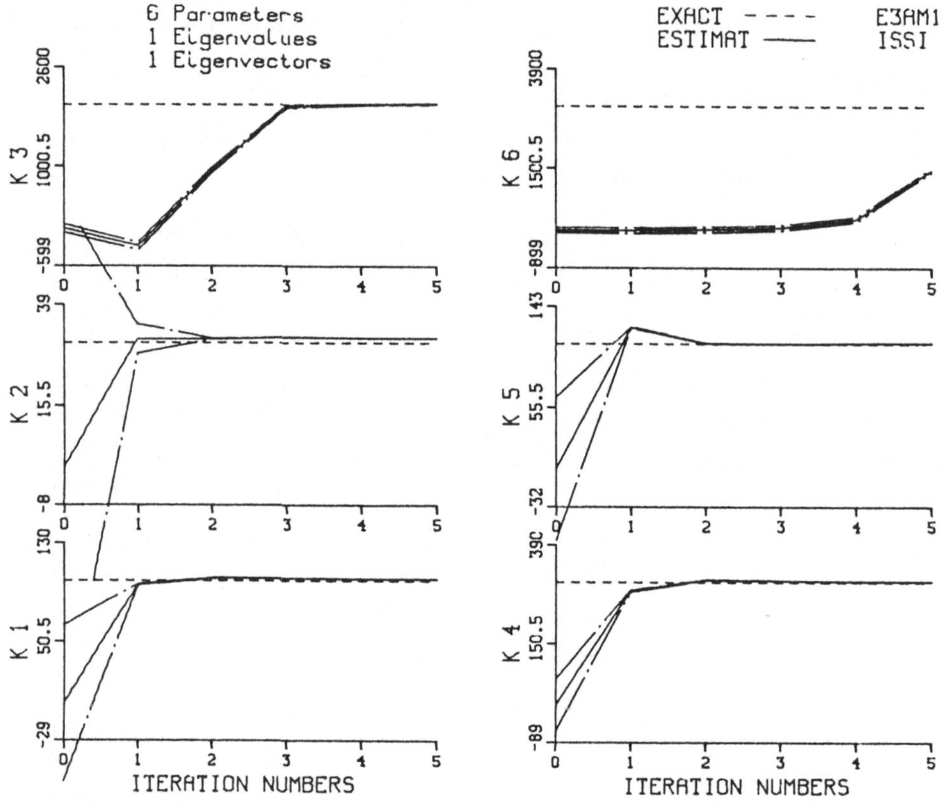

Figure 2:   Estimated Parameters by ISSI, Using One Eigenvalues and Eignevectors with Variable $\beta(p)$

4. DESCRIPTION OF BUILDING

The building selected for this study is the Six PPG (Pittsburgh Plate Glass) Place building having fifteen stories above ground and three underground garage levels. It is located in downtown Pittsburgh, Pennsylvania. The exterior claddings of the building are made of laminated glass called "Natural Silver Solarban 550 Clear Reflective Glass". It consists of two layers of laminaed glass each .25 inches (6.35 mm) thick with .5 inches (12.70 mm) distance between them.

The above ground portion of the building is a steel frame which is 204.75 feet (62.41 m) tall and includes a mezzanine level, thirteen regular floors and a mechanical floor. The under-ground portion of the building is a three-story concrete frame parking garage extending 34 feet (10.36 m) below ground. A typical plan of these floors and the elevation view are shown in Fig. 3.

Considering both major and minor axes of the building there are thirty five plane frames in the building. Some of these frames terminate at the ground floor. Three of these frames, which continue up to the roof, have "K" bracing.

4.1 BUILDING MODEL

The type of mathematical model, which is used to analyze the Six PPG Place, is a pseudo three-dimensional model. The model consists of an assemblage of two-dimensional frames and shear walls that may be arbitrarily oriented in plan. The frames are connected at each floor level by a diaphragm which is assumed rigid in its own plane. The model has three dynamic degrees of freedom at each floor; two translations and one rotation about the mass center. Further discussion including the assumptions involved in this type of modeling procedure, advantages, disadvantages, and i.e., are given in Ref. [6].

South and west sides of the garage levels are filled by soil. The soil's effects are modeled by discrete external spring stiffness. To find the modulus of subgrade reaction K_S of the soil, it is assumed that the soil from ground level to garage level one is loose sand, and from garage level one to garage level three is medium sand. In this analysis, the lower limit of the soil properties is used to calculate K_S [3].

The garage levels on the east and part of the north sides are continuous. This continuation is modeled a rigid links connected to discrete springs. K_S for these two sides is computed as for the south and west.

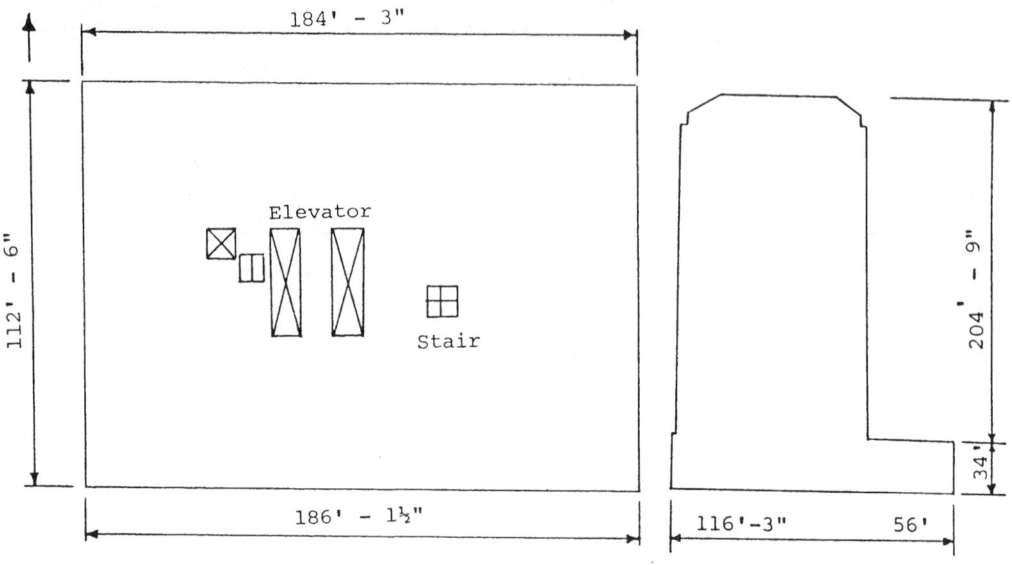

FIGURE 3:   A Typical Plan of Floors and Elevation View of the Building

4.2 AMBIENT VIBRATION TESTS

Three ambient vibration tests at different stages of construction were performed on this building. The first test was conducted when the main frames, floors, one stair case, and freight elevator were completed. The elevators (totally six of them), another stair case, and the exterior claddings of the building were in place prior to the second test. The third vibration test was performed after the partition walls were added. The data acquired from these tests was processed to obtain the mode shapes and frequencies of the building at three stages of construction.

Eight Ranger Seismometers (Model SS-1, Kinemetrics) were used to measure the responses simultaneously at different locations. The seismometer signals, after amplification and filtering by two Signal Conditioners (Model SC-1, Kinemetrics), are recorded on magnetic tape using an eight channel FM analog tape recorder. Figure 4 shows the general layout of ambient response measurement instrumentation.

As shown in Fig. 5 seismometers 1-4 were set and remained on the 14th floor during the vibration test, while the other four seismometers were moved to different floors. Fifteen minutes of normal signals (velocities) were recorded for each set up. Throughout the testing, Seismometers 1 and 5 were used to obtain information about the N-S direction, while 2 and 6, obtained information about the E-W direction. Torsional information was obtained by diminishing the effects of N-S modes. This was accomplished by algebraically subtracting the signals of seismometers 3 from 4 and 7 from 8 at each end of the floor.

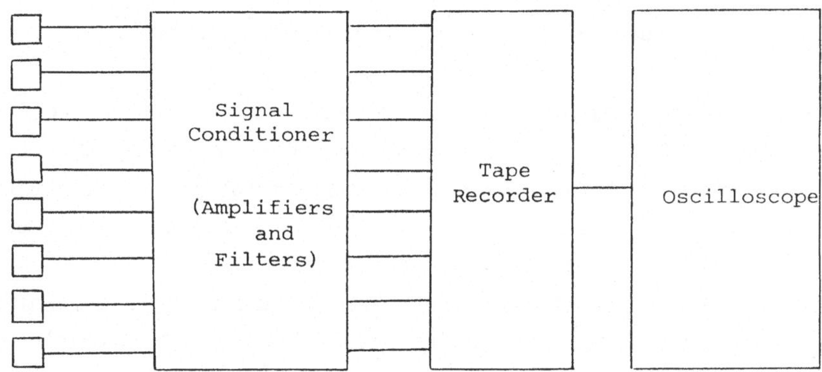

Figure 4: Ambient Response Measurement Instrumentation

In the laboratory 102.4 seconds of each recorded signal was digitized at a rate of 40 samples per second resulting in a Nyquist frequency of 20 Hz. All digitized data were passed through the Hanning time window to minimize the spectral spreading effect caused by the finite record length [11].

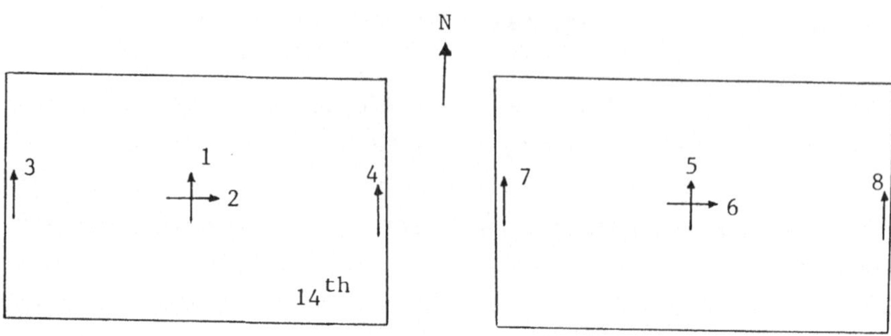

Figure 5: Lay out of Ranger Seismometers

A Fast Fourier Transform was computed for 4096 points of each record resulting in 2048 spectral ordinates and phase angles. The Fourier spectra are smoothed using five point averaging to increase the accuracy of the spectral estimates. Figure 6 shows the smoothed Fourier spectra of the 14th floor of the third vibration test in both N-S and the E-W directions. Fourier spectra of the responses are used to determine the frequencies, mode shapes, and damping ratios (half-power bandwidth method is used to find damping ratios) of vibration of Six PPG Place. The first nine frequencies and mode shapes of vibration of the building are easily detected. Frequencies resulting from data processing along with the first nine frequencies of the analytical model are shown in Table 1. Mode shapes obtained are only representative of the dominant directions (one dimensional).

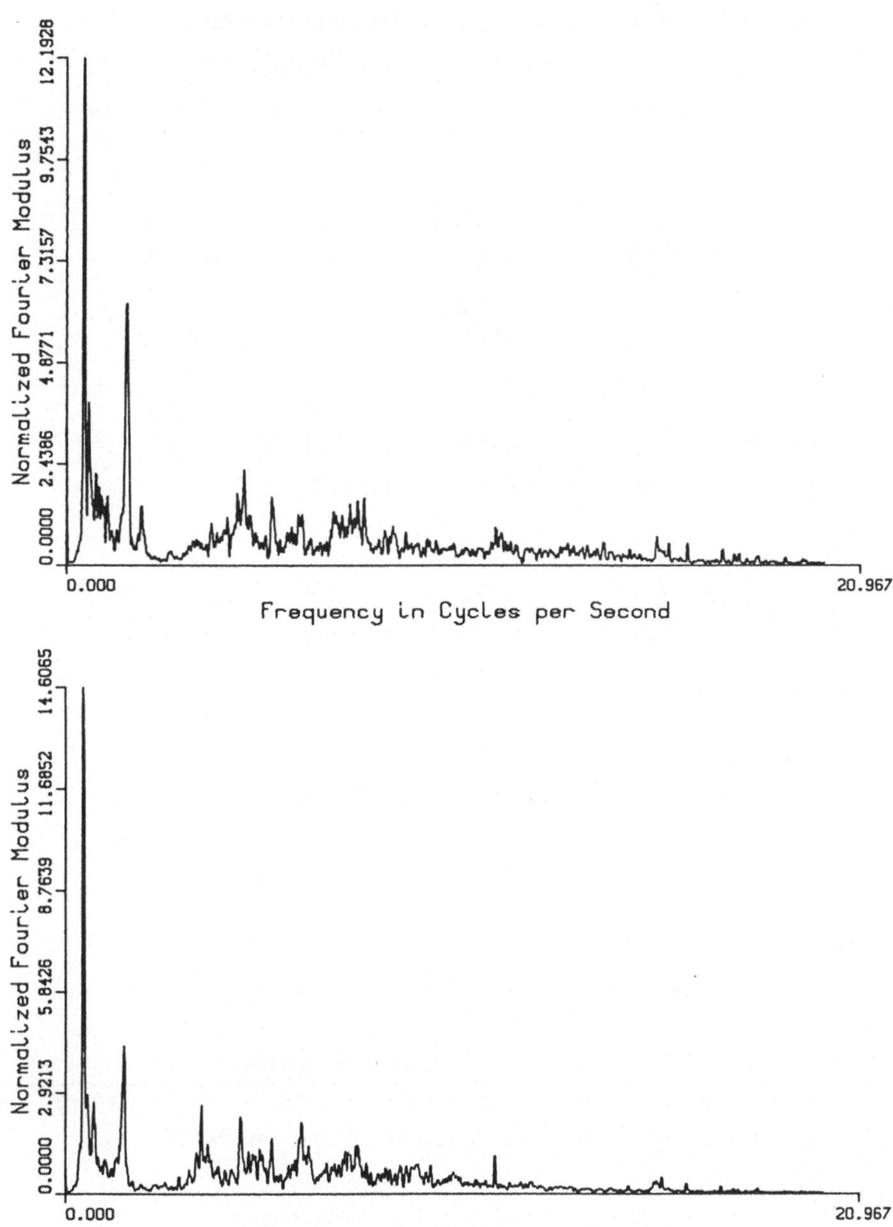

Figure 6:   Smoother Fourier Amplitude Spectra of 14th Floor Response;
(a) N-S Direction, (b) E-W Direction

Table 1:   Modal Frequencies of Mathematical Model and
Three Vibration Tests (Hz)

| | | Mode | | |
Directions		1	2	3
	E-W	.32089	.94571	1.63045
Analytical	N-S	.47875	1.57720	2.14984
	Tor	.41570	1.25126	2.02402
	E-W	.40039	1.15230	1.96289
Test No. 1	N-S	.48828	1.58203	2.73437
	Tor	.42969	1.25976	2.14844
	E-W	.44922	1.45508	2.54883
Test No.2	N-S	.49805	1.63086	3.14453
	Tor	.55664	1.70898	2.83203
	E-W	.46875	1.49414	2.55859
Test No.3	N-S	.50871	1.64062	3.24219
	Tor	.57617	1.94336	2.83203

4.3 PARAMETER ESTIMATION

There are three steps in a parameters estimation :
1. determination of the form of the model and isolation of the unknown
parameters, 2. selection of a criterion function (error function), and
3. selection of an algorithm for adjustment of the parameters [2].

The model considered for this building is a pseudo three-dimensional
model with 3nx3n (n is the number of floors) stiffness and mass matrices.
To obtain an accurate mass matrix, the mass of all structural elements,
systems and components are considered. To study the sensitivity of the
modal responses to changes in the mass of the structure, the computed
mass of each floor is increased by 5%, 10%, 15%, 20%, and 25%. The
sensitivity study indicates that the range of frequency change is

between 1% to 10% while the variation in mode shapes is negligible. However, in the estimation of the parameters considering tests number two and three, the mass of the mechanical floor and all other floors is increased by 50% and 5%, respectively, in the analytical model. These changes are due to construction changes, i.e., addition of non-structural elements, and installation of heavy machinery at the mechanical floor.

The Improved Statistical Structural Identification is used to improve the stiffness matrix of the Six PPG Place at different stages of construction. All measured eigenvalues and eigenvertors (nine frequencies and mode shapes) are used in these identification processes to estimate the 54 parameters of the building. Because, one dimensional mode shapes are obtained from Ambient vibration test, system identification is carried out with 1/3 of entries of each analytical mode shape. In the computation of $[C_r]$, it is assumed that all measured responses have a standard deviation less than 4% of their value. Frequencies of improved mathematical model, resulting after three iterations, are shown in Table 2.

Figures 7 through 9 are the first three, E-W direction, mode shapes of the analytical model, vibration tests, and improved models. Figure 7 shows a very good agreement between the test results (test number one) and the mode shapes of analytical model of the main frames, floors and one staircase. The mode shapes of the improved model and the test results have an excellent agreement. Figures 8 and 9 show that the agreement between the test results and analytical model is not as good as in test number one. The mode shapes of the improved mathematical model shows a good agreement for the first mode and revised stiffness matrix contributed more in matching for the higher modes. Tables 1 and 2 are modal frequencies of the analytical model, test results and improved analytical model. It shows a good agreement between the frequencies of the test results and the improved analytical model. Revised stiffness matrices are reported in Ref. [1].

Table 2: Modal Frequencies of Six PPG Place from
Improved Mathematical Models (Hz)

| | | Mode | | |
Directions		1	2	3
	E-W	.38562	1.12857	1.94006
Fir. Imp Mod.	N-S	.49053	1.58254	2.56854
	Tor	.42297	1.25239	2.14251
	E-W	.42484	1.19161	2.08325
Sec. Imp.Mod.	N-S	.51033	1.60416	2.97280
	Tor	.54255	1.74588	2.43492
	E-W	.44264	1.20355	2.07885
Thd. Imp. Mod	N-S	.50935	1.56481	3.15151
	Tor	.58361	1.93231	2.41916

5. SUMMARY AND CONCLUSIONS

A pseudo three-dimensional mathematical model is used to model an 18-story building. Computation of the mass matrix, which is based on the masses of all structural elements, systems and components, is more accurate than the stiffness matrix. Typical structural elements such as columns, beams, panels, and diagonal elements are considered in the assembly of the stiffness matrix.

Three ambient vibration tests at different stages of construction were performed on this building. The first nine frequencies resulting from the analytical model and three vibration tests are shown in Table 1. Mode shapes identified from ambient vibration tests are one dimensional. Each mode shape represents the dominant direction of vibration of the building.

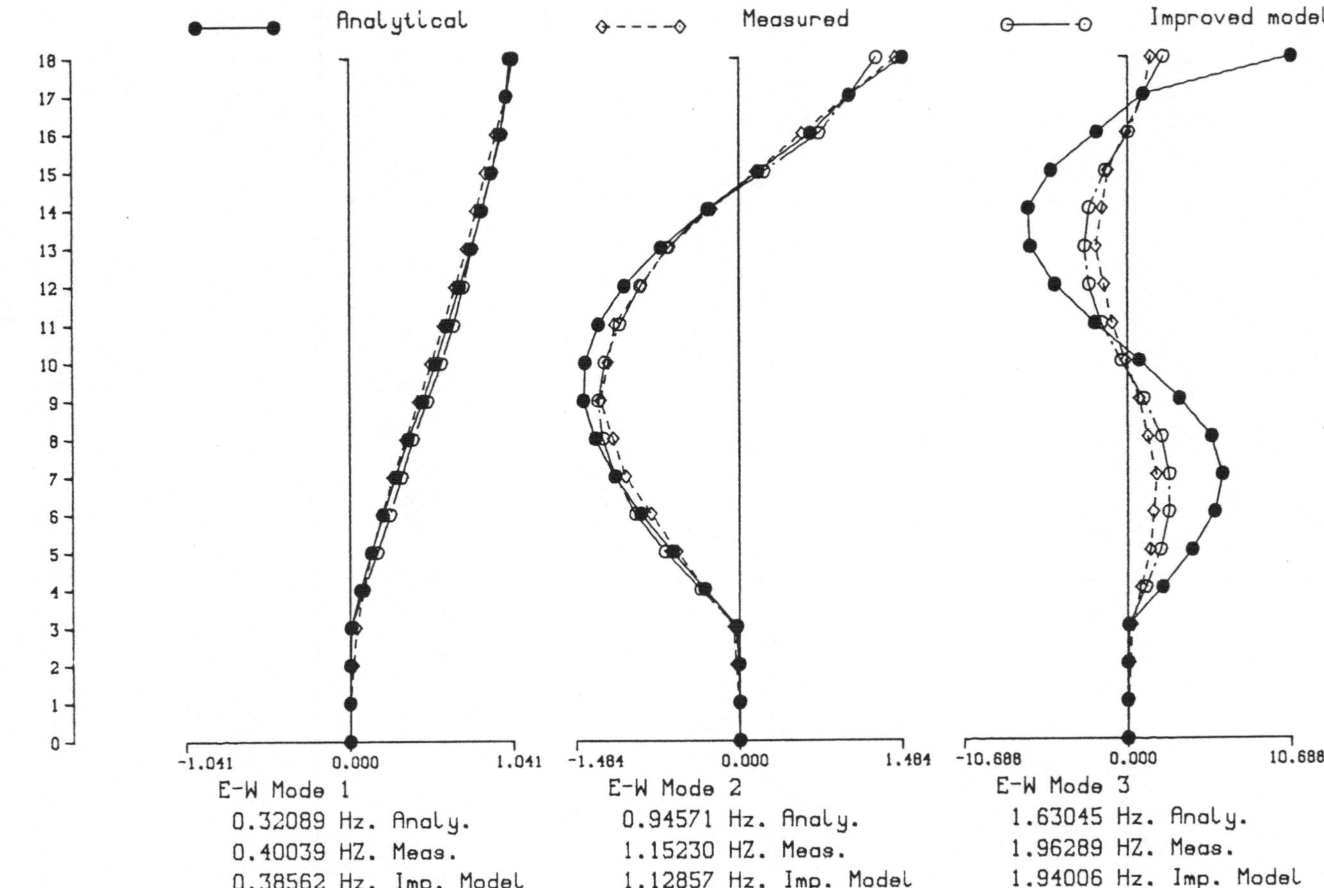

Figure 7 : Comparison of Analytical, Measured, and Improved Model Mode Shapes, E-W Test No. 1

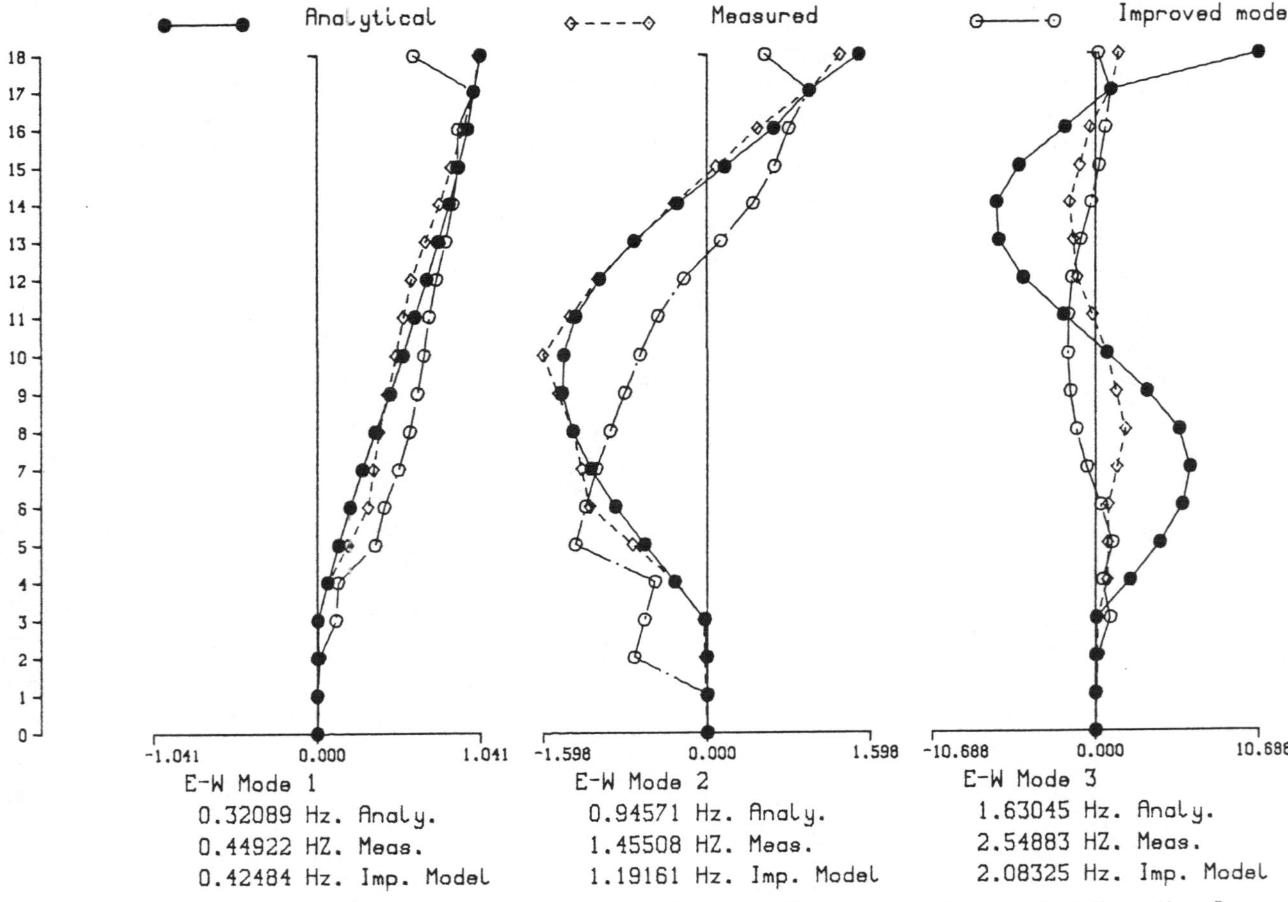

Figure 8 : Comparison of Analytical, Measured, and Improved Model Mode Shapes, E-W Test No. 2

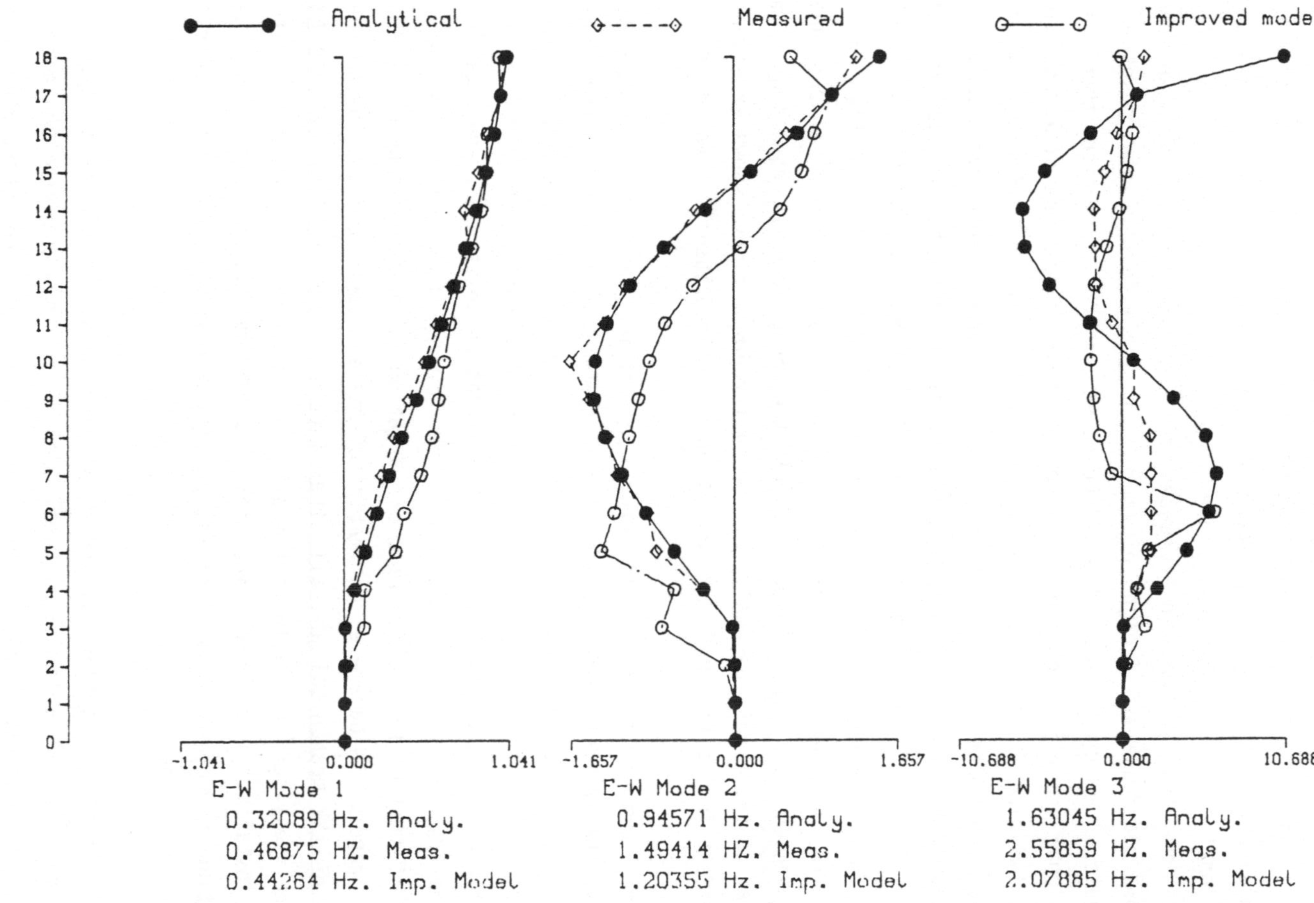

Figure 9 : Comparison of Analytical, Measured, and Improved Modal Mode Shapes, E-W Test No. 3

The following conclusions are drawn from a comparison of these mode shapes and frequencies: 1. Stairs, elevators and partition walls have significant effect on the frequencies of the building. 2. Frequencies resulting from the analytical model and first vibration test for N-S and torsional directions are very close. The results of the second vibration test show that a small difference resulted from the addition of the second staircase and freight elevator. 3. The effect of staircases and elevators is greater on the frequencies of E-W direction than the torsional and N-S directions. Since the direction of the staircases in this building is N-S, their major axes are in the E-W direction which causes a greater effect in this direction. Also, elevators doors are facing east and west directions. 4. Because the exterior cladding of this building is made from glass, it does not have a significant effect on frequencies and mode shapes of the building. 5. The partition walls in this building also have a small effect on the mode shapes and frequencies ^{3}. 6. Comparison of the mode shapes resulting from vibration tests and the analytical model for lower floors shows excellent agreement indicating that the assumption of a fixed base is a valid one.

The Improved Statistical Structural Identification formula is used to study the effects of non-structural elements, at three stages of construction, for Six PPG Place. Nine frequencies and corresponding mode shapes, with only 1/3 of the entries (dominant directions) which identified from data processing of the vibration tests data, are used for system identification. Also, a mass sensitivity study for this building is performed to find the effect of variation of mass on frequencies and mode shapes. In parameter estimation of tests numbers two and three, the mass of the mechanical and all other floors is increased by 50% and 5% respectively for the analytical model. These changes in mass are due to the installation of heavy machinery at the mechanical floor and the addition of elevators, staircases, partition walls, etc., which occurred after test number one.

The following conclusions are drawn from this part of the study: 1. The Improved Statistical Structural Identification along with the model, which is presented in the preceding paper, for non-structural and structural elements can be used for parameters identification of large degrees of freedom models without the ill-condition problem. 2. Because the triple matrix product $[S]_n[C_p][S]_n^T$ is singular, the Least Square and Weighted Least Square methods can not be used for identification of three-dimensional problems. It should be noted that as β approaches to zero, the matrix that must be inverted in Eqs. 2 and 3 approaches the above singular matrix, therefore the final value of β is small but never equal to zero. 3. The rate of convergence of frequencies is very good. See Tables and 1. The rate of convergence of mode shapes is slower, which could result from using 1/3 of each mode shape for system identification. 4. The effects of non-structural elements on the stiffness matrix of this building are considerable. However, the change in stiffness for the first few floors is not significant. This may be the result of considering the fixed base in the analytical model, which effects the degrees of freedom of the first few floors. 5. To study the effect of the variation of mass on frequencies and mode shapes, the mass matrix of the building is increased by 5%, 10%, 15%, 20%, and 25%. This study shows that the frequencies of this building are not very sensitive to variation of mass. The range of change of frequencies is between 1 to 10%, with negligible change for mode shapes.

6. ACKNOWLEDGMENT

The writers are grateful for partial support from National Science Foundation Grant CEE-8206909 and to Dr. John B. Scalzi, program director.

7. REFERENCES

[1] A. K. Ahmadi, "Application of System Identification in Mathematical Modeling of Buildings", February, 1986, Ph.D. Thesis in Progress at University of Pittsburgh, Pittsburgh, Pennsylvania.

[2] G. A. Bekey, "System Identification-An Introduction and a Survey", Simulation: 151-166, October, 1970.

---

3 The partition walls are few in number and are located at the core of he building.

[3]   J. E. Bowles, "Foundation Analysis and Design", McGraw-Hill Book Company, New York, 1982.

[4]   J. D. Collins, J. P. Young and L. A. Kiefling, Methods "and Applications of System Identification in Shock and Vibration", In Pilky, W. D. and R. Choben (editor), System Identification of Vibrating Structures, Mathematical Models from Test Data, pages 45-71. ASME, New York, NY, 1972.

[5]   D. A. Foutch, "A Study of the Vibrational Characteristics of Two Multistory Buildings", Technical Report EERL 76-03, California Institute of Technology, Pasadena, California, September, 1976.

[6]   W. E. Gates, "The Workshop on Earthquake-Resistant Reinforced Concrete Building Construction", Vol. 11, pages 857-886.   University of California, Berkeley, California, July 11-15, 1977.

[7]   ASTRO/Move; A Finite Element Program for Performing System Identification Analysis, GELMANCO, Culver City, California, 1983.

[8]   G. C. Hart and M. A. M. Torkamani, "Structural System Identification, Stochastic Problems in Mechanics", In Proceedings of the Symposium on Stochastic Problems in Mechanics, pages 207-228, University of Waterloo, Waterloo, Ontario, Canada, September 24-26, 1973.

[9]   P. C. Jennings, R. B. Matthiesen, and J. B. Hoerner, "Forced Vibration of a 22-Story Steel Frame Building",Technical Report EERL 71-01, California Institute of Technology, Pasadena, California and University of California at Los Angeles,Los Angeles, California, February, 1971.

[10]   J. Petrovski, R. M. Stephen, E. Gartenbaum. and J. G. Bauwkamp. "Dynamic Behavior of a Multistory Triangular-shaped Building", Technical Report EERC 76-3, College of Engineering University of California Berkeley, California, October, 1976.

[11]   D. S. Samuel, "Digital Signal Analysis", Hayden Book Company, Inc., New Jersey, 1975.

[12]   M. A. M. Torkamani and A. K. Ahmadi, "Stiffness Identification of Frames Using Simulated Ground Excitation", ASCE, Journal of Engineering Mechanics, Accepted for Publication, August, 1987.

Interaction between System Identification and Damage

Reliability Based Factor of Safety for Unmanned Spacecrafts
T. De Mollerat, C. Vidal, M. Klein

Parameter Identification for Reliability in Markov Cumulative Damage Processes
P. Voltz, F. Kozin

A System Identification Approach to the Detection of Changes in Structural Parameters
M. S. Agbabian, S. F. Masri, R. K. Miller, T. K. Caughey

Time Domain Identification of Linear Structures
V. C. Matzen

Fuzzy Data Processing in Damage Assessment
H. Furuta, M. Shiraishi

Reliability Based Factor of Safety for Unmanned Spacecrafts

by T.De Mollerat *
   C.Vidal   *
   M.Klein   **

*   Societe Nationale des Industries Aerospatiales
        Cannes-la-Bocca   (France)
**   European Space Agency - ESTEC
        Noordwijk (The Netherlands)

Study performed under ESA-ESTC contract nr.6370/85/NL/PB
        "Evaluation of Test and Design Factors"

Presented by M.Klein at the "Workshop on Structural
Safety Evaluation Based on System Identification Approach". LAMBRECHT FRG June 29-July 1, 1987.

TABLE of CONTENTS

1 ABSTRACT

A method is presented for the definition of Factor of Safety ensuring structural reliability. The proposed procedure encompasses both theoretical reliability and practical aspects, combining design requirements, qualification and acceptance tests. Design criteria in terms of reliability have been defined based on previous spacecraft projects and then translated into structure design requirements that can be met in a practical design process: this was performed using a rational approach in terms of statistical variation coefficient of both strength and loads. As a consequence, a set of Factor of Safety (FOS) is proposed which is valid for unmanned spacecrafts up to 2000 kg. The methodology can easily be extended to larger masses; work is in progress for manned spacecraft aspects and computer implementation.

2 FACTOR of SAFETY : PRESENT STATUS

2.1 FACTOR of SAFETY USED IN VARIOUS PROJECTS

A general survey of safety factor practice has been performed for airplanes, lauchers and spacecraft projects :

AIRPLANES The 1.5 factor of safety for ultimate in airplane design practice is fundamental and has become an accepted standard. However, good correlation (at least 10%) between static test results, analysis and flight load measurements is required. Fracture control plan is always implemented.

LAUNCHERS The typical values for the ultimate factor are: 1.25 for unmanned systems and 1.40 for manned systems. For the latter a fracture control plan is also implemented. In both cases, the yield factor of safety is typically 1.1.

SPACECRAFTS Table 2.1-1 summarizes the safety factors used in several significant projects. The qualification test factor varies from 1.2 to 1.5, but as additional factors are specified, the ultimate factor lies between 1.5 and 1.95 for the primary structure (up to 3 for the secondary structure). The definition of each safety factor (qualification, yield, ultimate, additional) is often not very clear (excepted ERS-1 and RO-SAT).

QUALIFI-CATION (3)	TYPE (2)	LAUNCHER (1)	SAFETY FACTORS EMPLOYED IN SPACECRAFT PROJECTS AS RELATED TO FLIGHT LIMIT LOADS				
			Structure of satellite project	Qualification (static test level)	Ultimate (Safety factor at qualif. test level)	Yield (Safety factor at qualification test level	A – Additional factor R – Resulting factor for ultimate loads
SM FM1	S	ARIANE IV	ERS-1	1.45	Primary 1.25 (1.45 x1.25)=1.8 Secondary 2.0 (1.45 x 2) = 2.9	Primary 1.1 1.45 x 1.1) = 1.6 Secondary 1.4 (1.45 x 1.4) = 2.0	A 1.25 (higher risks elements) R Primary 2.25 = (1.25 x 1.8)
SM PM FM	S	DELTA	GEOS	1.5	1.25 (1.5x1.25) = 1.88	1.1 (1.5 x 1.1) = 1.65	A 1.15 R 2.2 (= 1.15 x 1.88)
STM PFM	S	DELTA	GIOTTO	1.5	1.75 (1.5x1.75) = 2.63	1.3 (1.5 x 1.3) = 1.95	A 2.0 R 3.0 (= 2x1.5)
SM PFM	S	DELTA	ISEE-B	1.25	1.25 (1.25x1.25) = 1.56	1.1 (1.25 x 1.1) = 1.38	A = 1 R 1.56
STM PFM FM	T	ARIANE	TV-SAT	1.25	1.25 (1.25x1.25) = 1.56	1.1 (1.25 x 1.1) = 1.38	A = 1 R 1.56
STM	S	SCOUT	AEROS	1.65	1.1 (1.65x1.1) = 1.82	1.1 (1.65 x 1.1) = 1.82	A -1.5, 1.15, 1.25 in different parts R -2.7, 2.1, 2.3
STM PFM	S	SHUTTLE	RO-SAT	1.2	1.4 Ultimate factor 1.25 buckling factor	1.1	A -1.25, 1.15 in different parts R -1.75, 1.61
STM PFM FM	S	ARIANE IV	PFM-SPOT	1.5	1.25, 2.0 in different parts (1.5x1.25) = 1.87 (1.5 x 2.0) = 3.0	1.1, 1.4 in different parts (1.5 x 1.1) = 1.65 (1.5 x 1.4) = 2.1	A -1.25 R -2.34, 3.75
PFM	S	SHUTTLE	IPS	Derived enveloppe of flight loads from different configurations	1.4	1.1	1.15 for rivets, fatigue/fracture required load cycles: 4 x nominal cycles
STM PFM FM	T	ARIANE III + SHUTTLE	ARABSAT	1.5	1.1(buckling) 1.0	1.0	A = 1 R = 1.65
STM PFM	S	DELTA	EXOSAT	1.5	1.25	1.1	A = 1 R = 1.27

(1) S: Scientific
   T: Telecom

(2) SM: structural model STM combined structural and thermal model
   PM: QM = prototype model for qualification
   PFM: Protoflight model
   FM: Flight model

TABLE 2.1-1   SAFETY FACTORS EMPLOYED IN SPACECRAFT PROJECTS

FACTOR of SAFETY for PAYLOADS required by launcher authorities

ARIANE requirements are summarized in table 2.1-2.

<table>
<tr><td colspan="3" align="center">Table 2.1-2

ARIANE requirements for payloads</td></tr>
<tr><td align="center">Requirements</td><td align="center">FOS</td><td align="center">Comments</td></tr>
<tr><td>Main structure design</td><td>1.25</td><td>at ultimate wrt flight limit loads</td></tr>
<tr><td align="center">Secondary structure design</td><td>1.5</td><td>at ultimate wrt flight limit loads</td></tr>
<tr><td align="center">Static qualification test</td><td>1.25</td><td>* flight limit loads without rupture</td></tr>
<tr><td align="center">Sinusoidal qualifica-tion test</td><td>1.50</td><td>* acceptance level (ARIANE 1 and ARIANE 3)</td></tr>
<tr><td align="center">- id-</td><td>1.25</td><td>* acceptance level (ARIANE IV)</td></tr>
</table>

STS requirements

The stuctural design shall provide ultimate factors of safety equal to or greater than 1.4 for all STS mission phases except emergency landing (FOS=1.). All playloads structures shall be demonstrated to be safe for flight prior to flight. The available options for structural qualification are listed in Table 2.1-3 in order or preference.

STS Loads definition: In the preliminary design phase, conservative load are used for sizing the structural members and developing the analytical model. Confidence that the payload mathematical model represents its dynamic behaviour has a direct bearing on the adequacy of safety factors used in the verification of the structure: if the structural design loads are obtained using a space vehicle system transient loads analysis and if minimum safety factors are used, then the correlation requirements are quite stringent, more than if conservative quasi-static load factors are used.

<table>
<tr><td colspan="5" align="center">Table 2.1-3

STS QUALIFICATION OPTIONS</td></tr>
<tr><td align="center">OPTION</td><td align="center">TEST LEVEL</td><td align="center">FOS Yield</td><td align="center">FOS Ultimate</td><td align="center">TEST SUCCESS CRITERIA</td></tr>
<tr><td>Test Dedicated Article to Ultimate (Baseline)</td><td align="center">1.4</td><td align="center">1.0</td><td align="center">1.4</td><td>No detrimental deformation at yield load. No failure at ultimate load.</td></tr>
<tr><td>Reduced Level Test on the First Flight Article</td><td align="center">1.25</td><td align="center">1.25</td><td align="center">1.4</td><td>No detrimental deformation</td></tr>
<tr><td>Proof Test Each Flight Article</td><td align="center">1.1</td><td align="center">1.1</td><td align="center">1.4</td><td>No detrimental deformation</td></tr>
<tr><td>No System Static Qualification</td><td align="center">NA</td><td align="center">1.6</td><td align="center">2.25</td><td>Positive margins of safety and fail-safe design</td></tr>
</table>

For large and complex playloads which used a transient system loads analysis, the goal is to achieve after a modal test a cross orthogonality check of 0.95 for all significant normal modes. NASA recognized that there are no well defined requirements for the degree of correlation and the relationship between correlation and safety factor.

A fracture control plan must also be implemented. JSC exempts composites from fracture control if initial defects that exist in the structure are minor and if structure is intended for single-mission applications with a low number of load cycles.

CONCLUSION

This survey has shown that safety factors are used in each aeronautical system in the same purpose (cover all the possible uncertainties). But no evidence of consideration of correlation between these safety factors and the real scatter of the applied loads and of the structure properties has been found. The spacecraft projects appear to use safety factor higher than the requirements of the launcher.

These two conclusions reinforce the necessity of a clear definition of the objectives searched by the safety factors.

2.2 HARDWARE FLOW PHILOSOPHY

One of the dominant questions in a spacecraft program with respect to cost, schedule and verification is the amount and quality of hardware required for verification testing. The different philosophies are presented in this paragraph.

"OLD" APPROACH

STM   Qualification levels (static and dynamic
                 tests)

QM   Qualification levels (dynamic tests)

FM1, 2,... Acceptance levels

This approach was used until METEOSAT (1977) in the ESA projects. The additional QM (qualification model) was found necessary because the confidence in analytical development of design loads was insufficient. The qualification is performed independent from flight hardware (lower risk).

"CURRENT" APPROACH

STM   Qualification levels (static and dynamic
                 tests)

PFM/QM Qualification level but flight time (dynamic
= FM1   tests)

FM2, 3,... Acceptance level

The first FM (FM1) is used for qualification. So the risk of failure is higher and a late refurbishment can be necessary. This approach is based of sufficient confidence in analytical development of design loads. This approach is generally used for the present spacecrafts (GIOTTO, EXOSAT, TV-SAT...).

"PROTOFLIGHT" APPROACH

PFM/QM Qualification level (static and dynamic
= FM1   tests; dynamic tests only, in some cases)

FM2, 3,... Acceptance level

In this approach the STM is not used. The qualification is demonstrated directly on the PFM which increases the risk of failure. For the qualification of the primary structure two approaches are possible:
- the primary structure is qualified by a previous static

test with an STM (the primary structure is common to a family of spacecrafts), this is a "BUS" approach.
- local tests are performed at large component level, before the assembly, for the critical parts.
This approach is more and more used because it allows to decrease the development costs (the cost of the STM is supported by several projects instead by one only in the current approach). The french Direct Broadcast Satellite TDF-1 is a good example of this approach (ref (14)). As a matter of fact the STM was common to TV-SAT (German D.B.S.) and TDF-1 which are very close. The Swedish D.B.S. TELE-X will also profit of this STM (a STM model was nevertheless necessary for the antenna tower due to a new design).

CONCLUSION

Although the confidence in analytical development of design loads is increasing, the new test philosophies increases the risk of flight hardware failure during qualification tests. Consequently we have to define safety factors adapted to each approach.

2.3 RELIABILITY ASPECTS

The results of the survey on satellite projects may be summarized as follows :
- very often, no reliability calculation for structural aspects
- when a reliability calculation is implemented, two methods are used :
    . the constant failure rate method
    . the "stress-strength" method
- the method used (or the assumptions) may be modified as the project proceeds (for instance constant failure rates at the beginning, "stress-strength" method when the safety margins on the parts are known).
- fracture mechanisms aspects, when considered (launch by shuttle, or safety requirements), are never considered in a probabilistic way and thus not integrated in the reliability estimation process.
- the satellite structural reliability is assessed to be the product of the reliabilities of each part (no degraded mode, loads applied to the parts assessed statistically independent).
- the definition of the word "part" is not always clear (is a screw a part or a junction by screw (several items)a part ?).
- the satellite structural reliability is 1.0 after launch (only risk during launch).
- in all projects reviewed, test and reliability activities are in practice disconnected (determination of test levels, processing of tests results).

- when the "stress-strength" method is used for reliability estimations, the parameters of Figure 2.3-1 are used. Typical values for these parameters are given on Table 2.3-1.

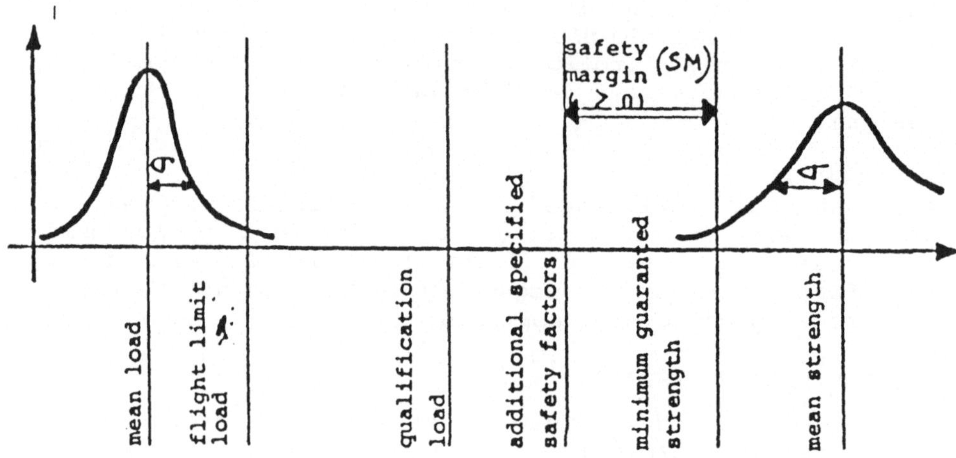

FIGURE 2.3-1   "STRESS-STRENGTH" METHOD PARAMETERS

<table>
<tr><td colspan="2" align="center">

Table 2.3-1

TYPICAL VALUES OF STRENGTH-STRESS METHOD

USED FOR CONVENTIONAL ANALYSIS

</td></tr>
<tr><th align="center">PARAMETERS</th><th align="center">VALUES IDENTIFIED</th></tr>
<tr><td>Load distribution</td><td>. Always assessed as normal
. Standard deviation equal to 10% of the mean value (in most cases this value is specified)
. No distinction between the various load cases</td></tr>
<tr><td>Flight limit load</td><td>. 90% value or
. 97% value (2σ value)</td></tr>
<tr><td>Qualification load

Additional speci-
fied safety fac-
tors</td><td>Specified values for each project</td></tr>
<tr><td>Safety margin</td><td>Specific to each component. Extracted from the dimensioning analysis (in preliminary reliability assessment, assessed to be zero)</td></tr>
<tr><td>Minimum guaranteed strength</td><td>It is the strength with regard to which the safety margin is calculated in the dimensioning analysis. According to the projects :
. is considered as a mean value
. is considered as a 99% value
. is considered as a 97% (2σ) value</td></tr>
</table>

Strength distribu-
tion
	. always assessed as normal
	. Standard deviation, equal to 10% of the mean value (in most cases this value is specified) or to 8% of this mean value
	. No distinction between :
	  - the strength modes (shearing, tension, yield strength,...)
	  - the material (metals, composite, adhesives)

2.4 LOAD and TESTING ASPECTS

2.4.1 LOADS

A spacecraft structure can achieve successfully its mission with a high reliability but without excessive conservatism only if the design process takes account of the real environmental conditions during the whole lifetime of the hardware. In the case of this paragraph only the "low frequency" portion of the mechanical load environment will be considered (this is usually the design driver for the primary and secondary structures). A spacecraft structure is designed to survive :
- the real flight load environment.
- the simulation of the predicted flight environment by test.
In the ideal case both would be identical (optimal design with low risks). However in the actual life this is not true, due to :
- incomplete measurements of flight loads
- incorrect prediction of flight load environment.
- uncertainties in the simulation of flight environment.
These three aspects have been considered in our study.

MEASUREMENTS OF FLIGHT LOADS
Generally speaking, a few number of measures are available, and generally limited at the measurement of the interface accelerations. For typical spacecraft structures launch (or landing if any) generate maximum design loads rather than on orbit events. The nature of the launch loads is multiple :
- steady state loads (thrust, aerodynamic, wind...)
- transient (thrust build-up and decay,wind shear and gust, stage separation, landing...)
- quasi harmonic (engine chugging, pogo instability...).
- random
- high frequency (acoustic noise at lift off and transonic events)
- thermal

A work package of our study was dedicated to the analysis of the interface accelerations during the sixteen first flights of ARIANE (ref (3)). The conclusions of this study have been compared with two recent presentations of ARIANESPACE (ref (5) and (6)).

ARIANE I TO III LAUNCH LOADS
The results of an analysis are :
- the maximum quasi-static loads at the launcher interface are due to the first stage thrust (4.5 g with a low standard deviation (1% to 2%)).
- the maximum dynamic loads are due to the flame-outs of the first and second stages and have to be combined only with a residual quasi-static acceleration (1 to 0.5 g) as follows.
- load statistics have been drawn-up assuming a law of normal distribution. Maximum loads have been obtained with a 95% probability and a 90% confidence level. The effect of the limited sampling, 15 flights maximum, has been taken into account in the evaluation of those loads. For the first stage burn-out several propellant depletion are possible: N depletion (best performance but severe loads), simultaneous N/U depletions, U depletion (lower loads). In fact great variations of the cut-off severity are possible. As also a few number of flights with the same configuration is available (ARIANE I or III, with or without SYLDA), the statistical treatment leads to a high coefficient of variation.
A comparison with some ARIANE III coupled analysis showed that these predictions are situated between the average and the maximum load (95%).
As a matter of fact, the coupled analyses predicts the dynamic loads that the payload would have undergone during a previous flight considered as the most severe of all known flights, and not maximum loads (see tables 2.5.1. to 2.5.4).This remark may lead to reconsider the definition of the coupled analysis process using synthetic thrust transient corresponding to maximum loads on the playloads (a Monte-Carlo method could be interesting).

STS LAUNCH LOADS
From ref (22) containing measurements of flight 1 to 5 we have verified that the preliminary design load factor has a probability of no exceedance higher than 2σ. If we except the severe landing of STS-2 flight, this probability is close to 5σ which justifies lower design loads based on coupled analyses.

PREDICTION OF LAUNCH LOADS
The first spacecrafts (launch mass , 100 kg) were considered as the electronic boxes of the aircraft industry. As the influence of the light and very rigid

spacecraft on the launch vehicle was disregarded, and the vibrations measured at the interface were assumed to be the same for all possible payloads. When the spacecraft became heavier and more elastic the necessity of prediction methods included the spacecraft feedback effects was discovered. A rough evaluation of the various methods (spectral method, shock-spectra method, generalized modal shock-spectra method, transient response method) has been performed during our study, (based on several papers of M.TRUBERT and WADA from the JPL (ref (12), (18), to (21)). A more detailed evaluation of the JPL methods can be found in a paper written by MM. HORNUNG-BREITBACH-ORY (ref 10)). The most accurate flight load prediction method is obviously the transient analysis of the composite structure launch vehicle+spacecraft, even if this method is the most expensive.

CONCLUSION

Obviously different types of loads required different Safety Factors according to the way the loads have been produced. This aspect is almost not considered in the actual practice, but has been accounted for in our study.

2.4.2 TESTING

STATIC LOADS

A structure is nowadays divided in two parts :
- the primary structure sized and tested by the quasi-static loads
- the secondary structure sized and tested by vibration loads.

This partition was not considered for the very first spacecrafts which were tested like electronic boxes on uniaxial electrodynamic shakers. When the spacecrafts became heavier and more elastic the notion of primary structure was introduced. Then static tests (at component, subsystem or system level) were used to qualify the primary structure, and the vibrations tests were controlled to have the same stresses on the primary structure than during static tests. The main difficulty in establishing the test procedure is to load all the members which have to be qualified to the proper load without overloading other members which may lead to an unexpected failure. At system level the static load test is performed with a whiffle tree (like an aircraft), or an centrifugal machine when the size of a tested structure can be accepted on this type of machine (tests on a centrifugal machine are less questionable).

When the margin of safety is greater than a specified value a qualification by analysis is accepted. A summary of the safety factors required for this "no-test" option is given hereafter:

```
GALILEO      SF > 2.0
SPACELAB     SF > 2.0   (yield)
             SF > 3.0   (ultimate)
ROSAT        SF > 1.875 (yield)
             SF > 2.5   (ultimate)
```
(the reference is the "flight limit load" level)

VIBRATIONS LOADS

These tests are performed to verify and qualify the secondary structures and can be split in low and high frequency environment.

Low frequency environment

COS-B was the first ESA project which authorizes notchings for the primary structure during these tests. Nowadays it is also authorized for the elements of secondary structures which are well defined in the launcher/spacecraft coupled analysis. For the large structures which begin to be tested, this procedure of notching become questionable (it seems that some parts can be undertested, see ref (10) and (13)). Notching based on coupled analysis results is compulsory, since otherwise using the test environment specified by the launcher authorities (ARIANE) the large spacecrafts would be mainly (over)sized for test survival. Due to this, there is a tendency to envisage replacement of some low frequency vibration tests by transient tests.

High frequency environment

Until a few years ago, both sine and random vibration test were performed on the structures. Nowadays random vibration tests are used only at subassembly level. At spacecraft level they have been replaced by acoustic tests (the launcher manuals indicate now that a properly performed acoustic test is the best simulation of the high frequency launch environment).

Transient loads

Because the launch loads on the modern launchers are mainly of transient nature, it is now foreseen to simulate them by transient tests rather than by sine vibration loads. Several studies and tests have been conducted in this area. The details of three significant experiences can be found in papers by HUGHES/COMSAT (ref.7) , MBB/ERNO/INTELSAT (ref.10) and CNES/ESA (ref.13 evaluation in progress) during which respectively INTELSAT IV, MAROTS, and ARABSAT qualification models were tested. All these experiences have concluded that transient testing is well feasible, but some test facility problems had to be solved (see ref.15). Depending on the expected nature of the loads, two approaches are considered:
- single axis testing, adapted to the expendable launcher (like ARIANE IV) where the transient vibrations are mainly axial and applied at a single quasi non-redundant interface.

- multi-axes testing adapted to the spacecrafts launched
by STS (or similar type of carriers) and also for
subsystems with redundant interface(s).

2.5 SYNTHESIS

The assessment of the spacecraft design loads can be
summarized by the following schema:

A   Flight measurements
       during /prior to launches
              |

B   Flight loads predictions -------   Design loads
                                               spacecraft's FOS
              |

C   Simulation of flight loads
       during ground based tests

All the steps of this chain have to be homogeneous. For
example sophisticated load predictions and test methods
are only justified if the flight measurements are accurate
enough. Safety factors have to be associated with this
scheme:
- low safety factors will be required for well known
launchers and accurate flight loads predictions.
- higher safety factors will be necessary to cover the
uncertainties of the other cases.
Test loads have to be of the same nature as the real
flight loads (static, sine, random, or transient). As a
matter of fact, a good simulation during ground based
tests increases confidence in the structure. On the other
hand, specifications more severe than the reality increase
the mass of the structure and also the risk of unjustified
failure during test.

3 DERIVATION OF FACTOR of SAFETY

3.1 PROPOSED METHODOLOGY

3.1.1 STRESS-STRENGTH APPROACH

The method selected for the definition of the Factor of
Safety for each spacecraft component considers a
reliability objective at component level based on the well
known "stress-strength" method. This method assesses the
structural reliability of a part by the statistical
comparison between it strength and stress applied to it.
Various statistical distributions of stress and strength
are possible with this method which covers in practice a
wide area : Normal, Log-normal, Weibull, truncated,
experimental data... However,the classical case of Normal
law distribution has been considered for both stress and

strength. Indeed, it would not be realistic today to consider other sophisticated distributions since the database are of limited importance except for particular cases. Consequently for each combination of load type-material the necessary design Safety Factor is defined with normal law as shown in Figure 3.1.1-1 .

3.1.2 DESIGN and TESTING

The definition of the Factor of Safety has also to consider the test aspects, ie Qualification and Acceptance. On one hand, the Qualification test must be sufficiently high to prove a good structural reliability. On the other hand, knowing that both Qualification and Acceptance tests will actually be performed, a margin wrt the test levels is necessary to limit the risk of failure during test. This latter risk is itself depending on the test philosophy: the risk which can be accepted on flight models (PFM) is lower than for the test models (STM).

3.1.3 RELIABILITY PROVED BY TEST

The main idea to assess a reliability proved by test (with an associated confidence level) is to make the following assumption:
- qualification tests are performed to detect potential design problems which consist of underdimensioning of some elements.
- So the mean strength of a part can be different from the predicted value, but the strength distribution remains normal and the coefficient of variation $v = \sigma/m$ is unchanged (due only to the material).

Consequently, after a successful completion of a qualification test, it is possible to define a "proved strength" and a proved failure probability. These considerations are illustrated in Figure 3.1.3-1.

3.1.4 IN-TEST FAILURE RISK

The risk of failure during test is given by the probability of failure at test level considering that the stress distribution is nominal. See Figure 3.1.4-1 .

Figure 3.1.1-1 Stress-strength method (normal law)

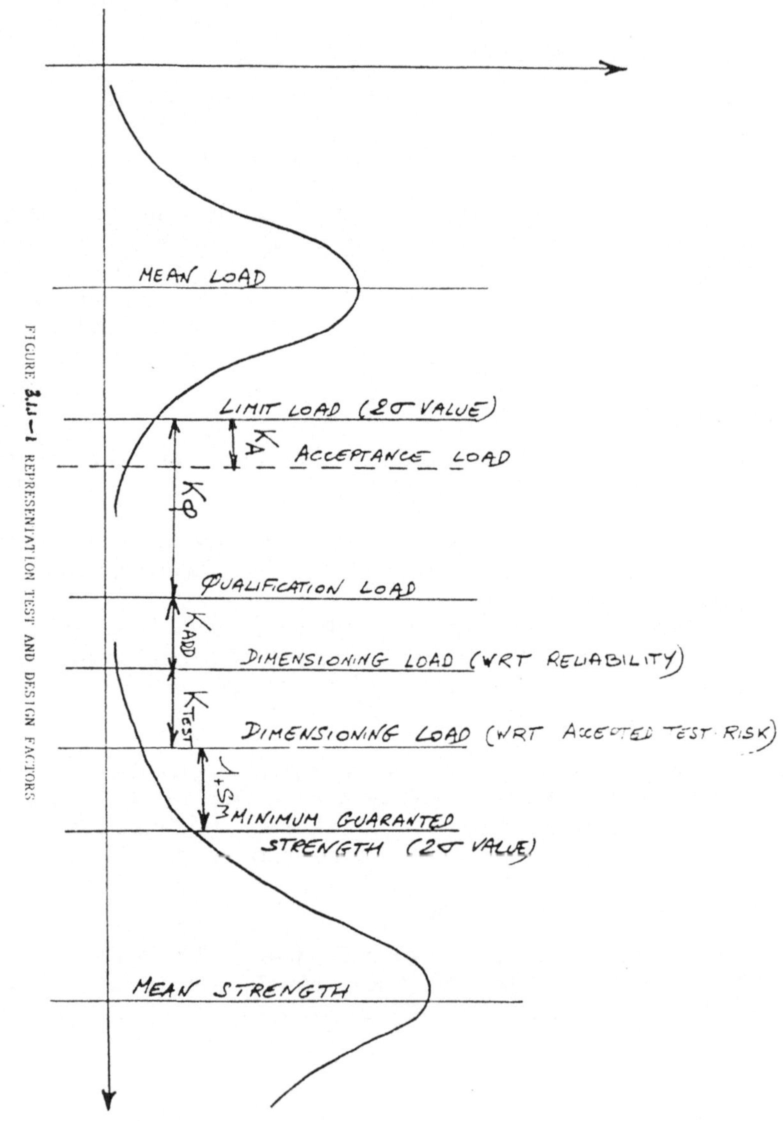

Figure 3.1.3-1 Reliability proved by Qualification

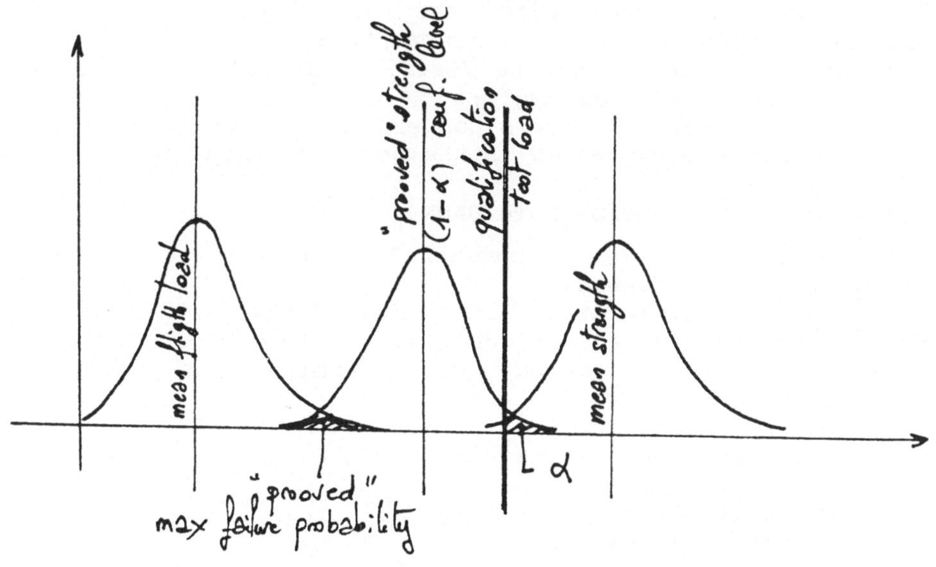

Figure 3.1.4-1 In-test failure probability

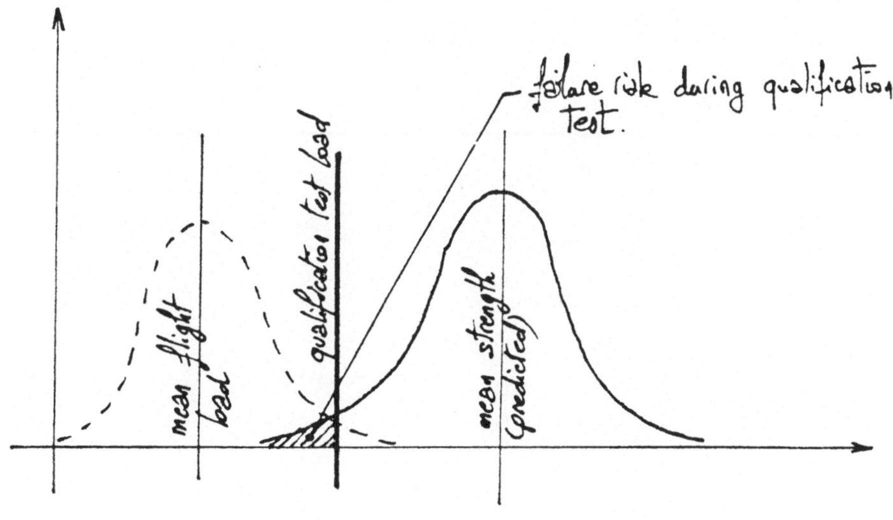

3.2 INPUTS FOR THE METHODOLOGY

The method presented in the previous chapters requires 3
types of inputs:
- a data-bank concerning the distributions of the
different loads applied to the structure;
- a data-bank concerning the distributions of the strength
of each materials or assembly;
- the reliability objectives concerning the project;
These inputs are presented in the next sections.

3.2.1 LOAD DISTRIBUTION DATA-BANK

The loads acting on a spacecraft structure have been
classified in 8 classes:
- a - launch vehicle static loads due to the thrust
- b - other launch vehicle quasi-static loads (POGO,
                        aerodynamics, landing, gust...)
- c - transient loads
- d - thermal loads
- e - deployment shocks (regulated mechanisms)
- f - thruster loads
- g - acoustic loads
- h - vibration loads.
For each class of loads a significant documentation has
been collected. In many cases a statistical treatment of
tests or analyses results has been performed. This
synthesis is presented in the Table 3.2.1-1.

<table>
<tr><td colspan="4" align="center">

Table 3.2.1-1

VALUES OF σ/m
for LOADS

</td></tr>
<tr><td>

TYPE of LOADS

</td><td>

PROBABILITY LEVEL

</td><td>

σ/m

</td><td>

Reference of the phase II report

</td></tr>
<tr><td>Launch vehicle : thrust</td><td>2σ</td><td>5%</td><td>Conservative value from table 1.1.1</td></tr>
<tr><td>Launch vehicle : other static loads</td><td>2σ</td><td>30%</td><td></td></tr>
<tr><td>Transient loads</td><td>2σ</td><td>50%</td><td>Chapter 1.2 and 1.3(a)</td></tr>
<tr><td>Thermal loads :</td><td></td><td></td><td></td></tr>
<tr><td>correlated temperatures</td><td>2σ</td><td>7.5%</td><td></td></tr>
<tr><td>uncorrelated temperatures</td><td>2σ</td><td>20%</td><td></td></tr>
<tr><td>Deployment shocks</td><td>2σ</td><td>10%</td><td>Chapter 1.5</td></tr>
<tr><td>Thruster loads</td><td>3σ</td><td>2%</td><td>Chapter 1.6</td></tr>
<tr><td>Acoustic loads</td><td>2σ</td><td>40%</td><td>Chapter 1.7</td></tr>
<tr><td>Vibration loads</td><td>2σ</td><td>20%</td><td>Chapter 1.8(b)</td></tr>
<tr><td colspan="4">

note: $2\sigma = 97.7\%$ probability
$3\sigma = 99.3\%$ probability

a. The nominal loads must be defined with a realistic launch vehicle excitation model, not a pessimistic one.
b. The nominal loads must be computed with a damping factor of 2%, the loads at 2σ with 1.4%.

</td></tr>
</table>

Only one value σ/m has been defined for each class of loads because a great number of coefficients would not have been applicable in the frame of a real project.

To establish this table many sources have been searched and processed as for example :
- a - STS and ARIANE ascents
- b - STS, ARIANE and DELTA flight events
- c - STS and ARIANE transient loads based on CNES, JPL
     and ARIANE events analysis performed in this study
- d - AEROSPATIALE test results for various spacecraft
                                                  projects
- e -   id.
- f - SEP thrust measurement, and others.
- g - STS NASA flight measurement analyses and ARIANE mock up tests
- h - Various spacecraft projects
        See references : - study final report
                         - study phase II report

3.2.2 STRENGTH DISTRIBUTION DATA-BANK

The purpose of this task was to propose realistic statistical distributions of the strength properties of materials commonly used in the design of spatial structures and mechanisms. In this area, the situation at this day may be summarized in the following way :
- reliability calculations, when performed, are conducted assuming a normal distribution for strength with a standard deviation of 10% of the mean value (8% in some limited cases)
- no distinction is made between the kinds of materials and the kinds of strengths.
This study, due to its limitations considering the amount of data to collect and process, does not intent to establish, for each couple material/strength, a very accurate statistical distribution modelling. The aims are :
- to establish if it is necessary to associate to each couple material/strength specific assumptions
- to propose these assumptions, in a realistic and conservative way.
The first step of this study, that we think to be very useful for improvement of mechanical reliability assessment methods, was to establish a preliminary data-bank. A further step should be to extend this data-bank, involving manufacturers of materials and components, to increase the confidence level in the proposed assumptions.
Table 3.2.2.-1 gives an overview of all strength item surveyed, after a first simplification in classes. Data scanned originated from various sources AEROSPATIALE Central Test Laboratory, CEA, BRUHN, ARBOCZ, BUSHNELL, ERNO, etc...but correspond all to actual hardware/test samples measurements.

Table 3.2.2-1 MATERIALS STATIC STRENGTH DISTRIBUTIONS

MATERIAL	STRENGTH CHARACTERISTIC	RECOMMENDED ASSUMPTION	
METALLIC MACHINED PART	RUPTURE	Normal law, $\dfrac{\sigma}{m} = 8\%$	
	YIELD STRENGTH	Normal law, $\dfrac{\sigma}{m} = 8\%$ if R/yield $\leqslant$ 1.2 $\dfrac{\sigma}{m} = 15\%$ if R/Yield >1.2	
METALLIC CONICAL SHELL	BUCKLING	Normal law, $\dfrac{\sigma}{m} = 16\%$ for axial loading $\dfrac{\sigma}{m} = 12\%$ for bending moment $\dfrac{\sigma}{m} = 14\%$ for combined loading	
METALLIC CYLINDRICAL SHELL	BUCKILING	$R/e < 10^3$	$R/e > 10^3$
		Normal law axial: $\sigma/m = 16\%$ bending: $\sigma/m = 12\%$ combined: $\sigma/m = 14\%$	Normal law axial $\sigma/m = 22\%$ bending $\sigma/m = 12\%$ combined $\sigma/m = 17\%$
CARBON FIBER COMPOSITE PARTS	RUPTURE	- if the characteristic is individually checked on each lot: normal law, $\dfrac{\sigma}{m} = 10\%$ - if not: normal law, $\dfrac{\sigma}{m} = 17\%$	

TABLE 3.2.2-1 SUMMARY OF MATERIALS STATIC STRENGTH DISTRIBUTIONS

Table 3.2.2-1 contd. MATERIALS STATIC STRENGTH DISTRIBU-
TIONS

MATERIAL	STRENGTH CHARACTERISTIC	RECOMMENDED ASSUMPTION
CARBON FIBER COMPOSITE PARTS	MODULUS	- if the characteristic is individually checked on each lot: normal law, $\dfrac{\sigma}{m} = 6\%$ - if not, normal law, $\dfrac{\sigma}{m} = 13\%$
SCREW, RIVET WELDING	RUPTURE	Normal law, $\dfrac{\sigma}{m} = 8\%$
BONDING	RUPTURE	- if the characteristic is individually checked on each lot: normal law, $\dfrac{\sigma}{m} = 12\%$ - if not, normal law, $\dfrac{\sigma}{m} = 16\%$
HONEYCOMB	RUPTURE	Normal law: $\dfrac{\sigma}{m} = 16\%$ for tension $\dfrac{\sigma}{m} = 10\%$ for shear, compression
	FACE WRINKLING (Honeycomb plates)	Normal law $\dfrac{\sigma}{m} = 8\%$
INSERT (in honey-comb)	RUPTURE	Normal law $\dfrac{\sigma}{m} = 16\%$

TABLE 3.2.2-1 SUMMARY OF MATERIALS STATIC STRENGTH DISTRIBUTIONS (CONT'D)

3.2.3 RELIABILITY OBJECTIVES

If Factors of Safety have to be based on reliability, reliability objectives are needed on spacecraft level, component level and at test level (proved reliability and risk of failure during test). These allocations have been determined from the following steps :
- determination of a structural reliability allocation at system level
- determination of a realistic number of low margin elements at satellite level
- deduction of a reliability allocation at elementary level.

STRUCTURAL RELIABILITY ALLOCATION AT SYSTEM LEVEL

No structural reliability allocation at spacecraft level has been evidenced during the investigations. Available allocations found include in fact the failure probabilities associated to structural problems, but also the functional failure probabilities, the distinctions being not made between the two cases.

Thus, two satellite projects (ARABSAT and TV-SAT) for which reliability calculations had been performed on a majority of subsystems with the stress-strength method have been selected. The cumulated available structural reliability estimations are the following:

TV-SAT .... 0.9973   ARABSAT 0.9995

Consequently a requirement of 0.999 at spacecraft level seemed realistic and is proposed.

NUMBER OF LOW MARGIN ITEMS AT SPACECRAFT LEVEL

Is defined as an item a component or a structural interface between components even if it involves many parts. These items are located in structure subsystem but also in mechanism subsystems, propulsion subsystem, experiments... Is considered as a structural item each item requesting a mechanical dimensioning with regards to flight or functionally induced loads.

Are considered as low margin items the items having a safety factor (ratio between mean strength and mean stress) lower than 3. This figure has been chosen because, with the classical reliability assumptions of stress and strength normally distributed with a variation coefficient $\sigma/m = 10\%$, it leads to a failure probability of 10^{-10}, thus considered as negligible.

The documentation available concerning the structural dimensioning of several satellite subsystems has been processed (GIOTTO, INTELSAT V, ARABSAT, SPOT, TV-SAT,

TDF-1, TELE-X) and the results are summarized in Table 3.2.3-1. A realistic number of 1000 low margin items at satellite level is proposed.

<table>
<tr><td colspan="2" align="center">Table 3.2.3-1

TYPICAL NUMBER OF LOW MARGIN ITEMS FOR CLASSICAL SATELLITE SUBSYSTEMS</td></tr>
<tr><td align="center">SUBSYSTEM</td><td align="center">NUMBER OF LOW MARGIN ITEMS</td></tr>
<tr><td align="center">Structure subsystem</td><td align="center">50</td></tr>
<tr><td align="center">Solar arrays</td><td align="center">400</td></tr>
<tr><td align="center">Antennas</td><td align="center">200</td></tr>
<tr><td align="center">Propulsion</td><td align="center">50</td></tr>
<tr><td align="center">Other mechanisms (optics, AOCS, scientific payload)</td><td align="center">300</td></tr>
<tr><td align="center">Electronics (structural parts)</td><td align="center">0</td></tr>
<tr><td align="center">TOTAL</td><td align="center">1000</td></tr>
</table>

RELIABILITY ALLOCATION AT ITEM LEVEL
A reliability allocation of 1.10^{-10} is proposed at item level. This figure is deducted from the spacecraft allocation (0.999) and the number of critical items (1000 for spacecrafts in the range 1000 to 2000kg) assuming an equal allocation for each item.

RULES FOR EVOLUTION OF ITEM RELIABILITY ALLOCATION
The proposed allocation has been established based on the processing of data of "classical" satellites (earth observation, telecommunication, scientific) in the range of 1000 to 2000 kg. The rules proposed to allow the evolution of this figure according to specific project requirements is the following:

$$\text{allowed structural failure item level} = \frac{\text{allowed failure probability at system level}}{\text{number of low margin items}}$$

The number of low margin items may be assessed at the beginning of a project by two methods:
- If the overall number of structural items may be assessed by similarity with previous projects, a number of low margin items corresponding to 15 to 20% of this overall number is realistic.
- If no information is available (no comparison elements), a rough estimation of the number of low margin items may be done with the formula :

$$\text{nb.of low margin items} = 1000 * \frac{\text{system weight (kg)}}{1500.}$$

RELIABILITY PROVED BY TEST / TEST RISK
These figures could not be found explicitly in any satellite documents. Processing of ARABSAT and TV-SAT data, for which reliability data and analysis exist, typical figures have been computed. These figures have then been compared to statistics drawn from the literature on qualification and acceptance tests and failures during tests.

PROPOSED RELIABILITY OBJECTIVES
Finally, following numerical values have been defined based on values currently accepted for unmanned spacecraft project:
- at spacecraft level   0.999
- at component level $1-1.10^{-6}$
  (based on 1000 critical items per spacecraft
  and same reliability objective for each for them)
- reliability proved by Qualification test
  at spacecraft level   0.950
- risk of failure during Qualification test
                        STM approach   10%
                        PFM approach   1%
- reliability proved by Acceptance test   0.950

3.3 DERIVATION OF THE FACTOR of SAFETY

The factors used in this study and their definitions are summarized on Figure 3.1.1-1 . The steps that have been followed for their evaluation are the following(*):
- Determination of a reliability objective,for structural

aspects, at satellite level: 0.999 has been selected.
- Determination of a reliability objective, for structural aspects, at part level: 0.999999 is chosen, after the determination of a realistic number of critical components at satellite level(1000).
- Determination of typical scatters on:
  . the various load cases
  . the various materials
- definition of a limited number of load classes
- definition of a limited number of strength classes (materials)
- For each combination load case/material, determination, using the "stress-strength" method, of the necessary safety factor to achieve the reliability objective at part level :

$$K = \frac{\text{Mean strength}}{\text{Mean stress}}$$

$$K1 = \frac{\text{Minimum guaranteed strength } (2\sigma\,value)}{\text{Limit load } (2\sigma\,value)}$$

Tables 3.3-1 and 3.3-2 show typical figures obtained after this processing for the classes of loads and materials previously defined.

Table 3.3-1   K FACTORS

	METALLIC MATERIAL $\sigma/m = 8\%$	METALLIC MATERIAL: YIELD STRENGTH WHEN R/YIELD> $1.2\ \sigma/m = 15\%$	BUCKLING STRENGTH OF CONICAL OR CYLINDRICAL METALLIC SHEELS $\sigma/m = 14\%$	CARBON FIBER COMPOSITES $\sigma/m = 10\%$	JUNCTION BY SCREW, RIVET WELDING $\sigma/m = 8\%$	BONDING STRUCTURAL IN-SERT (AXIAL LOADING) $\sigma/m = 12\%$	HONEYCOMB: TENSION $\sigma/m = 16\%$	HONEYCOMB: SHEAR, COM-PRESSION $\sigma/m = 10\%$	HONEYCOMB: FACE WRINKL-ING $\sigma/m = 10\%$	EQUIPMENT INSERT (IN HONEYCOMB AXIAL LOADING) $\sigma/m = 16\%$
LAUNCH VEHICLE THRUST $\sigma/m = 5\ \%$	1.69	3.52	3.03	1.96	1.69	2.38	4.21	1.96	1.69	4.21
LAUNCH VEHICLE OTHER STATIC LOADS $\sigma/m = 30\ \%$	2.77	4.53	4.04	3.02	2.77	3.41	5.21	3.02	2.77	5.21
TRANSIENT LOADS $\sigma/m = 50\ \%$	3.78	5.72	5.19	4.06	3.78	4.50	6.44	4.06	3.78	6.44
THERMAL LOADS (CORRELATED) $\sigma/m = 7.5\ \%$	1.76	3.57	3.08	2.03	1.76	2.43	4.25	2.03	1.76	4.25
DEPLOYMENT SHOCK $\sigma/m = 10\ \%$	1.85	3.63	3.15	2.11	1.85	2.50	4.32	2.11	1.85	4.32
THRUSTER LOADS $\sigma/m = 2\ \%$	1.63	3.49	2.99	1.92	1.63	2.34	4.17	1.92	1.63	4.17
ACOUSTIC LOADS $\sigma/m = 40\ \%$	3.27	5.10	4.60	3.54	3.27	3.95	5.80	3.54	3.27	5.80
VIBRATION LOADS – THERMAL LOADS (UNCORRELATED) $\sigma/m=20\%$	2.29	4.02	3.54	2.54	2.29	2.92	4.70	2.54	2.29	4.70

Failure probability = 10^{-6} --> $K = \dfrac{\text{Mean strength}}{\text{Mean stress}}$ = safety factor $(1.63 < K < 6.44)$

TABLE 3.3-1 K FACTORS

Table 3.3-2 K1 FACTORS

	METALLIC MATERIAL $\sigma/m = 8\%$	METALLIC MATERIAL: YIELD STRENGTH WHEN R/YIELD > $1.25\ \sigma/m = 15\%$	BUCKLING STRENGTH OF CONICAL OR CYLINDRICAL METALLIC SHEELS $\sigma/m = 14\%$	CARBON FIBER COMPOSITES $\sigma/m : 10\%$	JUNCTION BY SCREW, RIVET WELDING $\sigma/m = 8\%$	BONDING STRUCTURAL IN-SERT (AXIAL LOADING) $\sigma/m = 12\%$	HONEYCOMB: TENSION $\sigma/m = 16\%$	HONEYCOMB: SHEAR, COM-PRESSION $\sigma/m = 10\%$	HONEYCOMB: FACE WRINKL-ING $\sigma/m = 8\%$	EQUIPMENT INSERT (IN HONEYCOMB AXIAL LOAD-ING) $\sigma/m = 16\%$
LAUNCH VEHICLE THRUST $\sigma/m = 5\ \%$	1.29	2.24	1.98	1.42	1.29	1.64	2.60	1.42	1.29	2.60
LAUNCH VEHICLE OTHER STATIC LOADS $\sigma/m = 30\ \%$	1.45	1.98	1.82	1.51	1.45	1.62	2.21	1.51	1.45	2.21
TRANSIENT LOADS $\sigma/m = 50\ \%$	1.59	2	1.87	1.62	1.59	1.71	2.19	1.62	1.59	2.19
THERMAL LOADS (CORRELATED) $\sigma/m = 7.5\ \%$	1.29	2.17	1.93	1.41	1.29	1.61	2.51	1.41	1.29	2.51
DEPLOYMENT SHOCK $\sigma/m = 10\ \%$	1.29	2.11	1.89	1.41	1.29	1.58	2.45	1.41	1.29	2.45
THRUSTER LOADS $\sigma/m = 2\ \%$	1.31	2.34	2.07	1.48	1.31	1.71	2.73	1.48	1.31	2.73
ACOUSTIC LOADS $\sigma/m = 40\ \%$	1.53	1.98	1.84	1.57	1.53	1.67	2.19	1.57	1.53	2.19
VIBRATION LOADS – THERMAL LOADS (UNCORRELATED) $\sigma/m = 20\%$	1.37	2.01	1.82	1.45	1.37	1.59	2.28	1.45	1.37	2.28

Failure probability $= 10^{-6} \rightarrow K_1 = \dfrac{\text{Minimum guaranted strength } (2\sigma \text{ value})}{\text{Limit load } (2\sigma \text{ value})}$ ($1.29 < K_1 < 2.73$)

TABLE 3.3-2 K1 FACTORS

- Simplification of this safety factor matrix, expressing the safety factor as the product KQ x KADD , with :
    . KQ depending only of the load case
    . KADD depending only of the material
- Determination of the acceptance test factor KA in a way to "prove" a sufficient reliability by this test (the figure selected is # 0.98 at a confidence level estimated as being 95 %).
- Determination of the qualification test factor KQ in a way to "prove" a sufficient reliability by test (the figure selected is # 0.95 at a confidence level estimated as being 99.5 %). In our application the determination on this factor has conducted to partially modify the KQ factors evidenced during step 5.
- Determination of the qualification test risks acceptable according to the test philosophy (STM or PFM approach, tests at subsystem or system level). Deduction of an additional factor KTEST to respect the accepted test risk.

NOTA   : The structural reliability objective (0.999 at satellite level) represents the "theoretical" reliability (achieved by design, but not considering potential realization problems as manufacturing defects or poor workmanship). This objective is verified by the qualification test: "proved theoretical reliability".
        The reliability proved by acceptance, to the contrary, integrates manufacturing and integration aspects specific to each flight model: "proved practical reliability".

REMARK : (*) For another application (with different inputs) the same process would be applied.

3.4 PROPOSED FACTOR of SAFETY

CALCULATION ASSUMPTIONS :
The establishment of the design and test factors made during this study has needed a certain number of assumptions. They are summarized here below, and their validity must be verified (or required) before the application of the proposed set of factors in a project:

- Application to a satellite in the "classical" range: weight between 1000 and 2000 kilos, with an average number of mechanical parts (parts = elementary part of junction between part) of 10000 (the application can be extended to the spacecrafts with a weight between 500 and 3000 kg).
- Structural reliability objective of 0.999 at satellite level
- Flight limit load and minimum guaranteed strength to consider for dimensioning are 2σ values
- The scatters considered for composite and bonding

materials are applicable only if each lot of these materials is acceptance tested with regard to the strength characteristic used
- For the determination of the reliability proved by qualification, the parts having an important strength scatter (honeycomb in tension, equipment insert in honeycomb ($\sigma/m = 16\%$) and yield strength of metallic materials with R/Yield > 1.2 ($\sigma/m = 15\%$)) have not been considered. It means that either they have an important margin, or they are individually tested (DVT).

PROPOSED DESIGN FACTORS
The safety factor between the minimum guaranteed strength and the limit loads is given by :

$$K = KQ * KADD * KTEST * (1 + SM)$$

with KQ = qualification test factor, depending on the load case according to Table 3.4-1.

KADD = additional factor depending on the material according to Table 3.4-2 .

KTEST = additional factor depending on the accepted failure risk during test according to Table 3.4-3 .

SM = safety margin, required to be > 0.

This set of factor allows to achieve a reliability with regard to structural aspects greater than 0.999 at satellite level (in practice, near from 0.99995).

PROPOSED TEST FACTORS
The Qualification test, conducted at KQ level, "proves" a theoretical reliability of # 0.95 with an estimated confidence level of 99.5%.

The Acceptance test, conducted at KA = 1.1 level, "proves" a practical reliability of # 0.95% with a 95% confidence level.

Table 3.4-1	
QUALIFICATION TEST FACTORS : KQ	
LOADS CASES	**KQ**
Launch vehicle thrust	1.50
Other static loads	1.50
Transient loads	1.58
Thermal loads	1.50
Deployment shocks	1.50
Thruster loads	1.50
Acoustic loads	1.58
Vibration loads	1.50

Table 3.4-2	
ADDITIONAL FACTOR : KADD	
MATERIALS	**KADD**
Metallic material, screw,rivet, welding, honeycomb (face wrinkling)	1.0
Carbon fibber*,honeycomb (shear,compression)	1.0
Bonding*,structural insert(axial loading)	1.1
Buckling strength	1.3
Honeycomb (in tension), equipment insert in honeycomb (axial loading)	1.65 if specific tests to verify the margin 3.60 if not
Metallic materials for which R/Y>1.2 sized by YIELD	1.45 if specific tests to verify the margin 2.45 if not
* with acceptance test for each batch	

<table>
<tr><td colspan="5" align="center">Table 3.4-3

ADDITIONAL FACTOR : KTEST</td></tr>
<tr>
<td>TEST</td>
<td>INDIVIDUAL SUBSYSTEM TEST

RISK 10%

STM</td>
<td>INDIVIDUAL SUBSYSTEM TEST

RISK 1%

PFM</td>
<td>TEST AT SYSTEM LEVEL (NO TEST AT SUBSYSTEM LEVEL)
RISK 10%

STM</td>
<td>TEST AT SYSTEM LEVEL (NO TEST AT SUBSYSTEM LEVEL)
RISK 1%

PFM</td>
</tr>
<tr><td>STATIC</td><td>1.05</td><td></td><td></td><td></td></tr>
<tr><td>VIBRATION</td><td>1.05</td><td>1.10</td><td>1.10</td><td>1.20</td></tr>
<tr><td>TRANSIENT</td><td></td><td></td><td>1.10</td><td>1.20</td></tr>
<tr><td>ACOUSTIC</td><td>1.0</td><td>1.05</td><td>1.0</td><td>1.10</td></tr>
<tr><td>THERMAL</td><td>1.0</td><td>1.05</td><td>1.0</td><td>1.10</td></tr>
<tr><td>DEPLOYMENT</td><td>1.05</td><td>1.15</td><td></td><td></td></tr>
<tr><td>ORBITAL LOADS</td><td>1.0</td><td>1.10</td><td></td><td></td></tr>
<tr><td>PRESSURE or INITIAL TENSION</td><td>1.0</td><td>1.10</td><td></td><td></td></tr>
</table>

4 HUMAN ERRORS ASPECTS

The set of design factors proposed in this study has been based on a structural reliability objective at satellite of 10^{-3} (probability to failure). This reliability figure represents the theoretical probability of the stress exceeding the strength. But this figure doesn't consider the items failure probabilities due to incorrect design, realization or assembly (generally speaking "human errors"). Qualification tests allow the detection of the design errors. Acceptance tests allow to verify the good workmanship of each model. Due to the cost of such tests, the tendency is now to reduce the quantity of qualification / acceptance tests on the future projects. So one task of our study was to evaluate the efficiency of the check procedures (text or inspection), and then to evaluate a realistic failure probability and to compare it with the theoretical value.

FAILURE PROBABILITIES AT ITEM LEVEL

A simplified FMECA (failure mode, effects and critical analysis) has been performed in order to define general types of causes. The probability of failure has been defined for each identified failure cause, based on an AEROSPATIALE internal research completed by data found in several documents. The results of this task are summarized in Table 4-1.

Then the efficiency of each type of control has been defined with the same procedure; see Tables 4-2 and 4-3.

Finally the probabilities of an incorrect realization not detected by any control is given in the Table 4-4 .

PROBABILITY OF FAILURE BY INCORRECT REALIZATION AT SATELLITE LEVEL

Considering that...
- a satellite has nearly 1000 critical elements,
- a typical failure probability at item level due to incorrect realization is nearly 6.10^{-5} (majority of metallic critical items in present spacecrafts),
    ...the probability of failure at satellite level, due to incorrect realization of an item is near to 6.10^{-2}.

CONCLUSION

This failure probability at satellite level may seem to be very high (6.10^{-2}).
The processing of in flight data given by the BOEING study leads nevertheless to a figure of the same order of magnitude (7 structural failures in flight for 80 launched spacecrafts, giving an "experimental" failure probability of 9.10^{-2}). A figure of 5.10^{-2} is thus probably realistic. It is on one hand necessary to notice that these failures (predicted or detected) are very probably not of the same criticality with regard to the satellite mission. But, on the other hand, it means that it would not be realistic, with currently applied quality provisions on structural aspects, to require at satellite level, a reliability objective greater than 0.999. This figure corresponds in fact to a good compromise between the theoretical structural reliability (governed by the selection of the design factors) and the practical one (governed by the probability of realization errors not detected by implemented quality provisions).
Considering these results, with the safety factors defined

in our study at least one qualification test is required or an acceptance test on each model, the best thing being to have both tests.

<table>
<tr><td colspan="3" align="center">Table 4-1

SUMMARY OF FAILURE CAUSE PROBABILITIES</td></tr>
<tr><td>TYPICAL FAILURE CAUSE</td><td>ASSESSED PROBABILITY OCCURRENCE</td><td>REMARKS</td></tr>
<tr><td>Incorrect design</td><td>$5.\ 10^{-4}$</td><td>Issued from experimental data</td></tr>
<tr><td>Incorrect machining</td><td>$3.\ 10^{-3}$</td><td>"</td></tr>
<tr><td>Metallic material non compliant with its definition</td><td>$3.\ 10^{-3}$</td><td>"</td></tr>
<tr><td>Incorrect assembly</td><td>$5.\ 10^{-4}$</td><td>"</td></tr>
<tr><td>Degradation during handling, storage, transportation...</td><td>$7.7\ 10^{-4}$</td><td>"</td></tr>
<tr><td>Non-metallic material non compliant with its definition</td><td>$1.5\ 10^{-2}$</td><td>Estimated</td></tr>
<tr><td>Incorrect application of a manufacturing process (composite, bonding, welding...)</td><td>$8.\ 10^{-3}$</td><td>"</td></tr>
</table>

<table>
<tr><td colspan="2" align="center">

Table 4-2

INSPECTION EFFICIENCY

</td></tr>
<tr><td align="center">

TYPE OF INSPECTION

</td><td align="center">

ASSESSED EFFICIEN-CY

</td></tr>
<tr><td align="center">Dimensional (AEROSPATIALE-CANNES data on 793 dimensions out of tolerances)</td><td align="center">96%</td></tr>
<tr><td align="center">Surface treatment (evaluated from AEROSPA-TIALE-CANNES data)</td><td align="center">>99%</td></tr>
<tr><td align="center">Heat treatment (insufficient data)</td><td align="center">95 to 99%</td></tr>
<tr><td align="center">Welding</td><td align="center">90 to 95%, if the inspection method is adapted to the size of the cracks to be detected</td></tr>
<tr><td align="center">Efficiency for the establishment of a procedure (literature, from NUREG CR3688, comparing the figures proposed by "human error" with adapted procedure by "human error" without procedure</td><td align="center">98%</td></tr>
</table>

<table>
<tr><td colspan="3" align="center">Table 4-3

SUMMARY OF CHECK EFFICIENCIES</td></tr>
<tr><td align="center">CHECK TYPE</td><td align="center">ASSESSED EFFICIENCY</td><td align="center">REMARKS</td></tr>
<tr><td>Qualification test</td><td align="center">0.995</td><td>Experimental data + as-sumptions (probably pessimistic)</td></tr>
<tr><td>Acceptance test</td><td align="center">0.950</td><td>Experimental data</td></tr>
<tr><td>Manufacturing or inte-gration inspection</td><td align="center">0.950</td><td>Order of mag-nitude issued from experi-mental data</td></tr>
</table>

<table>
<tr><td colspan="2" align="center">Table 4-4

ITEM   FAILURE PROBABILITY DUE TO INCORRECT REALIZATION</td></tr>
<tr><td align="center">ITEM TYPE</td><td align="center">ASSESSED PROBA-BILITY</td></tr>
<tr><td align="center">Metallic parts</td><td align="center">$5.6 \ 10^{-5}$</td></tr>
<tr><td align="center">Composite parts</td><td align="center">$9.8 \ 10^{-5}$</td></tr>
<tr><td align="center">Bonding, welding, inserts</td><td align="center">$9.8 \ 10^{-5}$</td></tr>
<tr><td align="center">Bolted junction</td><td align="center">$5. \ \text{to} \ 7. \ 10^{-5}$</td></tr>
</table>

5 PROPOSED DESIGN/TEST PROCEDURES

As a synthesis for the study, a draft design and test procedure is proposed based on the reliability figures outlined in the previous chapters.

DESIGN PROCEDURE
- Flight limit loads defined at 2σ
- Minimum guaranteed strength defined at 2σ
- Minimum thickness for isostatic parts of the structures
    (nominal thickness when fail-safe design).
- Qualification factor KQ = 1.5
    (excepted acoustic and transient test KQ = 1.58

(4dB).
- Design with K = KQ x KADD x KTEST x Flight limit loads
     (KADD and KTEST defined in Tables 4-2 and 4-3).
- Margin of safety as defined in Figure 3.3-1 has to be
     positive on null.
- Allowable strength should be :
          . Ultimate stress when possible,
          . Yield limit when the same structure is used
            for several tests, or functional aspects are
            driving the design.

TEST PROGRAMME

- **General case :**   Qualification test at KQ level
                          Acceptance test at level KA = 1.1

In order to have a significant qualification test, the
properties of the test model must be known and close to
the minimum values used in the analyses.

- **Particular cases :** During the design, the appropriate
qualification and acceptance test must be defined for each
element. Special care has to be taken for the elements
which are not really tested during qualification or (and
may be) acceptance tests; 4 such cases can be found:

	QUALIFICATION TEST	ACCEPTANCE TEST
CASE 1	YES	YES
CASE 2	YES	NO
CASE 3	NO	YES
CASE 4	NO	NO

A very high practical reliability is only ensured in CASE
1: both design and manufacturing errors can be discovered
with a high efficiency.

In CASE 2, the manufacturing control procedures have to
ensure the reliability. Some tests (at subassembly,
component or element level) might be necessary for the
critical elements (low margin or high strength variation
coefficient).

In CASE 3, the quality of the design is not really
verified. However all the "low strength" structures are
eliminated before the flight. So an inflight failure can
occur only if the flight loads are higher than expected.

In CASE 4, nothing is verified before flight, neither the design, nor the manufacturing errors. Special procedures must be defined for this case in order to avoid a very high safety factor which is very penalizing for a space structure; the current practice is to multiply the usual factors of safety by 1.5 or 2. These special procedures could be, for example:
- independent verification of the stress analysis reports (by a certification authority);
- two independent controls for each manufacturing step.

RECOMMENDATIONS CONCERNING EACH TYPE OF TESTS
Static test
- Qualify the primary structure (on a qualification model)
- For a modular concept, qualify by static test at large component level, reproducing for each test the flexibility of the remainder of the system at the interfaces.
Vibration test
-perform qualification test by correct simulation of the load environment on a qualification model: sine, random, transient (single or multiaxis). The choice of the appropriate excitation will depend on the confidence in the dynamic environment of each launcher. Initially, for safety reasons, transient might only be applied to non-flight items.
Acoustic test
-perform on any model without restriction
-for a large model, if acoustic loads are significative, a single acoustic test might be used as acceptance test
Thermal test
- The test factor 1.5 has to be understood as applied to the stresses due to the thermal loads.
-As it can be difficult to increase the temperature during a qualification test (the material allowables might vary, for example) it can be better have an acceptance test on each model.
- if thermal loads are combined with static loads, the static test factor has to be increased in order to reproduce the total stress if the thermal environment is not reproduced (assuming the material data do not vary).
Deployment shock test
for the mechanisms using a regulation (i.e. almost all in space application) it is not possible to have a "qualification" deployment. So each mechanism has to be properly deployed, this being considered as an acceptance test. Qualification tests can be performed separately on specific structural parts using deployment loads times the qualification factor (hinges...).

REMARKS

- Properties of qualification models: before the qualification, the properties of the test model should be known as far as possible(thicknesses, strengths...). If they deviate from the minimum values used for design, the applied test loads can be corrected using the so call "$J_{corrected}$" method"; example: $J_{corrected}$(membrane) = measured thickness/minimum thickness.

- Flight loads probabilities: the qualification load factors have been defined for 2σ limit loads (97.7%). It can be reduced if these loads are specified with a higher probability level; for example KQ=1.4 for loads specified at 2.3σ (99.0%), KQ=1.25 for loads specified at 3σ (99.8%).

- Coupled analysis: Loads obtained from a coupled analysis might be multiplied by an uncertainty factor KU. This factor could be function of the accuracy of the launcher model and the one of the spacecraft model (correlated or not). A statistical approach of coupled analysis including sensitivity aspects could be considered to obtain 2σ loads (see ref.16).

6 PROPOSED STRESS SUMMARY STANDARD

In order to have an overview of all the subsystems stress analysis reports in a project we have defined a standardized form containing all the necessary informations : Table 6-1. The aims of this procedure are :

- facilitate the detailed reliability calculations
- detection of the critical elements
- detection of the elements not tested
- assessment of the criticity of each test (failure risk, reliability proved by test).

Each element analysed in the stress analysis reports constitutes one line of this table.

KQ, KADD and SM are the informations contained in these reports : KQ qualification factor, KADD specified additional factor, SM margin of safety achieved after dimensioning.

σ/m on stress, σ/m on strength, KL, KS, SF, are the data generally added by reliability people for the reliability analysis (realistic values of σ/m have been defined in our study).

The three last columns are results from these reliability calculations.

Depending on the application the whole table can be necessary or only a part.
Results concerning each element can be combined to obtain many types of outputs:
- failure probability at subsystem level
- failure probability at system level
- risk of failure during a given test (concerning one or several subsystems)
- reliability proved by the qualification tests (at subsystem or system level).

An example of treatment is given in the Table 6-2: application on ARABSAT telecommunication spacecraft.

Remark: Due a great quantity of data to be handled and to the type of these data, a computer implementation of this procedure is very interesting. So it is foreseen to develop a software in 1987 on a personal computer which appears well adapted to this application.

ITEM	MATERIAL	LOAD CASE	QUALIF. TEST	FAILURE CASE	$\sigma_{/m}$ ON STRESS	$\sigma_{/m}$ ON STRENGTH	K_L	K_Q	Kadd	SM	K_S	K	FAILURE PROBABILITY (FLIGHT)	FAILURE PROBABILITY (QUAL. TEST)	PROVED RELIABILITY	QTY
- Part	- Metallic	- Vibration loads	Vibration Test	- Rupture (tension)												
- Junction between parts	- Composite	- Deployment shock	nu	- Buckling												

[DRAFT]

K_L = factor between limit loads and mean load

K_Q = factor between qualification loads and limit loads

Kadd = additional specified safety factor giving the minimum required strength

SM = [factor between the minimum guaranted strength and the minimum required strength] −1

K_S = factor between the mean strength and the minimum guaranted strength

SF = safety factor = $K_L \times K_Q \times$ Kadd $\times [1 + SM] \times K_S$

Table 6-1 STRESS-RELIABILITY SUMMARY TABLE

| SUBSYSTEM | RELIABILITY | QUALIFICATION TEST CHARACTERISTICS | | | |
		LOAD CASE CORRESPONDING TO IDENTIFIED CRITICAL ITEM	IMPLEMENTED QUALIF. TEST	PROVED RELIABILITY AFTER QUALIFICATION	QUALIFICATION TEST RISK
STRUCTURE	$0.9_{(4)}69$	Thermal loads	No test *	Not applicable	N.A
		Vibration loads	Vibration test	0.99	$2.4.10^{-2}$
		Launch loads	Static test	0.9996	7.10^{-3}
ANTENNA (mechanisms + reflectors)	$0.9_{(5)}4$	Safe loads	No test *	N.A	N.A
		Deployment shock	No test *	N.A	"N.A
		Vibration loads	Vibration test	$0.9_{(4)}6$	$2.6.10^{-4}$
SOLAR ARRAY (mechanisms + panels)	$0.9_{(4)}84$	Vibration loads	Vibration test	0.973	$1.84.10^{-5}$
		Deployment shock	No test *	N.A	N.A
		Safe loads	No test *	N.A	N.A
PROPELLANT TANKS	$0.9_{(6)}85$	Vibration loads	Vibration test	$0.9_{(4)}74$	$4.6.10^{-2}$
RESULTS AT SATELLITE LEVEL	0.99995	—		< 0.963	> 7.6 %

* No test = no qualification test to "verify" a margin w.r.t limit loads
  (acceptance tests are peformed on each model)

Table 6-2 RELIABILITY SUMMARY TABLE: ARABSAT S/C

7 CONCLUSION

A survey of the safety factors has been firstly performed, showing that various safety factors were employed without a rational basis for their definition.
In a second step, a methodology has been established, based on reliability aspects with realistic inputs. Particular attention was devoted to data search and definition of these inputs. Subsequently, reliability allocations and factors of safety have been outlined. Although these factors are close to the one generally used, an application on an actual satellite (ARABSAT) has shown that this methodology improves the design process without impact on the mass (example not shown in this paper). Finally, recommendations have been issued, including a set of reliability design allocations, design and test factors of safety, valid for standard unmanned spacecrafts. A mean to extend its validity has been proposed.

Presently, work is going on following this study. Running actions cover:
- manned spacecraft aspects
- generation of a PC computer tool (including a load - strength - allocation database) allowing to envisage impact of different variation coefficients on the factors of safety, and vice-versa.

8 REFERENCES

(1)   Evaluation of design and tests safety factors phase I report (WP E11-21)

(2)   Evaluation of design and tests safety factors phase II report (WP E12-E22)

(3)   Evaluation of design and tests safety factors WP E13 report (ARIANE launcher loads analysis)

(4)   Evaluation of design and tests safety factors synthesis of the phase III (WP E41-E42)

(5)   Resultats des vols du lanceur ARIANE et leurs effets sur le dimensionnement des structures satellites.
       M.VEDRENNE, JP DULOUT and JL CANEILLES, ARIANESPACE, EVERY, proceedings of an international conference held at CNES TOULOUSE, FRANCE, on 3-6 December 1985 on the spacecraft structures

(6)   Flight loads and test requirements for ARIANE launcher users M. VEDRENNE, ARIANESPACE   ESA/ESTEC structure presentations, June 11th and 12th 1986, NOORDWIJK, NL

(7)   Alternatives to notched swept sine testing for
spacecrafts and other large systems. MM.KOWLER/ZUZIAH/DES-
FORGES (HUGHES) M.SWRANTZ (COMSAT) TES, 1877, p351 to 361

(8)   Logique de developpement et performances du lanceur
ARIANE par M.DESLOIRE Revue l'Aeronautique et l'Astronau-
tique numero 77, 1974

(9)   Initial results from multi-axes transient testing K.
ECKHARDT MBB-ERNO
Proceedings of a conference spacecraft structures CNES
TOULOUSE 3.6 December 1985 (ESA SP.328 April 1986)
AGARD Conference proceeding N.397 mechanical qualification
of large flexible spacecraft structures (10) to (15)

(10)   Recent developments and future trends in structural
dynamic design verification and qualification of large
flexible spacecraft by E.HORNUNG, E. BREITBACH and H.ORY

(11)   Qualification of the faint object camera by
P.AMADIEU

(12)   Structural qualification of large spacecraft by
BH.WADA

(13)   Essais de satellites en vibrations basse frequence
par A.GIRARD, A.MAMDE et F.MERCIER

(14)   Dynamic analysis of direct television satellite
TV-SAT/TDF-1 by Y.PLUMAT

(15)   Multi-axis vibration tests on spacecraft using
hydraulic exciters by H.HAMN and W.RAASCH

(16)   Development of experimental /analytical concepts
for structural design verification
ESTEC study presentation NOORDWIJK 11-12 th June 1986 by
E.HORNUNG, K.ECKHARDT, E.ERBEN, E.BREITBACH, H.HUNERS,
H.ORY, H.GLASER

(17)   AEROSPATIALE internal document : ARIANE 4 systeme
vehicule; synthese des marges de securite A4.DE.1.C.02.

(18)   Structural load prediction methods for space
payloads by WADA proceedings congres TOKYO techniques
spatiales 28/06/82

(19)   Design of space payloads for transient environments
by WADA ASME 1979 winter annual meeting, NEW-YORK December
2-7 1979

(20)   Generalized modal shock spectra method for
spacecraft loads analysis by M.TRUBERT AIAA journal vol.
18 numero 8 August 1984 pp 982-994

(21)   Design and flight loads comparison by CHEN and
GARBA Journal of spacecraft and rockets vol. 16 numero 1,
January, February, 1979 pp 27-34

(22)   JSC 20052 Volume numero 8 Space shuttle payload
design and development structural mechanical interfaces
and requirements Rev 1 9/84.

Parameter Identification for Reliability in Markov Cumulative Damage Processes

by
P. Voltz and F. Kozin
Polytechnic University, Route 110, Farmingdale, N.Y. 11735

ABSTRACT

This paper is concerned with the identification of parameters in cumulative damage processes whose underlying statistics are Markov. We consider the problem of parameter estimation during normal field operation and maintenance, and discuss almost-sure convergence of the estimates in both single and multiple parameter situations.

We present theorems on convergence which are valid under a very general class of maintenance policies, and show computer simulation results to demonstrate the theorems.

1. INTRODUCTION AND PROBLEM STATEMENT

The engineering problem motivating this paper is the maintenance of a system of components which are subject to a cumulative damage phenomenon, the statistics of which depend upon unknown parameters. Our concern will be the estimation of these unknown parameters while the system undergoes normal field operation and maintenance.

We begin this introduction with a description of the mathematical model of cumulative damage to be employed, and a discussion of system maintenance when the parameters are known. We specify the maintenance objective and a method of controlling the process (via inspections and re-placement of worn parts) to achieve this objective. We then pose the problem of estimation when the parameters are unknown.

In our mathematical discussions we shall make use of a Markov model of cumulative damage which first appeared in a four part paper by Bogdanoff, Krieger and Kozin [1], and was expanded and refined by Bogdanoff and Kozin in a series of papers [2-4] and finally a book [5]. A review of the main features of the model follows.

To begin, we break the time axis up into discrete periods (duty cycles) of usage over which the damage occurs. The model also assumes a discrete set of damage states, say 1 through b, where state b represents failure of a component.

The Markov nature of the model is described as follows. Let the damage state of a given component just after the nth cycle of usage be denoted by D_n. Then D_n is assumed to be a discrete time, discrete state Markov process. D_n represents cumulative damage, and must be non-decreasing, so that the transition matrix P_n (in general time dependent) is upper triangular. As usual P_n is defined by

$$P_n = [p_{ij}^{(n)}] \qquad (1\text{-}1)$$

where

$$p_{ij}^{(n)} = \text{Prob}\{D_{n+1} = j \mid D_n = i\}. \qquad (1\text{-}2)$$

Throughout this paper we shall assume a stationary model with transition matrix P and shall concentrate on a one jump model with general form

$$
P = \begin{bmatrix}
\alpha_1 & 1-\alpha_1 & & & & \\
 & \alpha_2 & 1-\alpha_2 & & & \\
 & & \cdot & \cdot & & \\
 & & & \cdot & \cdot & \\
 & & & & \alpha_{b-1} & 1-\alpha_{b-1} \\
 & & & & & 1
\end{bmatrix}
\qquad (1-3)
$$

Thus, the damage jumps at most one state over each cycle and the probability of a jump depends on the state of damage just prior to that cycle. When the damage reaches state b (the absorbing state) the component is failed and it remains in the failed state thereafter. In what follows we shall consider the case when $\alpha_1 = \alpha_2 = \ldots = \alpha_{b-1}$ as well as the case that all of the α_j may be distinct.

With the basic damage model thus described let us now turn to the problem of maintenance. Suppose a fleet of N_c identical independent components is put into operation at time zero. As the number of operating cycles increases, each of the components will accumulate damage in a random fashion according to (1-3). If left in operation, all of the components would eventually fail. In order to avert a system failure therefore, we carry out routine inspections on some prescribed schedule to replace failed or excessively worn components. Since inspections can be costly however, we do not wish to inspect the fleet more often than necessary.

Since the cumulative damage process is stochastic in nature, we must rely on some sort of probabilistic or average criterion for determining the appropriate times between inspections based upon our observed information. But to explain the entire maintenance procedure more thoroughly we must now make some definitions and set up a precise mathematical formulation.

The next important procedure to quantify is that of the inspection. Ideally, during the inspection process the exact state of damage of a given component would be determined. This may be unrealistic in many cases, however, and we shall incorporate into the model the possibility of inaccurate measurements. At this point let us do it in the following natural way. We define

τ_{ij} = Prob{damage perceived in state j|damage in state i}

and obtain thereby a b x b observation probability error matrix. Now in certain specific cases the information related to the central processor by the inspector has only to do with whether or not a part or parts was replaced. Then as far as observation error is concerned, what is important to us is not τ_{ij} as defined above, but rather

$$\tau_i = \text{Prob}\{\text{part is replaced}|\text{damage in state i}\}.$$

It should be clear that once the replacement policy has been defined the τ_i can be computed from the τ_{ij}. We shall assume throughout that the inspection process is well understood, so that the τ_{ij} are known, and also that (as seems quite reasonable) $\tau_j \geq \tau_i$, if $j > i$.

Now suppose the nth inspection is performed and a perceived state of damage $\hat{d}_n$ is obtained (we use the notation d_n to denote damage at the nth inspection as opposed to the notation D_n which denotes the damage after the nth cycle of operation between inspections). This is the only indication of the actual damage d_n, and therefore the component should be replaced if $\hat{d}_n = b$, and left in place if $\hat{d}_n < b$ (the component is perceived to be operational). This is the physical aspect of maintenance. The system aspects of maintenance shall be discussed shortly. In this paper we shall assume that new parts start out undamaged (they begin in state 1).

Now let us return to the system problem and describe in detail the flow of a particular maintenance procedure. At time zero a fleet of N_c identical new components is put into operation. Suppose these components are all allowed to operate for Δ_1 cycles before the first inspection is made (we inspect all components at the same time). Then if we assume P (the damage transition matrix) known, we could theoretically compute such statistical quantities as $\text{Prob}\{n_f^{(1)} \geq k\}$ where k is some integer and $n_f^{(1)}$ is the number of failed components at the first inspection time. If $n_f^{(1)}$ is large we may risk system failure, and we therefore want to keep $n_f^{(1)}$ small. The way we do this is to choose Δ_1 small enough, since it so happens that the above quantities are monotonically increasing functions of Δ_1. We shall choose as our maintenance objective to keep the quantity $\text{Prob}\{n_f^{(1)} \geq 1\}$ below a given acceptable value, say γ. Thus Δ_1 is chosen to be the largest (because we do not want to inspect too often) integer value such that

$$\text{Prob}\{n_f^{(1)} \geq 1\} < \gamma \; . \qquad\qquad (1\text{-}4)$$

The above quantity is of course the probability that one or more components has failed by the first inspection time.

The actual value of $n_f^{(1)}$ is random and in fact may be any integer up to N_c. Also, since there is observation error present, the number of parts perceived to need replacing (and hence, the number that are replaced) may be different than $n_f^{(1)}$. Thus let $n_r^{(1)}$ denote the actual number of parts replaced at the first inspection. We shall take $n_r^{(1)}$ as the bare minimum of information gathered at the first inspection. In some cases we may take more information, such as the perceived state of damage of each component in the fleet. In the first instance the information obtained is a scalar, and in the second it is a vector of dimension N_c. In any case let us denote by y_1 (scalar or vector) the totality of observed information at the first inspection. We shall use this information in choosing Δ_2, the number of cycles of operation between the first and second inspections, in order to keep the probability of failure within the prescribed bounds. In this paper we shall assume for simplicity that $\Delta_n \geq b\text{-}1$. This eliminates degenerate cases in which a component is inspected before it is possible for it to have failed.

We must bear in mind as we continue that each inspection performs two basic functions. The first is that components which are perceived to need replacing (due to damage in state b) are culled out and replaced with new ones, and the second is that whatever information, y_1, is obtained is relayed to some central processing unit.

After the first inspection we then restart the process with the required new parts in place and use the information y_1 to determine Δ_2 as follows. Just as before, we now choose Δ_2 as the largest integer value such that $\text{Prob}\{n_f^{(2)} \geq 1 | y_1\} < \gamma$. After Δ_2 cycles have passed we inspect a second time obtaining the information y_2, and restart the process with damaged parts having been replaced once again.

As this process continues we obtain at time (inspection time) n the nth observation, y_n, which gives us as a total observation history the sequence $y_1, y_2,\ldots,y_n$. We then choose Δ_{n+1} to be the largest integer value such that

$$\text{Prob}\{n_f^{(n+1)} \geq 1 | y_n, y_{n-1},\ldots,y_1\} < \gamma \qquad\qquad (1\text{-}5)$$

and replace all damaged parts.

We next state the basic problem of concern to us in this paper. Suppose we know that our damage process may be modelled as in the above, but that the parameters α_1, $\alpha_2,\ldots,\alpha_{b-1}$ of the probability transition matrix P in (1-3) are initially unknown to us. Then the maintenance policy described above cannot be carried out because the statistical quantities involved are unknown.

One approach in a situation like this is to attempt to learn the parameters of the model in an adaptive fashion and to apply the maintenance procedure based on our best estimate at any given time. To be more specific let y_1, $y_2,\ldots,y_n$ be the total history of observations up to time n. Since we have no other information as to the true parameter values, we should base our estimate of P_0 (the actual probability transition matrix) upon the observed sequence y_1, $y_2,\ldots,y_n$. We shall use the maximum likelihood estimate. Thus, if we denote by P_n the estimate of P_0 at time n, then P_n is chosen to be the matrix of the form (1-3) such that

$$\mathrm{Prob}\{y_1,\ y_2,\ldots,y_n\}$$

is maximum. We must keep in mind here that the above probability depends implicitly on the matrix P. In order to actually compute the maximum likelihood estimate at time n we must of course make this dependence explicit for a given observed sequence y_1, $y_2,\ldots,y_n$ and then choose the maximizing value of P.

Now once P_n is determined, it is our best estimate of P_0 at time n. It seems natural that to choose Δ_{n+1} we should apply the policy expressed in equation (1-5) under the assumption that P_n is the true transition matrix (since we have no better alternative). With this choice we allow the system to operate for Δ_{n+1} cycles until the next inspection, at which time y_{n+1} is observed and the above procedure is repeated. Thus we are adaptively estimating and maintaining the system. The only special case to note is at the start of the procedure. At that point, since we have not yet observed anything, we choose our initial estimate based upon our available past information or experience (it is deterministically chosen).

Our question is this: Does P_n converge to P_0 as $n \to \infty$?

In the following sections we consider two specific situations in which the answer to this question is yes.

The first is the single component-single parameter case wherein $N_c = 1$, and $\alpha_1 = \alpha_2 = \cdots = \alpha_{b-1} = \alpha$ in (1-3). In this case we take as our observation only whether or not the component is replaced at each inspection.

The second case is the multiple component-multiple parameter situation wherein $N_c > 1$, and $\alpha_1 \neq \alpha_2 \neq \cdots \neq \alpha_{b-1}$. In this case our observation is the vector of perceived states of damage for each component at inspections.

In Section 2 we outline the estimation procedures and state the convergence results. In Section 3 we illustrate the results with several computer simulation examples.

BIBLIOGRAPHY

1. J.L. Bogdanoff, et al., A New Cumulative Damage Model, Parts 1-4, Journ. Appl. Mech., 45(2), (June 1978-March 1980).

2. F. Kozin and J.L. Bogdanoff, A Review of the B-Model Approach to Cumulative Damage Processes, Proceedings of the 5th National Congress on Pressure Vessel and piping technology, ASME, San Antonio, Texas, June 1984.

3. J.L. Bogdanoff and F. Kozin, On Nonstationary Cumulative Damage Models, Journ. Appl. Mech., Vol. 104, March 1982.

4. F. Kozin and J.L. Bogdanoff, Cumulative Damage: Reliability and Maintainability, in Probabilistic Fracture Mechanics and Fatigue Methods: Applications for Structural Design and Maintenance, ASTM STP 798, J.M. Bloom and J.C. Ekvall, Eds., American society for testing and Materials, 1983, p. 131-146.

5. J.L. Bogdanoff and F. Kozin, Probabilistic Models of Cumulative Damage, John Wiley & Sons, 1985.

2. ADAPTIVE UPDATING OF PARAMETERS VIA THE MAXIMUM LIKELIHOOD METHOD

2.1 The Single Component-Single Parameter Case

In this section we assume that only a single component is in operation at any given time, and that the damage transition matrix depends upon a single parameter α. Thus the state transition matrix reduces to

$$P = \begin{bmatrix} \alpha & 1-\alpha & & & & \\ & \alpha & 1-\alpha & & & \\ & & & \cdot & \cdot & \\ & & & & \alpha & 1-\alpha \\ & & & & & 1 \end{bmatrix} \qquad (2-1)$$

We shall assume that the true parameter α_0 is neither 0 nor 1 since these cases are degenerate, and further we take α_0 to belong to some closed interval I_0 contained in $(0,1)$. Our observation here is the quantity $n_r^{(n)}$. Since we are dealing with only one component this quantity takes the value 1 or 0 depending upon whether or not the part has been replaced at the nth inspection. Thus our observation y_n at time n will be precisely

$$y_n = n_r^{(n)} \ . \qquad (2-2)$$

Now the process begins just as in the introduction with the component operating for a prespecified number of cycles Δ_1. After the first inspection, y_1 is obtained (it is 0 or 1) and we make the maximum likelihood estimate of α_0 based upon y_1. That is, we choose α_1 to be the α value that maximizes the function $\text{Prob}\{y_1\}$. More specifically, if $y_1 = 1$ we maximize $p\{y_1 = 1\}$, and if $y_1 = 0$ we maximize $\text{Prob}\{y_1 = 0\}$. For clarity we will at this point make the dependence on α explicit. For this purpose, it will be convenient to introduce some notation. Since at an inspection the component may be replaced, the damage state just prior to, and just after the nth inspection may be different. Therefore, let us denote by d_n^- the damage just prior to the nth inspection, and by d_n^+ the damage just after the nth inspection.

Now we may write that

$$\text{Prob}\{y_1 = 1\} = \sum_{j=1}^{b} \text{Prob}\{y_1 = 1 \mid d_1^- = j\}\text{Prob}\{d_1^- = j\}$$

$$= \sum_{j=1}^{b} \tau_j \text{Prob}\{d_1^- = j\} \; . \tag{2-3}$$

But (since $\text{Prob}\{d_o = 1\} = 1$)

$$\text{Prob}\{d_1^- = j\} = \sum_{i=1}^{b} \text{Prob}\{d_1^- = j \mid d_0 = i\}\text{Prob}\{d_0 = i\}$$

$$= P_{1j}^{\Delta_1}(\alpha)$$

where $P_{ij}^{\Delta_1}(\alpha)$ is the ij^{th} element of the matrix $P^{\Delta_1}(\alpha)$. Thus

$$\text{Prob}\{y_1 = 1\} = \sum_{j=1}^{b} P_{1j}^{\Delta_1}(\alpha) \, \tau_j \; . \tag{2-4}$$

Of course $\text{Prob}\{y_1 = 0\}$ is just $1 - \text{Prob}\{y_1 = 1\}$. It may readily be shown that $P_{ij}^{\Delta}(\alpha)$ is a polynomial in α for all i, j, and Δ.

Next we generate Δ_2 via equation (1-5) assuming that α_1 is the true parameter. Then after Δ_2 cycles of operation we observe y_2 and choose α_2 to maximize $P(y_1, y_2)$, and the process continues in this manner. Now let us turn to the estimate at the general time n.

Our total observation sequence up to time n is $y_1, y_2, \ldots, y_n$. At this time we choose α_n to be that value of $\alpha \in I_0$ such that $P(y_1, y_2, \ldots, y_n)$ is maximum. Now

$$P(y_1, y_2, \ldots, y_n) = \prod_{k=1}^{n} P(y_k \mid y_{k-1}, \ldots, y_1)$$

where for $k = 1$ we define $P(y_k \mid y_{k-1}, \ldots, y_1) = P(y_1)$.
The log likelihood function is then

$$L_n(\alpha) = \frac{1}{n} \sum_{k=1}^{n} \ln P(y_k \mid y_{k-1}, \ldots, y_1) \; . \tag{2-5}$$

We shall choose α_n as that value of $\alpha \in I_0$ which maximizes $L_n(\alpha)$ for the observed sequence $y_1, y_2, \ldots, y_n$. Recall that $y_1, y_2, \ldots, y_n$ is a sequence zeros and ones for this case. In order to compute α_n we should as before make the dependence on α in (2-5) explicit. We must therefore be prepared to compute as functions of α each of the $P(y_k | y_{k-1}, \ldots, y_1)$ for all possible sequences $y_1, y_2, \ldots, y_n$. This is a somewhat tedious task, and we merely state the results here.

Consider first the probability

$$P(y_k = 1 | y_{k-1} = 1, \ y_{k-2} = q_{k-2}, \ldots, y_1 = q_1)$$

where $q_1, \ldots, q_{k-2}$ take the values 0 or 1 arbitrarily. After some manipulations we find

$$P(y_k = 1 | y_{k-1} = 1, \ldots) = \sum_{j=1}^{b} P_{ij}^{\Delta_k}(\alpha) \tau_j \ . \tag{2-6}$$

Notice that the functional form depends only on Δ_k. If the part had not been replaced at the last inspection, but at the one before that, things get a little more complicated. We have in that case

$$P(y_k = 1 | y_{k-1} = 0, \ y_{k-2} = 1, \ y_{k-3} = q_{k-3}, \ldots, y_1 = q_1) =$$

$$= \frac{\displaystyle\sum_{j_k=1}^{b} \sum_{j_{k-1}=1}^{b} \tau_{j_k} (1 - \tau_{j_{k-1}}) P_{j_{k-1} j_k}^{\Delta_k}(\alpha) P_{1 j_{k-1}}^{\Delta_{k-1}}(\alpha)}{\displaystyle\sum_{j_k=1}^{b} \sum_{j_{k-1}=1}^{b} (1 - \tau_{j_{k-1}}) P_{j_{k-1} j_k}^{\Delta_k}(\alpha) P_{1 j_{k-1}}^{\Delta_{k-1}}(\alpha)} \tag{2-7}$$

Note that the probability in Eq. (2-7) depends only upon Δ_k and Δ_{k-1}, and not on anything before the last replacement.

The functional form in Eq. (2-7) can be generalized in a natural way to the case that the last replacement occured say, m inspections in the past. In that case the form depends on $\Delta_k, \Delta_{k-1}, \ldots, \Delta_{k-m+1}$. Of course in Eq. (2-5) we shall encounter conditional probabilities of the form

$$P(y_k = 0 | y_{k-1} = 0, \ y_{k-2} = 1, \ y_{k-3} = q_{k-3}, \ldots, y_1 = q_1)$$

as well, and this probability is computed by simply subtracting from 1 the expression in Eq. (2-7).

Having categorized the various terms that will appear in (2-5), it is, conceptually at least, straightforward to compute the maximum likelihood estimate α_n. Next we state without proof our first convergence theorem.

Theorem 1: Let α_n be a maximizing value of α in (2-5) at time n on the closed interval I_0.

Then $\alpha_n \to \alpha_0$ a.s. as $n \to \infty$.

Although this theorem has been stated in the context of the maintenance policy (1-5), it is in fact true under much more general conditions. It holds under any non-anticipating maintenance policy for which Δ_n is uniformly bounded in n. That is, for which

$$\Delta_n \leq \Delta_{Max}$$

for some $\Delta_{max} < \infty$.

It can be shown that the policy (1-5) satisfies this requirement for any fixed γ and I_0 as defined previously.

2.2 The Multiple Component-Multiple Parameter Case

Next consider a system with N_c components operating in parallel. Each component is governed by the damage state transition matrix of (1-3) where all the α_i are distinct. In this case the observation, y_n, at the nth inspection is taken to be the vector of perceived states of damage of each component. We shall assume here that the matrix $\tau = [\tau_{ij}]$ contains only positive elements and is non-singular.

Just as in Section 2.1, we choose as our estimate, α_n at time n the maximizing value of $L_n(\alpha)$ (2-5), except here the y_k are vectors and the conditional probabilities are more involved than in the single component situation. Since the components are assumed to decay independently of one another, however, the conditionals may be separated into factors due to each of the components, and the functional forms then become comparable in complexity to those of Section 2.1.

In the present situation α_0 refers to the vector of actual decay parameters in (1-3), and α_n denotes the maximum likelihood estimate of α_0 at time n. As before we assume that $\alpha_i \in I_0$ for $i = 1,\ldots,b-1$, where α_i is the ith component of α_0.

With regard to convergence in the present case we have Theorem 2.

Theorem 2: Let α_n be a maximizing value of $L_n(\alpha)$ on I_0^{b-1} at time n. Then for any maintenance policy such that $\Delta_n \leq \Delta_{max}$ for all n,

$$\alpha_n \to \alpha_0 \text{ a.s. } \quad \text{ as } n \to \infty .$$

3. COMPUTER SIMULATION RESULTS

In this section we present computer simulation results to illustrate the convergence properties of the estimation procedure.

3.1 The Single Parameter Case

Here a single component operates under the governing damage state transition matrix of equation (1-3) except that $\alpha_1 = \alpha_2 = \ldots = \alpha_{b-1} = \alpha$. We shall take the order, b, of the transition matrix to be 5.

Let us define the row vector τ as

$$\tau = (\tau_1, \tau_2, \ldots, \tau_b) \qquad (3-1)$$

where the τ_i are as defined in the introduction,

$$\tau_i = \text{Prob}\{\text{part is replaced} \mid \text{damage in state i}\}.$$

For the first example let

$$\tau = (0.0,\ 0.1,\ 0.2,\ 0.5,\ 1.0),$$

and $\gamma = 0.2$, where γ is the control objective quantity defined in (1-5). The true parameter is $\alpha_0 = 0.90$.

In this example the control procedure of (1-5) was carried out to determine the appropriate Δ's between inspections.

Table 3.1 shows the first several values of y_n, Δ_n, and α_n for this example under closed loop operation. Δ_1 was chosen to be 10 here, and $y_1 = 0$ indicates that the component was not perceived to require replacement after 10 cycles of operation. According to the discussion in Section 2.1 the first estimate, α_1, should be that value of α which maximizes $P\{y_1 = 0\}$. But it may be shown that $P\{y_1 = 0\}$ is a monotone function of α, so α_1 should be the largest allowable value in the parameter space. In this example we have chosen to restrict the domain of α to $0.03 \leq \alpha \leq 0.97$.

As the number of inspections progresses some general trends are the following. Whenever $y = 0$ (the part is not replaced) the parameter estimate increases, and when $y = 1$ the estimate decreases. This is natural because $y = 0$ tends to indicate a slower rate of decay and $y = 1$ indicates a faster rate of decay. Also the Δ's should generally be large when the component is just replaced and then should decrease as more inspections go by without replacing. This trend can be violated locally if it is outweighed by the fact that the parameter estimate increases, thereby requiring a larger Δ.

A graph of α_n vs. n for this example is shown in Fig. 3.1. We see that the estimate approaches the vicinity of $\alpha_0 = 0.90$ rather quickly, but that fine adjustment may take appreciably longer.

We have made the point that the parameter estimate should converge under any control law, and not only the one used above. Consider therefore the same example as above except that the Δ's are chosen at random each time with $5 \leq \Delta \leq 20$. Figure 3.2 is the resulting plot of α_n vs. n. Figure 3.3 shows the successive estimates when the true parameter is changed to $\alpha_0 = 0.80$. The convergence behavior here is very similar to the earlier situation in which the control is generated as a function of past data.

3.2 The Multiple Parameter Case

Recall the problem in which the damage state transition matrix contains distinct diagonal elements as in (1-3). We shall consider the multiple component situation in which the observed state of each component is used in updating the parameter estimate after each inspection. We will next give a few examples to illustrate the convergence in this case. At this time we are restricted by speed of computation to only a few parameters.

For the following discussion let τ denote the observation error matrix introduced in the introduction, and let α denote the row vector of unknown parameters

$$\alpha = (\alpha_1, \alpha_2, \ldots, \alpha_{b-1}) \ . \tag{3-2}$$

We shall consider a 4×4 damage state transition matrix ($b = 4$) and for the first example let

$$\tau = \begin{bmatrix} 0.70 & 0.20 & 0.05 & 0.05 \\ 0.12 & 0.70 & 0.12 & 0.02 \\ 0.02 & 0.12 & 0.70 & 0.12 \\ 0.05 & 0.05 & 0.20 & 0.70 \end{bmatrix} \ ,$$

$$\alpha_0 = (0.90, \ 0.85, \ 0.80) \ ,$$

and $N_c = 100$ components. The non-singularity of the matrix τ is a consequence of the fact that it is diagonally dominant (diagonal elements are greater than the sum of all other elements in each row).

For numerical convenience we shall use the following control law to generate the Δ's in this case. At each inspection we record the perceived state of damage of each component, and compute the number of components perceived to be in each damage state. Denote by N_i the number of components perceived to be in damage state i, and by pf_i the probability of failure for a component by the next inspection if it begins in state i after the present inspection (pf_i depends implicitly upon Δ and α).

The next Δ is then chosen to be the maximum integer such that

$$\frac{\sum_{i=1}^{b-1} N_i pf_i}{\sum_{i=1}^{b-1} N_i} < \gamma \qquad (3-3)$$

where γ plays the same role as in the single parameter–single component case. In this example $\gamma = 0.4$. Figure 3.4 shows a graph of Δ_n vs. n for this example and figure 3.5 shows α_n vs. n.

Notice from Fig. 3.4 that the Δ_n tend to cluster about the value $\Delta = 9$. This is not surprising since when a large number of components operates in parallel we would expect that once the system reaches steady-state the average number of components in each non-failure state should remain fairly constant over time. As a result, the Δ computed from (3.3) will not fluctuate much once the parameters have sufficiently zeroed in on α_0 so that $pf_i(\alpha) \approx pf_i(\alpha_0)$.

Turning to figure 3.5 we see that just as in the single parameter case, the estimate reaches the neighborhood of α_0 fairly rapidly, but fine adjustment is somewhat slower. In figure 3.6 we show α_n vs. n for the same example except that the observation error matrix is the identity. This corresponds to a noiseless state observation at each inspection. The estimate in this case appears to be somewhat better than in figure 3.5, as one might expect.

In figures 3.7 through 3.9 we illustrate the effect of reducing the number of components. In these examples τ is as in the first example but the order of the parameters in the state transition matrix has been reversed, so that $\alpha_0 = (0.80, 0.85, 0.90)$. The number of components drops from $N_c = 100$ in figure 3.7 to $N_c = 50$ and $N_c = 20$ in figures 3.8 and 3.9

respectively. The convergence apparently becomes more sluggish as N_c decreases, as we would expect, since a larger number of components implies a larger number of independent observations.

For our final example we again consider the system under a randomized inspection policy. In figure 3.10 Δ_n is chosen at random such that $8 \leq \Delta_n \leq 13$, and all else is as in the first example of this section. The convergence here is not substantially different from the non-random inspection case.

In order to increase the speed of computation the curves in section 3.2 were generated using a modified maximum likelihood method. This method includes in the likelihood function (2-5) only those terms corresponding to components which were replaced at most two inspections before. That is, each component contributes information until it has survived two inspections without replacement. If more information is used, the convergence would improve, and there is thus a tradeoff between computation complexity and convergence rate.

This modification was not employed in the single-parameter case. There we quantized the parameter to increments of 0.01 and the likelihood function at the quantized values of α were stored in an array. Then, with the arrival of new data, the likelihood function is incremented by the new log-conditional probability function in (2-5) instead of recomputing the entire function. In the multiple-parameter case this method becomes very slow for any reasonable degree of quantization.

n	Δ	y	α
1	10	0	0.97
2	48	1	0.88
3	18	1	0.77
4	9	0	0.81
5	5	0	0.84
6	5	1	0.80
7	11	0	0.82
8	4	0	0.84
9	5	0	0.85
10	4	0	0.87
11	5	0	0.88
12	5	0	0.89
13	6	1	0.87
14	16	0	0.88
15	7	0	0.89
16	6	1	0.88
17	18	0	0.89
18	7	0	0.89
19	6	0	0.90
20	6	0	0.90

Table 3.1

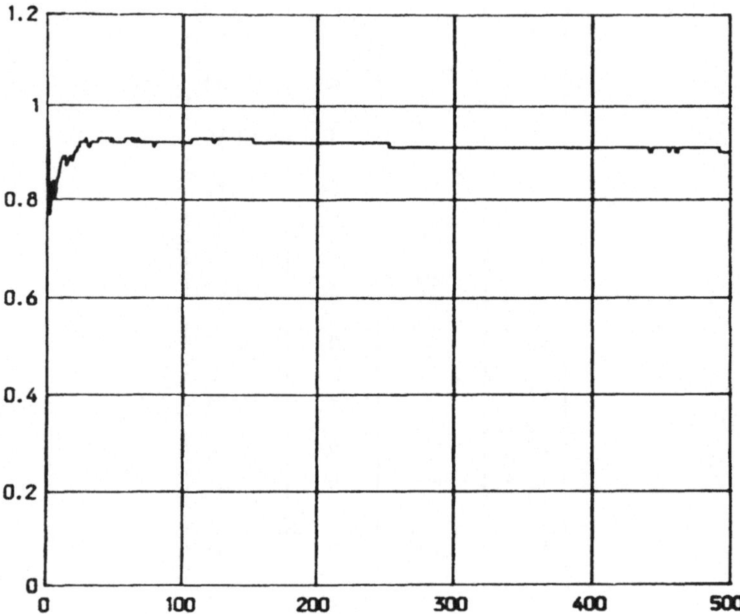

Figure 3.1 - α_n vs. n, $\alpha_0 = 0.90$

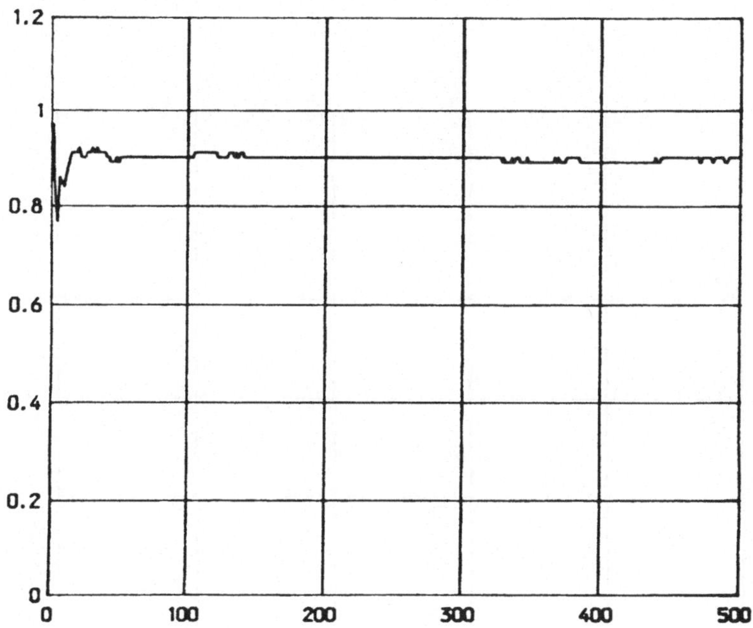

Figure 3.2 - α_n vs. n, $\alpha_0 = 0.90$, Δ random, $5 \le \Delta \le 20$

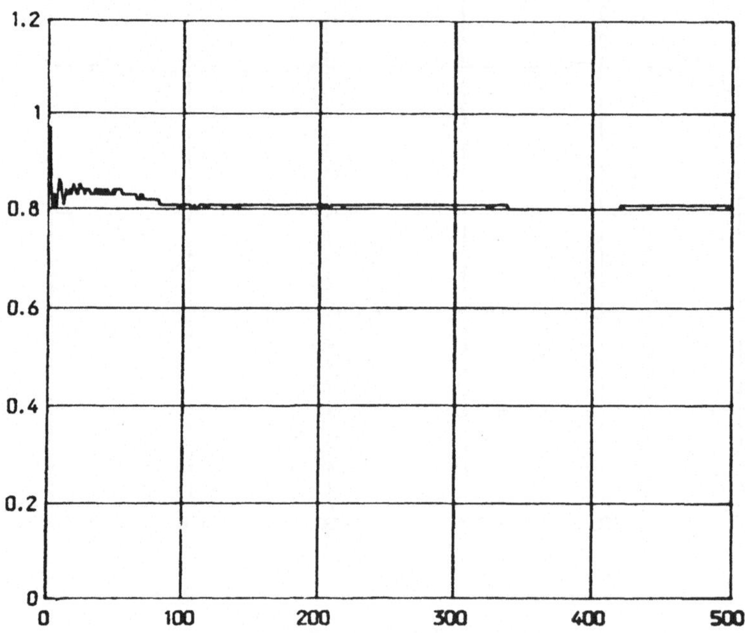

Figure 3.3 - α_n vs. n, $\alpha_0 = 0.80$, Δ random, $5 \leq \Delta \leq 20$

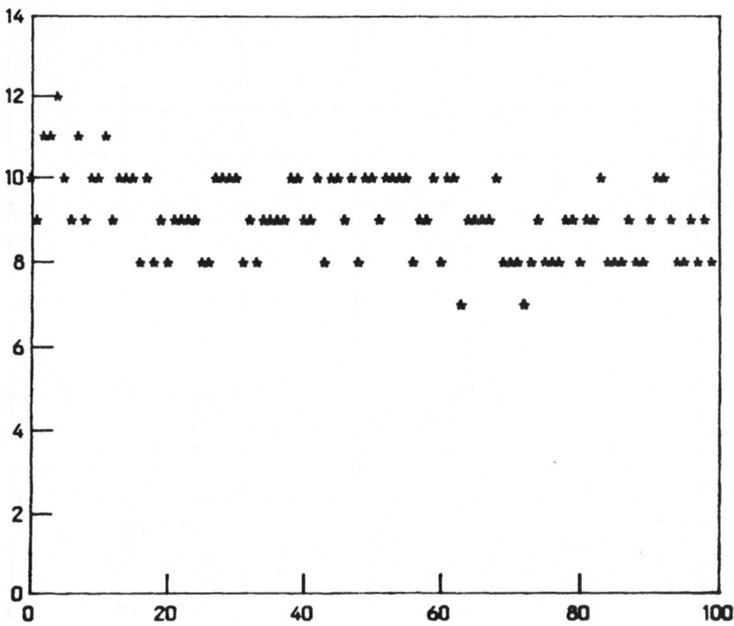

Figure 3.4 - Δ_n vs. n

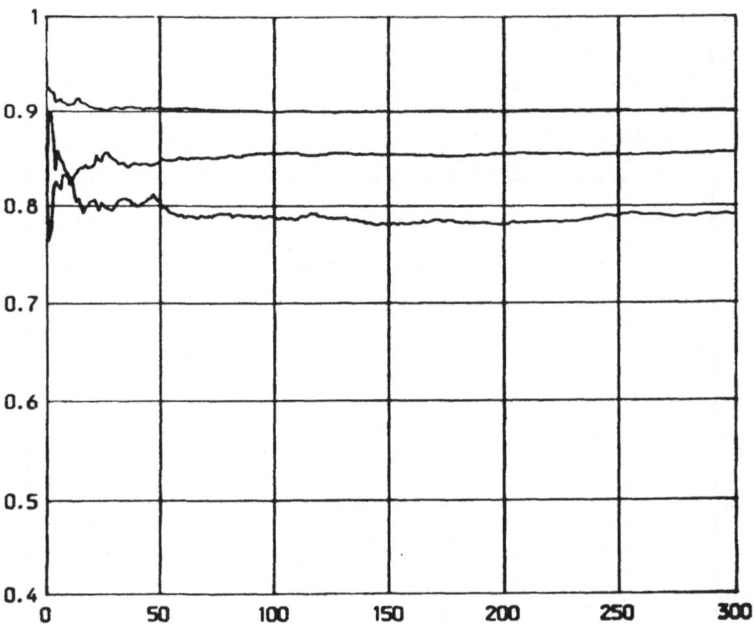

Figure 3.5 - α_n vs. n, $\alpha_0 = (0.90, 0.85, 0.80)$

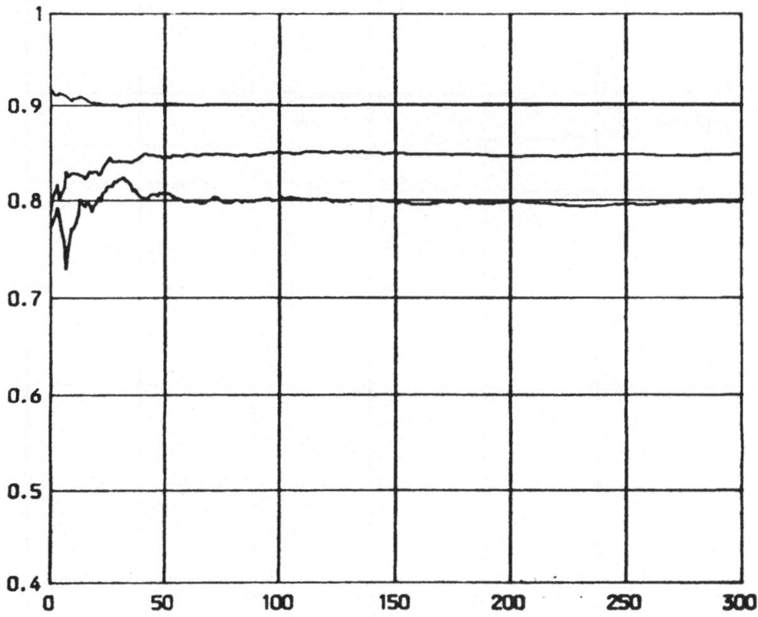

Figure 3.6 - α_n vs. n, α_0 = (0.90, 0.85, 0.80), τ = Identity Matrix

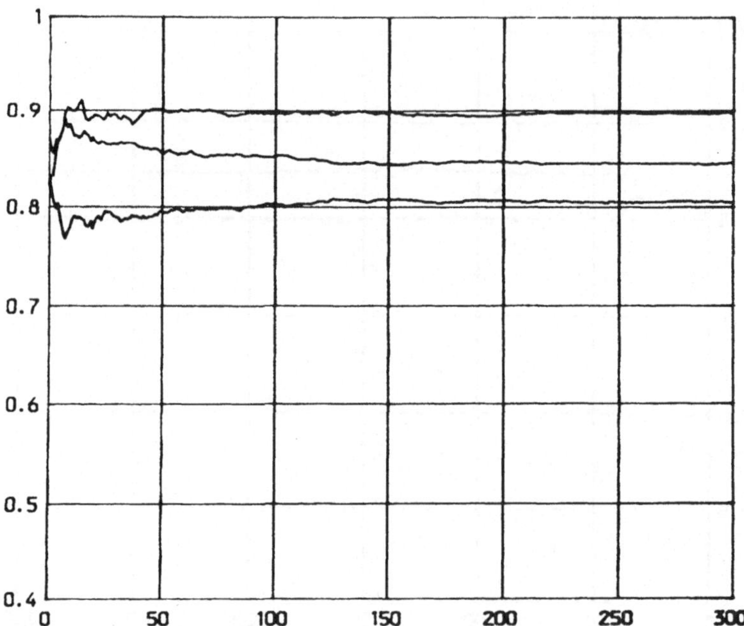

Figure 3.7 - α_n vs. n, α_0 = (0.80, 0.85, 0.90), N_c = 100

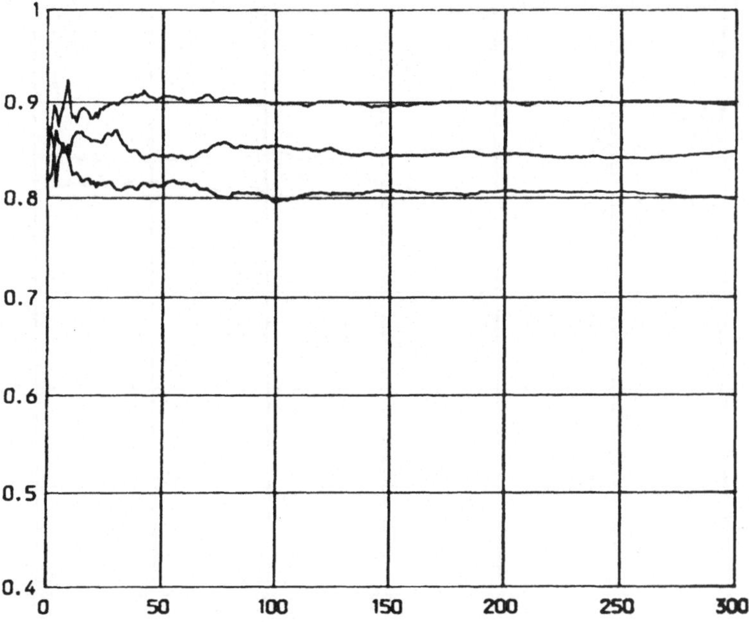

Figure 3.8 - α_n vs. n, $\alpha_0 = (0.80, 0.85, 0.90)$, $N_c = 50$

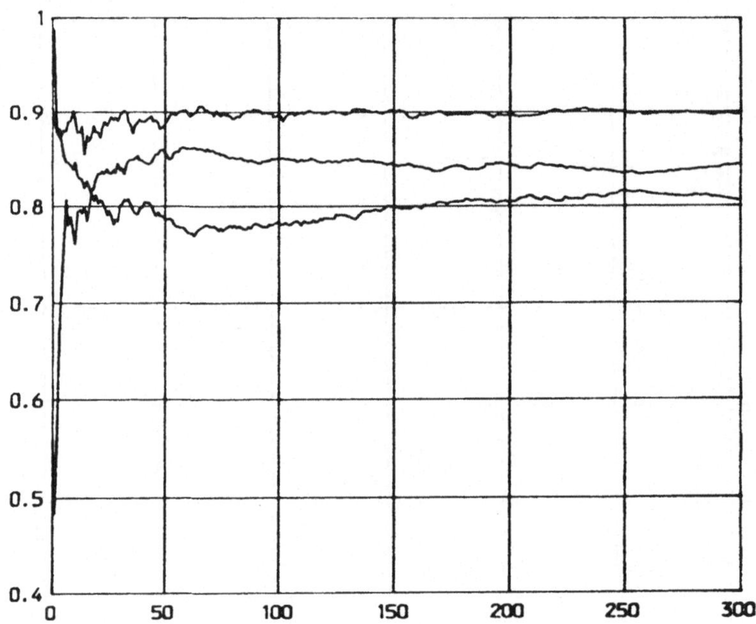

Figure 3.9 - α_n vs. n, $\alpha_0 = (0.80, 0.85, 0.90)$, $N_c = 20$

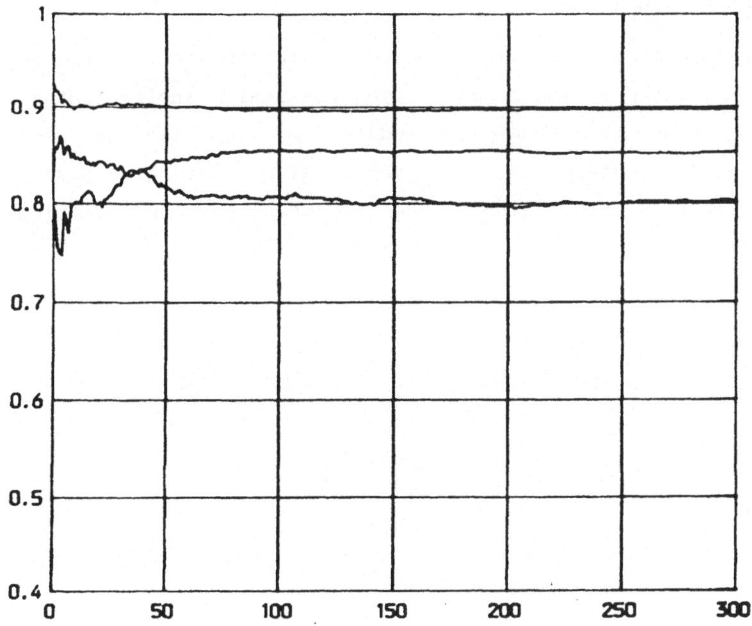

Figure 3.10 - α_n vs. n, α_0 = (0.90, 0.85, 0.80), Δ random, $8 \le \Delta \le 13$

4. CONCLUSION

We have considered the problem of parameter estimation in cumulative damage processes which are governed by Markov decay statistics. We have found that the maximum likelihood estimate, in the situations discussed, converges almost-surely to the true parameters under a very general class of maintenance policies.

We have demonstrated via computer simulation that, although the exact maximum likelihood estimate can become numerically inefficient as the number of parameters increases, a modified maximum likelihood method may still be used to obtain consistent parameter estimation. There is a tradeoff, in this case, between speed of convergence and numerical complexity.

A System Identification Approach to the Detection of Changes in Structural Parameters

M.S. Agbabian   S.F. Masri   R.K. Miller
Department of Civil Engineering
University of Southern California
Los Angeles, California 90089-0242

T.K. Caughey
Div. of Engineering and Applied Science
California Institute of Technology
Pasadena, California 91109

1 INTRODUCTION

1.1 Motivation

Nondestructive evaluation (NDE) methods for the detection of damage in structural systems have been receiving increasing attention in the recent past. Among the promising NDE methods are those based on the analysis of structural dynamic response measurements to identify a suitable mathematical model corresponding to the (changing) state of the physical structure.

The use of system identification (SI) approaches for NDE problems has expanded in recent years due to (1) development of practical analytical methods for the solution of inverse problems related to the identification of mathematical models on the basis of experimental response measures, and (2) advances in data processing and signal analysis capabilities brought on by developments in sensor technology and the continually increasing computational power and economy of microprocessors.

1.2 Background

The potential of using SI approaches for damage detection in structures has been recognized by many researchers. Similar approaches have been used extensively in other fields, including (among others) nonintrusive medical diagnostics using tomology and geophysical prospecting using seismology. Noteworthy recent efforts to apply formal SI concepts to the problem of damage detection in structures include the works of Yao (1985), Stephens (1985) and Natke (1986).

Several technical challenges play a major role in SI applications to structural damage detection. For example, uniqueness and observability problems are inherent in many structural systems due to the presence of redundant structural members and limited sensor locations. In addition, the available physical measurements from the sensors always contain small amounts of noise superimposed on the desired signal. The presence of such noise can lead to serious problems with the accuracy and reliability of various SI algorithms. Finally, the available response measures of some structural systems are inherently insensitive to changes in structural parameters of interest, making it difficult to devise an adequate test arrangement.

1.3 Scope of Paper

This paper explores the potential of a time-domain identification procedure to detect structural changes on the basis of noise-polluted measurements. The method of approach requires the use of excitation and acceleration response records, to develop an equivalent multi-degree-of-freedom (MDOF) mathematical model whose order is compatible with the number of sensors used. Application of the identification procedure under discussion yields the optimum value of the elements of an equivalent linear system matrices. By performing the identification task before and after potential structural changes (damage) in the physical system have occurred, (hopefully) quantifiable changes in the identified mathematical model can be detected.

Section 2 of this paper presents the formulation for the time-domain identification procedure under discussion. The usefulness of this approach for damage detection is demonstrated in Section 3 where an example 3 DOF linear system is used to conduct synthetic experiments to generate noise polluted "data" sets that are subsequently analyzed to determine the mean, variance, and probability density function corresponding to each element of the identified system matrices. To gain some insight into the "physical" model, different versions of the model were investigated in which the location as well as the magnitude of the "damage" was varied.

2 TIME-DOMAIN IDENTIFICATION PROCEDURE

Consider a discrete MDOF linear system whose motion is assumed to be governed by the set of differential equations

$$M\ddot{\underline{x}}(t) + C\dot{\underline{x}}(t) + K\underline{x}(t) = \underline{f}(t),\tag{1}$$

where

$$\begin{aligned}
\underline{x}(t) \quad &= \text{system displacement vector of order } n;\ (x_1, x_2, ..., x_n)^T, \\
M, C, K \quad &= \text{mass, damping and stiffness matrices, each of order } n \text{ by } n, \\
\underline{f}(t) \quad &= \text{excitation vector of order } n.
\end{aligned}$$

Assuming that experimental measurements of $\ddot{\underline{x}}(t)$ and $\underline{f}(t)$ are available for a given time span, then, by integrating the accelerations with respect to time and by taking a sufficient number of measurements of the system excitation and response, the following set of algebraic equations is obtained:

$$R\underline{\alpha} = \underline{b},\tag{2}$$

where R is a matrix whose entries correspond to the system response at various observation times, $\underline{\alpha}$ is a vector containing all the unknown influence coefficients of the system, and $\underline{b}$ is a vector containing the excitation measurements.

Using least-squares approximation methods, the parameter vector $\underline{\alpha}$ can be computed from

$$\underline{\alpha} = R^{\dagger}\underline{b},\tag{3}$$

where $R^{\dagger}$ is the pseudoinverse of R. Further details regarding this approach are available in the work of Masri et al (1987).

3. APPLICATION

3.1 Example System Characteristics

To illustrate the method of approach under discussion, consider the model shown in Fig. (1). This one-dimensional structure consists of three unequal masses m_i that are interconnected and anchored to their supports by six linear elements. Each of these elements, denoted by g_i, consists of a linear spring k_i and viscous dashpot c_i.

The magnitudes of the system masses as well as the material properties of the model elements are tabulated in Fig. 1(b). The "exact" system matrices for this example structure are shown in Fig. 1(c).

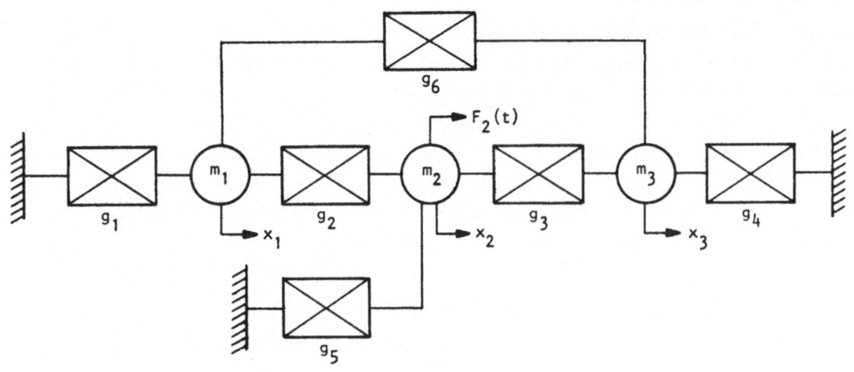

(a) Configuration

$$m_1 = 0.8 \qquad m_2 = 2.0 \qquad m_3 = 1.2$$

element number i	1	2	3	4	5	6
stiffness, k_i	2.0	1.0	1.0	2.0	2.0	1.0
damping, c_i	0.2	0.1	0.1	0.2	0.2	0.1

(b) Material Properties

$$M = \begin{pmatrix} m_1 & 0 & 0 \\ 0 & m_2 & 0 \\ 0 & 0 & m_3 \end{pmatrix} = \begin{pmatrix} 0.8 & 0 & 0 \\ 0 & 2.0 & 0 \\ 0 & 0 & 1.2 \end{pmatrix}$$

$$C = \begin{pmatrix} c_1 + c_2 + c_6 & -c_2 & -c_6 \\ -c_2 & c_2 + c_3 + c_5 & -c_3 \\ -c_6 & -c_3 & c_3 + c_4 + c_6 \end{pmatrix} = \begin{pmatrix} 0.4 & -0.1 & -0.1 \\ -0.1 & 0.4 & -0.1 \\ -0.1 & -0.1 & 0.4 \end{pmatrix}$$

$$K = \begin{pmatrix} k_1 + k_2 + k_6 & -k_2 & -k_6 \\ -k_2 & k_2 + k_3 + k_5 & -k_3 \\ -k_6 & -k_3 & k_3 + k_4 + k_6 \end{pmatrix} = \begin{pmatrix} 4 & -1 & -1 \\ -1 & 4 & -1 \\ -1 & -1 & 4 \end{pmatrix}$$

(c) Exact Matrices

Fig. 1 Characteristics of example 3 DOF system.

3.2 Generation of Synthetic Data Set

The method under consideration imposes no restrictions on the nature of the excitation source to be used as a probing signal. In the present "test", mass m_2 was subjected to a stationary, zero-mean wide-band random excitation.

For realistic simulation, the time histories of the applied excitation as well as the accelerations, velocities and displacements of the three masses were noise polluted. This was accomplished by adding to each data vector a corresponding noise vector whose rms level was equal to 5% of the rms of the unpolluted data vector. The components of all the noise vectors were uncorrelated, had a zero-mean and a Guassian distribution.

3.3 Identification of Original System Matrices (Case 1)

At this stage, the SI procedure outlined in Section 2 can be used to directly identify the system matrices corresponding to M, C and K without imposing any restrictions on the nature of these matrices. However, to simplify the illustration, it will be assumed that: (1) the mass matrix is known and need not be identified, and (2) matrices C and K are symmetric and need to be identified.

Re-writing Eq. (1) in the form

$$C\dot{\underline{x}}(t) + K\underline{x}(t) = \underline{f}(t) - M\ddot{\underline{x}}(t), \tag{4}$$

taking the parameter vector $\underline{\alpha}$ to be

$$\underline{\alpha} = (c_{11}, c_{12}, c_{13}, c_{22}, c_{23}, c_{33}, k_{11}, k_{12}, k_{13}, k_{22}, k_{23}, k_{33}), \tag{5}$$

and applying the SI procedure of Section 2, results in the optimum values of matrices C and K corresponding to the time segment used for identification purposes.

By separately applying the same SI procedure to different time segments of the "test" under discussion, an ensemble of matrices will be identified. Subsequent statistical analysis will yield the pertinent statistical measures such as mean, variance, and probability density function (pdf).

The SI procedure under discussion was applied to an ensemble of 600 data sets corresponding to distinct non-overlapping time segments of approximately two fundamental periods in duration. The results of these 600 independent tests were then used to find the mean, standard deviation, and the pdf associated with each element of matrices C and K.

For convenience, this data set will henceforth be referred to as Case 1. It will be used to establish the nominal (reference) values which will be used in three subsequent tests, to gauge the ability of the SI procedure under discussion to detect changes in the structural parameters. Furthermore, due to space limitations, this paper will focus the remaining discussion on the variation of the stiffness parameters (i.e., variations in the damping parameters will not be considered).

A summary of the identification results for this case is given in Fig. 2 in the form of 6 separate plots corresponding to the 6 distinct elements k_{ij} of symmetric matrix K. Noteworthy features of this composite figure are:

- The left-hand-side (LHS) columns of plots correspond to the diagonal elements k_{11}, k_{22}, k_{33}, while the three plots on the right-hand-side (RHS) correspond to the off-diagonal terms k_{12}, k_{13}, k_{23}.

- In each of the 6 individual plots, the abscissa corresponds to the sample (identification segment) number, and the ordinate corresponds to the identified magnitude of the indicated stiffness matrix element k_{ij}.

Fig. 2   Identification Results

Reference Case 1

Diagonal Elements

600 Individual SI Tests

Symmetry Assumption

Off−Diagonal Terms

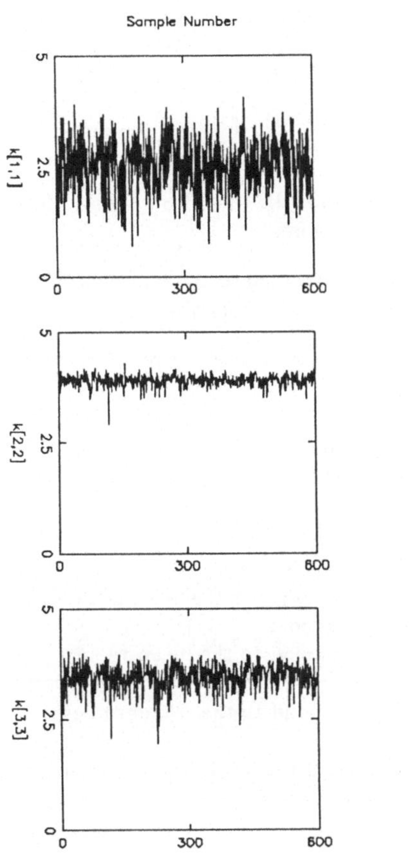

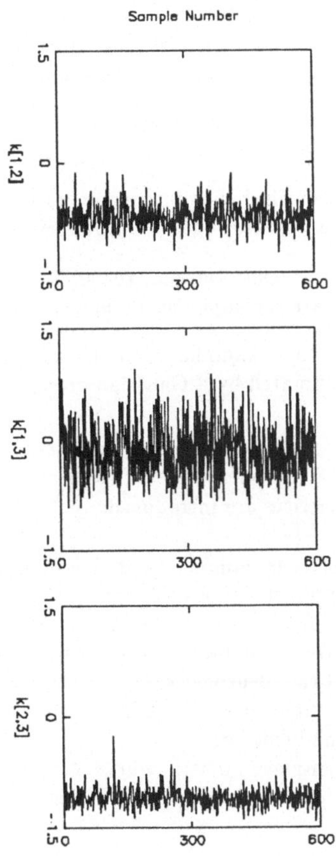

- The same range of abscissa (0 to 600) is used for all the plots.

- The same vertical sensitivity (0 to 5.0) is used for the LHS column of plots. Note that the "exact" values of the diagonal k_{ij} are all equal to 4.

- The same vertical sensitivity (-1.5 to +1.5) is used for the RHS column of plots. Note that the "exact" values of the off-diagonal k_{ij} are all equal to -1.

Fig. 3 shows a statistical analysis of the 600 identification tests carried on the noise-polluted data set corresponding to Case 1. Noteworthy features of this composite figure are:

- Each of the six individual plots shows the estimated probability density function of the indicated stiffness element, together with a superimposed curve corresponding to the pdf of a data set having the same mean and variance but with an exact Guassian distribution.

- For added resolution, each of the 6 individual figures shown has its own (different) horizontal and and vertical axis sensitivity. The horizontal axes correspond to the magnitude of k_{ij} while the vertical axes are a measure of the pdf.

- The horizontal axis in each of the 6 plots spans a range of three standard deviations ($\pm 3\sigma$) about the individual sample mean.

- The identified diagonal elements of K are shown in the LHS column plots, while the off-diagonal elements are plotted in the RHS column.

- Since, by definition, the area under the pdf curve is equal to unity, more sensitive horizontal scales are accompanied by less sensitive vertical scales, and vice versa.

- The actual randomness in the magnitude of the identification results can be reasonably approximated by a Gaussian process.

The mean values, standard deviations and percentage errors corresponding to each of the 6 elements of K associated with Case 1 are summarized in Fig. 4. To simplify the discussion, the following matrices are introduced:

K^* = exact (symmetric) stiffness matrix shown in Fig. 1(c),

ΔK = symmetric matrix whose entries correspond to percentage change in elements of K^* leading to the present K under discussion;

$\bar{K}$ = symmetric matrix whose elements $\bar{k}_{ij}$ correspond to the mean values of the identified k_{ij},

S = symmetric matrix whose elements s_{ij} correspond to the standard deviations of the identified k_{ij},

E = symmetric matrix whose elements correspond to the percentage error in the identified mean values relative to the exact values;

$$e_{ij} = \frac{\bar{k}_{ij} - k_{ij}}{k_{ij}} \times 100. \tag{6}$$

R = symmetric matrix whose elements correspond to the percentage change in the mean values of the identified results with respect to the corresponding mean values associated with the "undamaged" case (i.e.. Case 1)

Fig. 3 Probability Density Functions

Reference Case 1

Diagonal Elements

Analysis of 600 SI Tests

Symmetry Assumption

Off—Diagonal Terms

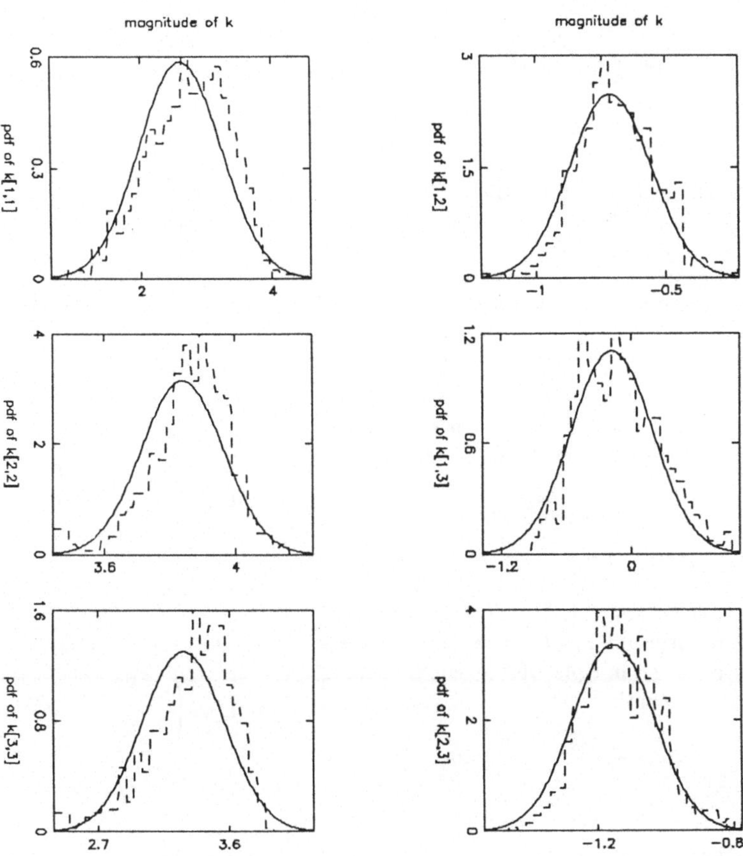

Fig. 4: Case Number: 1

Description: Reference case.

Formulation: $C\underline{\dot{x}}(t) + K\underline{x}(t) = \underline{f}(t) - M\underline{\ddot{x}}(t)$

Noise pollution level: 5%

Exact stiffness matrix for this case:

$$K^* = \begin{pmatrix} 4 & -1 & -1 \\ -1 & 4 & -1 \\ -1 & -1 & 4 \end{pmatrix}$$

Exact percentage change in elements of K^* relative to K corresponding to reference case shown in Fig. 1(c):

$$\Delta K = \begin{pmatrix} 0 & 0 & 0 \\ & 0 & 0 \\ & & 0 \end{pmatrix}$$

Mean value of identified K:

$$\bar{K} = \begin{pmatrix} 2.542 & -0.730 & -0.171 \\ & 3.898 & -1.090 \\ & & 3.443 \end{pmatrix}$$

Standard deviation of Identified K:

$$S = \begin{pmatrix} 0.611 & 0.160 & 0.366 \\ & 0.127 & 0.118 \\ & & 0.294 \end{pmatrix}$$

Percentage error in identified $\bar{K}$ relative to K^*:

$$E = \begin{pmatrix} -36.5 & -27.0 & -82.9 \\ & -2.6 & 9.0 \\ & & -13.9 \end{pmatrix}$$

Percentage change in $\bar{K}$ with respect to identified $\bar{K}$ corresponding to undamaged case:

$$r_{ij} \longleftarrow \frac{\bar{k}_{ij} - \bar{k}_{ij}^{(0)}}{\bar{k}_{ij}^{(0)}} \times 100$$

$$R = \begin{pmatrix} 0 & 0 & 0 \\ & 0 & 0 \\ & & 0 \end{pmatrix}$$

Percentage change in $\bar{K}$ with respect to identified $\bar{K}$ corresponding to undamaged case expressed as a multiple of the corresponding standard deviation.

$$d_{ij} \longleftarrow \frac{|\bar{k}_{ij} - \bar{k}_{ij}^{(0)}|}{s_{ij}} \times 100$$

$$D = \begin{pmatrix} 0 & 0 & 0 \\ & 0 & 0 \\ & & 0 \end{pmatrix}$$

Identified frequencies:

$$\underline{\omega} = \{1.172, 1.784, 1.855\}^T$$

Percentage change in frequencies relative to frequencies based on $\bar{K}$

$$\Delta\underline{\omega} = \{0, 0, 0\}^T$$

$$\Delta\omega \longleftarrow \frac{\omega_i - \omega_i^{(0)}}{\omega_i^{(0)}} \times 100$$

$$r_{ij} = \frac{\bar{k}_{ij} - \bar{k}_{ij}^{(0)}}{\bar{k}_{ij}^{(0)}} \times 100, \tag{7}$$

D = symmetric matrix whose elements correspond to the percentage change in $\bar{K}$ with respect to the identified $\bar{K}$ corresponding to the undamaged case expressed as a multiple of the corresponding standard deviation;

$$d_{ij} = \frac{|\bar{k}_{ij} - \bar{k}_{ij}^{(0)}|}{s_{ij}} \times 100, \tag{8}$$

$\underline{\omega}$ = vector of natural frequencies obtained by solving the eigenvalue problem associated with the exact mass matrix M and the identified stiffness matrix $\bar{K}$;

$$\underline{\omega} = (\omega_1, \omega_2, \omega_3)^T, \tag{9}$$

$\underline{\Delta\omega}$ = vector whose components represent the percentage change in frequencies relative to the frequencies corresponding to the "undamaged" (reference) case;

$$\Delta\omega_i = \frac{\omega_i - \omega_i^{(0)}}{\omega_i^{(0)}} \times 100. \tag{10}$$

It is seen from Fig. 4 that for the (reference) case under discussion, the noise pollution causes errors in the mean values of the identified k_{ij} ranging from -3% to -37% for the diagonal elements, and from -83% to +9% for the off-diagonal terms.

The identified natural frequencies for the reference case are:

$$\underline{\omega}^{(0)} = (1.172, 1.784, 1.855)^T. \tag{11}$$

The corresponding exact values based on M and K^* are:

$$\underline{\omega}^{(*)} = (1.173, 1.848, 2.354)^T. \tag{12}$$

Thus, the percentage error in the frequency estimates of $\underline{\omega}^{(0)}$ relative to $\underline{\omega}^{(*)}$ are:

$$\underline{\Delta\omega}^{(0)} = (-0.08\%, -3.46\%, -21.2\%)^T. \tag{13}$$

It is clear from Eq.(13) that the percentage error in the identified frequencies are substantially less than the percentage errors associated with the identified elements of the stiffness matrix $\bar{K}$.

3.4 Identification of Modified System (Case 2)

To get some insight into the detection capability of the identification technique under discussion, three separate cases were considered and are listed in Table 1.

The model considered in Case 2 was identical to the (reference) Case 1 model shown in Fig. 1, except that the stiffness of the element connecting m_1 to m_3 was changed from $k_6 = 1.0$ to $k_6 = 0.75$ (a reduction of 25% in the nominal value).

350

By again repeating the steps discussed in Section 3.2 (application of random excitation to m_2, measurements of the response, and noise-pollution of the data set), followed by the repetitive identification of 600 non-overlapping segments of the noisy measurements, as discussed in Section 3.3, the identification results summarized in Fig. 5 were obtained.

With reference to Fig. 5, it is seen that the 25% reduction in the stiffness of element k_6 from 1.0 to 0.75 contributes to changes in 4 elements of K: k_{11}, k_{13}, k_{31} and k_{33}. The exact percentage changes in K^* are given by matrix ΔK, where it is seen that elements k_{11} and k_{33} are reduced by 6.25%, while elements k_{13} and k_{31} are reduced by 25%, relative to the entries in the stiffness matrix shown in Fig. 1(c).

Analysis of the mean values in $\bar{K}$ and the standard deviations in S shows that the dispersion of the results for this case are similar to those of Case 1.

Table 1: Investigated Test Cases

Case Number	Changes Relative to Original Model Parameters in Fig. 1
1	None (reference case)
2	Reduce k_6 by 25% from 1.0 to 0.75
3	Reduce k_5 by 50% from 2.0 to 1.0
4	Reduce k_5 by 25% from 2.0 to 1.5

The percentage changes in the mean values of $\bar{K}$ with respect to the corresponding $\bar{K}$ elements associated with the reference Case 1, indicate a percentage change varying from -1.5% to +24%. Also, the percentage errors in the identification results as given by E are comparable to the corresponding values shown in Fig. 4.

A measure of the confidence level of the detected changes can be gleaned from an analysis of matrix D whose entries represent the quantitative change in $\bar{K}$ relative to the reference $\bar{K}$, expressed as a percent of the standard deviation of the corresponding element. It is seen that the range of the d_{ij} is from about 3% to 33% . This can be interpreted as saying that the largest observed change in magnitude in any of the elements of $\bar{K}$ is less than one third of the corresponding element standard deviation σ.

Note also that the structural changes in this case caused a change of about 2.5% in the identified ω_3 relative to $\omega_3^{(0)}$ corresponding to the reference case, and essentially no detectable changes in ω_1 and ω_2.

3.5 Identification of Modified System (Case 3)

The second "damaged" version of the model that was studied had the stiffness of element g_5 reduced from k_5 equal to 2.0 to 1.0 while all other parameters were identical to the model shown in Fig. 1. Notice that, unlike the previous Case 2 where the structural modification induced changes in 4 elements of the systems stiffness matrix, the present case involves a reduction of 50% in component k_5 thereby reducing the stiffness matrix element k_{22} by 25% from 4.0 to 3.0.

By repeating the testing, pollution, and identification procedures discussed in Sections 3.2 and 3.3, the results shown in Fig. 6 were obtained for this case.

Fig. 5: Case Number: 2

Description: Element k_6 changed from 1.0 to 0.75

Formulation: $C\underline{\dot{x}}(t) + K\underline{x}(t) = \underline{f}(t) - M\underline{\ddot{x}}(t)$

Noise pollution level: 5%

Exact stiffness matrix for this case:

$$K^* = \begin{pmatrix} 3.75 & -1 & -0.75 \\ & 4 & -1 \\ & & 3.75 \end{pmatrix}$$

Exact percentage change in elements of K^* relative to K corresponding to reference case shown in Fig. 1(c):

$$\Delta K = \begin{pmatrix} -6.25 & 0 & -25 \\ & 0 & 0 \\ & & -6.25 \end{pmatrix}$$

Mean value of identified K:

$$\bar{K} = \begin{pmatrix} 2.707 & -0.782 & -0.212 \\ & 3.902 & -1.051 \\ & & 3.390 \end{pmatrix}$$

Standard deviation of Identified K:

$$S = \begin{pmatrix} 0.535 & 0.161 & 0.304 \\ & 0.128 & 0.118 \\ & & 0.258 \end{pmatrix}$$

Percentage error in identified $\bar{K}$ relative to K^*:

$$E = \begin{pmatrix} -27.8 & -21.8 & -71.7 \\ & -2.5 & 5.1 \\ & & -9.6 \end{pmatrix}$$

Percentage change in $\bar{K}$ with respect to identified $\bar{K}$ corresponding to undamaged case:

$$r_{ij} \longleftarrow \frac{\bar{k}_{ij} - \bar{k}_{ij}^{(0)}}{\bar{k}_{ij}^{(0)}} \times 100$$

$$R = \begin{pmatrix} 6.5 & 7.1 & 24.0 \\ & 0.1 & -3.6 \\ & & -1.5 \end{pmatrix}$$

Percentage change in $\bar{K}$ with respect to identified $\bar{K}$ corresponding to undamaged case expressed as a multiple of the corresponding standard deviation.

$$d_{ij} \longleftarrow \frac{|\bar{k}_{ij} - \bar{k}_{ij}^{(0)}|}{s_{ij}} \times 100$$

$$D = \begin{pmatrix} 30.8 & 32.3 & 13.5 \\ & 3.1 & 33.0 \\ & & 20.5 \end{pmatrix}$$

Identified frequencies:

$$\underline{\omega} = \{1.172, 1.781, 1.901\}^T$$

Percentage change in frequencies relative to frequencies based on $\bar{K}$

$$\Delta\omega \longleftarrow \frac{\omega_i - \omega_i^{(0)}}{\omega_i^{(0)}} \times 100$$

$$\underline{\Delta\omega} = \{0, -0.2, 2.5\}^T$$

Fig. 6: Case Number: 3

Description: Element k_5 changed from 2.0 to 1.0

Formulation: $C\underline{\dot{x}}(t) + K\underline{x}(t) = \underline{f}(t) - M\underline{\ddot{x}}(t)$

Noise pollution level: 5%

Exact stiffness matrix for this case:

$$K^* = \begin{pmatrix} 4 & -1 & -1 \\ & 3 & -1 \\ & -1 & 4 \end{pmatrix}$$

Exact percentage change in elements of K^* relative to K corresponding to reference case shown in Fig. 1(c):

$$\Delta K = \begin{pmatrix} 0 & 0 & 0 \\ & -25 & 0 \\ & & 0 \end{pmatrix}$$

Mean value of identified K:

$$\bar{K} = \begin{pmatrix} 2.253 & -0.689 & -0.019 \\ & 2.888 & -1.086 \\ & & 3.292 \end{pmatrix}$$

Standard deviation of Identified K:

$$S = \begin{pmatrix} 0.622 & 0.150 & 0.376 \\ & 0.100 & 0.117 \\ & & 0.326 \end{pmatrix}$$

Percentage error in identified $\bar{K}$ relative to K^*:

$$E = \begin{pmatrix} -43.7 & -31.1 & -98.1 \\ & -3.7 & 8.6 \\ & & -17.7 \end{pmatrix}$$

Percentage change in $\bar{K}$ with respect to identified $\bar{K}$ corresponding to undamaged case:

$$r_{ij} \longleftarrow \frac{\bar{k}_{ij} - \bar{k}_{ij}^{(0)}}{\bar{k}_{ij}^{(0)}} \times 100$$

$$R = \begin{pmatrix} -12.8 & -5.6 & -88.9 \\ & -25.9 & -0.4 \\ & & -4.4 \end{pmatrix}$$

Percentage change in $\bar{K}$ with respect to identified $\bar{K}$ corresponding to undamaged case: expressed as a multiple of the corresponding standard deviation.

$$d_{ij} \longleftarrow \frac{|\bar{k}_{ij} - \bar{k}_{ij}^{(0)}|}{s_{ij}} \times 100$$

$$D = \begin{pmatrix} 46.5 & 27.3 & 40.4 \\ & 1000. & 3.4 \\ & & 46.3 \end{pmatrix}$$

Identified frequencies:

$$\underline{\omega} = \{0.997, 1.675, 1.790\}^T$$

Percentage change in frequencies relative to frequencies based on $\bar{K}$

$$\Delta\omega \longleftarrow \frac{\omega_i - \omega_i^{(0)}}{\omega_i^{(0)}} \times 100$$

$$\underline{\Delta\omega} = \{-14.9, -6.1, -3.5\}^T$$

Analysis of the summary results shown in Fig. 6 indicates that, due to the location and magnitude of the induced structural changes, more detectable changes can be observed, as can be seen from matrix R. Also, the confidence in the apparent changes is much higher, since as can be seen from element d_{22} of matrix D, the change in element $\bar{k}_{22}$ compared to its value in the reference Case 1 is 10 times the value of σ associated with the ensemble of 600 estimates of k_{22}.

Unlike Case 2, the changes made in component k_5 had a significant influence on the estimated value of ω_1, leading to about 15% reduction compared to ω_1 for the reference Case 1.

3.6 Identification of Modified System (Case 4)

This test case is similar to the previous Case 3 in all respects, except that the magnitude of the change in component k_5 was a 25% reduction from 2.0 to 1.5 leading to a 12.5% reduction in element k_{22} of the "undamaged" stiffness matrix.

The corresponding identification results for this case are summarized in Fig. 7. As would be expected, the statistical variation of the identification results are comparable to the previous case, while the magnitude of the observed changes have been reduced by about a factor of 2.

4 DISCUSSION

To simplify the discussion, the Gaussian distribution curves for the pdf associated with all 6 k_{ij} elements in each of the 4 test cases presented above, are shown in Fig. 8. As before, the LHS and RHS groups of the plots refer to the diagonal and off-diagonal elements respectively. In addition, the horizontal scale for each column of plots is constant for the different elements in the groups: the abscissa range is from 0 to 4.4 for the diagonal elements, and -2 to 1 for the off-diagonal elements. For added resolution, different vertical scales are used for the 6 plots in Fig. 8.

The following observations can be made in regards to Fig. 8:

- The wide difference in dispersion of identified parameters appears to be quantitatively related to the differences in known system masses, and therefore may be compensated by appropriate use of weighting matrices.

- The substantial errors in the mean values of the identified parameters are somewhat surprising, and are currently under investigation.

- The dispersions of the statistical variations in the estimates of the k_{ij} 's differ significantly from element to element.

- Superimposed pdf plots from different tests for elements that did not actually change (e.g., elements k_{12} and k_{23}) showed negligible quantitative variation in the shape of their pdf plots.

- A substantial shift appeared in the superimposed pdf plots for the k_{ij} elements that changed from test to test. Thus for Case 2, the corresponding pdf curves had a significant shift from the nominal case for elements k_{11}, k_{33} and k_{13}. This shift was substantially more than the scatter in the pdf curves associated with Cases 3 and 4, for the same elements. On the other hand, inspection of the pdf plots for k_{22} indicates that Case 2 did not result in any measurable change of the reference case, while the changes due to Cases 3 and 4 are very distinct.

- Even a relatively small change in the structural parameters (about 6% change in Case 2) can lead to detectable variations in the pdf curves of the influenced matrix elements, relative to their nominal values.

Fig. 7: Case Number: 4

Description: Element k_5 changed from 2.0 to 1.5

Formulation: $C\dot{\underline{x}}(t) + K\underline{x}(t) = \underline{f}(t) - M\ddot{\underline{x}}(t)$

Noise pollution level: 5%

Exact stiffness matrix for this case:

$$K^* = \begin{pmatrix} 4 & -1 & -1 \\ & 3.5 & -1 \\ & & 4 \end{pmatrix}$$

Exact percentage change in elements of K^* relative to K corresponding to reference case shown in Fig. 1(c):

$$\Delta K = \begin{pmatrix} 0 & 0 & 0 \\ & -12.5 & 0 \\ & & 0 \end{pmatrix}$$

Mean value of identified K:

$$\bar{K} = \begin{pmatrix} 2.421 & -0.715 & -0.105 \\ & 3.398 & -1.091 \\ & & 3.383 \end{pmatrix}$$

Standard deviation of Identified K:

$$S = \begin{pmatrix} 0.615 & 0.152 & 0.374 \\ & 0.108 & 0.112 \\ & & 0.303 \end{pmatrix}$$

Percentage error in identified $\bar{K}$ relative to K^*:

$$E = \begin{pmatrix} -39.5 & -28.5 & -89.5 \\ & -2.9 & 9.1 \\ & & -15.4 \end{pmatrix}$$

Percentage change in $\bar{K}$ with respect to identified $\bar{K}$ corresponding to undamaged case:

$$r_{ij} \longleftarrow \frac{\bar{k}_{ij} - \bar{k}_{ij}^{(0)}}{\bar{k}_{ij}^{(0)}} \times 100$$

$$R = \begin{pmatrix} -4.8 & -2.0 & -38.6 \\ & -12.8 & 0.1 \\ & & -1.7 \end{pmatrix}$$

Percentage change in $\bar{K}$ with respect to identified $\bar{K}$ corresponding to undamaged case: expressed as a multiple of the corresponding standard deviation.

$$d_{ij} \longleftarrow \frac{|\bar{k}_{ij} - \bar{k}_{ij}^{(0)}|}{s_{ij}} \times 100$$

$$D = \begin{pmatrix} 19.7 & 9.9 & 17.6 \\ & 463 & 0.9 \\ & & 19.8 \end{pmatrix}$$

Identified frequencies:

$$\underline{\omega} = \{1.091, 1.742, 1.822\}^T$$

Percentage change in frequencies relative to frequencies based on $\bar{K}$

$$\Delta\underline{\omega} = \{-6.9, -2.3, -1.8\}^T$$

$$\Delta\omega \longleftarrow \frac{\omega_i - \omega_i^{(0)}}{\omega_i^{(0)}} \times 100$$

Fig. 8 Probability Density Functions
4 Different Tests
Gaussian Distribution of k's
x range: 0 to 5
y range: variable

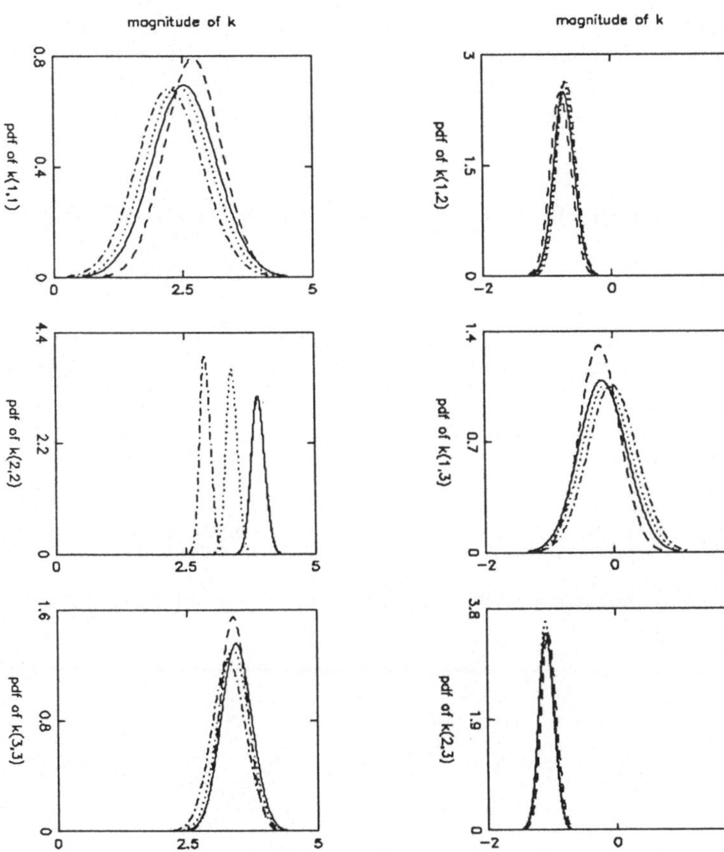

- Measures of the statistical variations of the identification results from test-to-test are more sensitive indicators of underlying changes in the test structure, than estimates of changes in the corresponding natural frequencies.

- By using the measures of change furnished by matrix R (which gives the percentage change from nominal conditions) and D (which denotes the significance of the changes relative to the statistical variation in the measurements) confidence limits may be established on observed changes in the system parameters.

5 SUMMARY AND CONCLUSIONS

As was stated earlier, this study was not meant to be a comprehensive treatment of the title subject, but rather an exploratory investigation of the usefulness of various indices, that can be conveniently extracted during an experimental test, to quantify changes in the characteristics of physical systems. On this basis it appears that computing the pdf of the identified quantities and subsequently constructing matrices R and D, discussed above, may furnish an effective tool to confidently gauge the location and extent of changes in a physical test article.

6 ACKNOWLEDGEMENTS

This study was supported by grants from the US National Science Foundation.

7 REFERENCES

Masri, S.F., Miller, R.K., Saud, A.F., and Caughey, T.K., (1987), "Identification of Nonlinear Vibrating Structures: Formulation", (to appear in the *Journal of Applied Mechanics. Trans. ASME*).

Natke, H.G., and Yao, J.T.P., (1986), "Research Topics in Structural Identification", Proc. Third Conference on Dynamic Response of Structures, EM Div./ASCE, UCLA, 31 March - 2 April.

Stephens, J.E., (1985), "Structural Damage Assessment Using Response Measurements", Ph.D. Thesis, School of Civil Engineering, Purdue University, West Lafayette, IN.

Yao, J.T.P., (1985), *Safety and Reliability of Existing Structures*, Pitman Advanced Publishing Program, Boston.

Time Domain Identification of Linear Structures

by

Vernon C. Matzen
Associate professor of Civil Engineering
North Carolina State University

1. ABSTRACT

A time domain system identification program that is able to determine the elements of the structural coefficient matrices in linear mathematical models is presented. A numerical investigation of a five story shear building is used to illustrate that the elements of the stiffness matrix can be correctly determined even when sensors are not placed at every degree of freedom and when single mode free vibration response data are used in the system identification procedure. Results from a laboratory experiment on a two story shear building show that the measured response at each degree of freedom can be matched almost perfectly using the parameters found in the identification process, but that the parameters found from one set of data are not always able to produce a response that matches well the response from another test on the same structure.

Information about damaged members can be obtained by studying the changes in the elements of the stiffness matrix. In structures more complicated than shear buildings, it will not usually be possible to define a "complete" model, and an "incomplete" model will have to be used. It is shown here that, when models are formulated using fewer DOFs than the number normally considered appropriate, damaged or modified members in a structure can not be precisely located. If the full set, or a more complete set, of DOFs is used on the other hand then, at least for the structures considered in this study, precise information can be obtained for individual elements of the coefficient matrices.

2. INTRODUCTION

System identification is a broad term designating that branch of numerical analysis that utilizes experimental input and output data to create or improve mathematicel models of "systems". For dynamic behavior of structural or mechanical systems, the mathematical models are sets of differential equations with coefficient matrices that describe the geometric and material properties of the systems. Professor Ewins calls models of this type _spatial_ models [1] (see the List of References in the Appendix).   If the behavior being modeled is linear or nearly so and there are certain restrictions on the type of damping, then according the Ewins there are two other types of models: _modal_ models and _response_ models. The overwhelming majority of applications today use modal models, and furthermore they utilize frequency domain methods to find the modal parameters.   There are times, however, when it may be desirable to use spatial models. Examples of such cases include the following: strongly nonlinear behavior, highly nonproportional damping, and isolated uncertainties in the mathematical model.

Besides the type of model, the user of system identification must select the type of excitation. For mechanical and small structural components, instrumented hammers are commonly used.   The use of these hammers and matching accelerometers with frequency domain methods are particularly well developed for determining modal parameters. Off-the-shelf hardware utilizing this approach is routinely used for machinery diagnostics, design verification, and system modification. For large structures, mechanical or hydraulic exciters can be used.

Even though frequency domain methods are undeniably valuable for many applications, they are not a panacea, - their use for nonlinear systems is not strictly appropriate, nonproportional damping cannot be handled by many systems, low frequency behavior created difficulties, and closely spaced modes cannot be sorted out, to name a few of the problems. In light of these shortcomings, I set out to investigate time domain methods to see if they might not be preferable in some applications.

The specific objective addressed in this paper is the following: Given a simple laboratory structure with well defined boundary conditions and an appropriate set of loadings and corresponding responses, formulate a spatial mathematical model that can accurately predict the structure's response to an arbitrary loading, not just the one used in the identification process.

3.   SYSTEM IDENTIFICATION ALGORITHM

Many approaches to system identification are described in the literature (see for example Hart and Yao [2] and Natke [3]). An early definition of system identification by George Bekey [4] creates a useful backdrop for a discussion of the method and how it is implemented in this work.

3.1 Form of the Model

The first step in Bekey's definition is the selection of the form of the model and of the parameters to be determined. In the current study, the structures are assumed to be linear, elastic, and viscously damped. The resulting discrete initial value problem takes the form:

$$[M]\{\ddot{x}\} + [C]\{\dot{x}\} + [K]\{x\} = \{P(t)\} \qquad (1)$$
$$\{\dot{x}\} = \{\dot{x}_o\}, \ \{x\} = \{x_o\}$$

The parameters to be determined are any combination of elements in the coefficient matrices. These "active" parameters are stored in a single vector designated $\{AP\}$. An alternative would have been to use member properties as parameters (see reference [5] for an example), but this approach was discarded as being too restrictive. If the structure is excited by an applied load, then all of the parameters (not really all, of course, since the matrices will be symmetric) are independent and hence can be obtained (in theory) through the identification process. If the response is caused by support excitation, or by an initial static loading and quick release, then the elements of the coefficient matrices will not be independent (in the system identification sense) and hence cannot be determined in the identification process.

3.2 Error Function

Bekey's second step is selection of a criterion for determining the goodness of fit between the measured response and the computed response. A weighted mean sum of the squared errors (WMSS) in accelerations was used in this work:

$$\text{WMSS(AP)} = \frac{1}{\text{NTS}} \sum_{j=1}^{\text{NDOF}} \sum_{i=1}^{\text{NTS}} W_j [\ddot{x}_{mj}(t_i) - \ddot{x}_{cj}(AP,t_i)]^2 \qquad (2)$$

where NTS is the number of time steps, NDOF is the number of degrees of freedom, $\ddot{x}_{mj}(t_i)$ is the measured acceleration at the jth DOF at the ith time step, $\ddot{x}_{cj}(AP,t_i)$ is the computed acceleration using the current set of parameters, and W_j is the weighting factor for the jth DOF. If an appropriate meaning were given to the weighting vector, then a statistical significance could in turn be given to the resulting error, but I have chosen not to do this. Instead, the weighting vector merely serves to indicate whether or not a sensor has been placed at a given degree of freedom (1 = sensor, 0 = no sensor).

3.3 Minimization Algorithm

The third step is the selection of an algorithm for minimizing the error function in terms of the parameters to be obtained. My expectation (which is sometimes unfounded!) is that I can always locate a set of initial parameters that is quite close to the minimizing set. To take advantage of this prospect, I use a modified Newton method [6] with cubic interpolation in the line searches. When the initial set of parameters is unusually bad, I resort to steepest descent for the first few iterations.

A computer program based on the system identification algorithm described above was written in Turbo Pascal on an IBM compatible personal computer with an 8087 math coprocessor. There are certainly limitations to working on a PC, but there

are advantages too, such as free computer time and easy access
to data from a PC-based data acquisition system.

There are many questions that can be asked about any particular application of system identification. A partial list
is given here:

1.   Are the test data reliable? (an entire paper could be
written on this topic!)
2.   Is the form of the model appropriate? (are the boundary
conditions properly accommodated? should joint flexibility be
considered? is the number of DOFs appropriate?)
3.   Are the parameters all identifiable? (are the parameters
independent? is the criterion function sensitive to changes in
the parameters?)
4.   Is the minimization algorithm sufficiently robust? (will
it converge in the presence of poor initial parameters, modeling errors, and noisy data?)
5.   If the program does converge, are the parameters unique?

Some of these questions are briefly addressed in this paper,
but the application here is straight forward enough that the
answers are not difficult.

4.   SIMULATED DATA

Verification of the computer program involved the use of
several sets of simulated data. In a simulated test presented
earlier [7], all elements in the mass, damping, and stiffness
matrices for a 2 DOF shear building (including an off diagonal
zero in the mass matrix) were found easily using data from a
forced vibration test.

Another simulated test was for a 5 story undamped shear
building.   The structure along with pertinent details is shown
in Figure 1 (all figures and tables are given in the appendix).
The excitation was from a pull-back-and-quick-release loading,
and the parameters obtained were the nine nonzero tridiagonal
terms of the stiffness matrix. When a single load was applied

at the top of the structure, and accelerations at each DOF were used in the error function, the program converged easily to the correct (assigned) set of stiffnesses.

A comment often made about pull-back loadings is that the resulting motion must contain contributions from each of the modes.   The reason for this presumably comes from the modal analysis approach - that is, if all of the modal parameters are available, then the mass, damping, and stiffness matrices can be reconstructed from the modal parameters using orthogonality and basic definitions of modal quantities. To investigate the effect of incomplete modal contribution in the measured re-sponse, an initial displacement equal to the first mode shape only was specified.   The mode shape was obtained to six-digit accuracy from the computer program CAL [8]. The response to this initial displacement was, as close as I could tell, exact first mode behavior, although the digits beyond six are clearly in error.

I looked at this identification problem from two points of view.   The first was to obtain the eigenvalues of the Hessian matrix of the error function at the minimum point on the error surface.   The eigenvalues of this matrix give the curvature of the error surface in the principal directions. If they are all positive, then the surface is convex and the point is at least a local minimum; on the other hand, if one or more of the eigenvalues is zero, then the minimum is not a point but a line and the minimum is not unique. When I tried this procedure for the five story structure, I found that the first 4 eigenvalues of the Hessian were on the order of E-3 to E-6 which were 9 to 12 orders of magnitude smaller that the 5th eigenvalue.   I considered these eigenvalues to be essentially zero, indicating that there was not enough information in single mode excitation to yield unique parameters.

The other way to look at this identification problem is less elegant, but no less useful - it is to pick an initial set of parameters and see what happens in the minimization process. The results from this study are given in Table 1. They show that, from a variety of starting points, the program converged to essentially the same set of parameters. The fact that this final set of parameters is somewhat different from the assigned set is a mystery that I have not yet solved. Recall that, for the more general initial displacement used in the previous example, the program did converge to the assigned set of parameters. As another point of reference, I have seen an identification problem where the minimum was not a point, but rather a valley of zero error. In that case, the program converged, but to different points in the valley for different starting points. The behavior of the minimization in the current problem indicates to me that the minimum is unique.

A second numerical study on this 5 story structure was to eliminate sensors at some of the DOFs. The first case was to use accelerations from stories 2, 3, 4, and 5 (by setting $w_1 = 0$). The program converged easily to the assigned set of parameters. Runs with accelerations from (2,4,5) and (3,5) were also successful. These results demonstrate that, at least in this simple example, the response need not be known at every degree of freedom. However, when accelerations were known at stories 4 and 5 only, the program still converged, but with great difficulty, and the stiffnesses for the lower stories were somewhat different from the assigned set.

4.1 Reduced Parameter Models

An alternative to finding all of the parameters using accelerations from only some of the DOFs, is to treat the structure as having the same number of DOFs as there are known accelerations. For example, a detailed finite element model may be needed to analytically model the elastic properties of a structure, but the mass may be physically lumped at only a few locations. In such cases (at least for dynamic testing) we might place the sensors at the dominant masses, and somehow "condense" out the remaining DOFs. Analytically, procedures such as Guyan Reduction [9] can be used. Experimental methods work well also when most of the mass is lumped at only a few locations (assuming high frequency response is not encountered). Using DOFs only at these locations, system identification will yield equivalent stiffnesses (substructure stiffnesses) for the structure in between the masses. If the mass is not physically lumped, then there are potential dangers in using either the analytical or the experimental methods. The nature of the response will dictate whether or not a useful reduced parameter model can be obtained.

To investigate this idea of reduced parameter models, accelerations from stories 1, 3, and 5 from the 5 DOF model were treated as if they were measurements from a 3 DOF model. The excitation was a transient harmonic forcing function applied at the top story. The frequency of the applied load was half that of the fundamental frequency of the 5 DOF model, thus virtually guaranteeing that the reduced parameter model would work well. Initial parameters were obtained in two ways: Guyan reduction and a method referred to as Incomplete Modes (described in references [10] and [1]). Initial and final sets of parameters from each method, as well as the frequencies and mode shapes of the resulting models are given in Table 2. Although not included here, a plot showed that the response of this reduced parameter model (using the final set of parameters) was virtually identical to the corresponding response from the 5 DOF model.

4.2 Damage Assessment

The potential of using a reduced parameter model approach as a tool in damage assessment was examined again using the 5 DOF model. This structure was modified in various ways, and then parameters were obtained for the 3 DOF reduced system for each case. (Had the parameters for the 5 DOF structure been obtained directly, the specific changes in the parameters would have been located exactly.) The modifications included increasing each story stiffness by 10 % one at a time, and then increasing masses at the first and second stories by 10 % one at a time. Accelerations at stories 1, 3 and 5 from the original structure were computed and used as measured quantities to find the parameters in the reduced system. Results from this part of the study are shown in Table 3. Final parameters in each case are shown, along with the percent difference from the original, undamaged structure. It can be seen that information about modified member properties is no longer distinct as it would be with a complete model, but is now smeared giving only a blurry view of the changes.

5.   LABORATORY EXPERIMENT

The objective of the laboratory tests was to compare properties of multistory shear buildings found using two different procedures. The first was to test _single_ story shear buildings experimentally, and then use this information to build up coefficient matrices for multi story buildings (see reference [11]).   The second procedure was to use system identification with data from the assembled structures. This second part is the one that will be described here.

5.1 Laboratory Procedure

The structure and test set up are shown in Figure 2. This particular structure has aluminum columns, but to the top set of columns had been applied a special tape (D-1-D) made for vibration damping applications. Standard laboratory tests were performed on the single story structures to determine their properties. (Masses and stiffnesses were determined analytically, the frequencies and damping ratios were found experimentally, and then the masses and stiffnesses were found a second time using a known added mass and measured frequencies.)

The assembled shear buildings were tested in free vibration. The initial displacements (measured using LVDTs) were obtained by hanging a weight on the structure at one of the stories with a string and then burning the string with a torch. Accelerations at each story were recorded on a Rapid System Digital Oscilloscope Peripheral [12] on a PC. The Rapid System is capable of recording 32,767 samples simultaneously on each of 4 channels, but we recorded only 2048 samples for each of the two accelerations. Two tests were performed on this structure — one in which the static load was hung on story 1 (file AD13), and the other in which the weight was attached to story 2 (file AD23). Figure 3 show the Rapid System display (showing only part of the data) for each of these tests.

Two problems became apparent during the testing. The first — finite rise time for the acceleration at the story where the load was applied — can be seen from the Rapid Systems displays. This problem is addressed below. The second was the rather poor repeatability of results in the early stages of the testing. This problem was almost entirely alleviated by using a torque wrench on all of the bolted connections.

5.2 Test Results and Interpretation

Although 2048 samples of data were available, the computer program is only able to use the first 300 because of current programing limitations. The problem of finite rise time of the measured accelerations was accommodated by skipping the first few milliseconds of response in the error function calculation. Using these measured accelerations and initial displacements, and initial parameters from models built up from the single story structures, the system identification program was able to determine the minimizing set of parameters easily. Initial and final sets of parameters as well as modal data for each test are shown in Table 4. Because pull-back-and-quick-release tests were used, only a subset of the parameters could be found. We chose to keep $M[1,1]$ and $M[1,2]$ fixed in both cases. The large variation between the final sets of properties from the two tests was not expected. The changes in the elements of the damping matrix are especially perplexing.

Figure 4 shows the measured response over 300 time steps for test AD13 as well as the simulated responses using both the original and the final sets of parameters. Except for the first few steps of response at DOF 1 (where the load had been applied) the measured and final simulated responses are virtually identical. To see how constant the behavior is over a longer time span, the response was integrated over 2000 time steps and compared to the measured response. This comparison for DOF 1 is shown in Figure 5. The comparison for DOF 2 was similar. The correlation remains quite good through 600 time steps, but it then begins to deteriorate. The reason for this deterioration is not clear, but it probably has to do with slight nonlinearities in the response.

The second test uses the same structure, but has the load applied at the second DOF. The final parameters from the previous test were used as initial parameters here. A plot of the measured response and the simulated response using this initial set of parameters is shown in Figure 6. The surprising thing about this figure is how poorly the responses match, since the structure has not been changed. The response using the final set of parameters is compared to the measured in Figure 7, and we see now that the correlation over 300 steps is now very good. It is not surprising that the two sets of parameters give almost identical frequencies, although the mode shapes do differ somewhat.

6.   CONCLUSIONS

This study showed that individual elements of the coefficient matrices could be obtained directly for small, well behaved structures, even when sensors are not placed at every DOF and when the excitation gives only single mode response. Additional research will be required to determine the behavior of the technique when it is used with more complicated structures and loadings. Perhaps the most significant result from this study is the difficulty, even for simple structural configurations, of extending the range of validity of a model from the set of data from which it was established to other sets of data. This, too, is an area that calls for further study.

7. APPENDIX

Figure 1. Five Story Simulated Shear Building

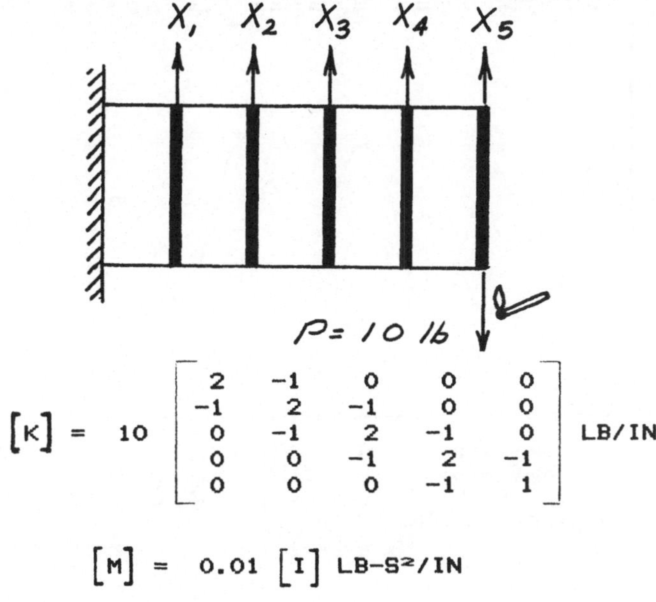

$$[K] = 10 \begin{bmatrix} 2 & -1 & 0 & 0 & 0 \\ -1 & 2 & -1 & 0 & 0 \\ 0 & -1 & 2 & -1 & 0 \\ 0 & 0 & -1 & 2 & -1 \\ 0 & 0 & 0 & -1 & 1 \end{bmatrix} \text{LB/IN}$$

$$[M] = 0.01\,[I]\ \text{LB-S}^2/\text{IN}$$

Figure 2. Two Story Structure Tested in the Laboratory

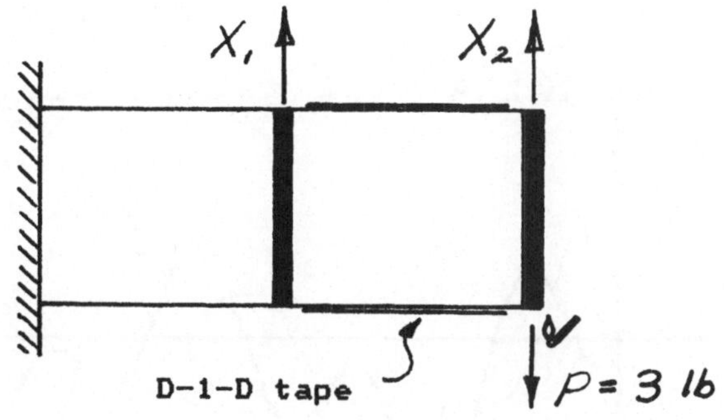

Initial displacements

AD13 (load at X_1): $\{X_o\} = \begin{Bmatrix} -0.1413" \\ -0.1413" \end{Bmatrix}$

AD23 (load at X_2): $\{X_o\} = \begin{Bmatrix} -0.1413" \\ -0.2832" \end{Bmatrix}$

Figure 3. Rapid Systems Displays for Tests AD13 and AD23

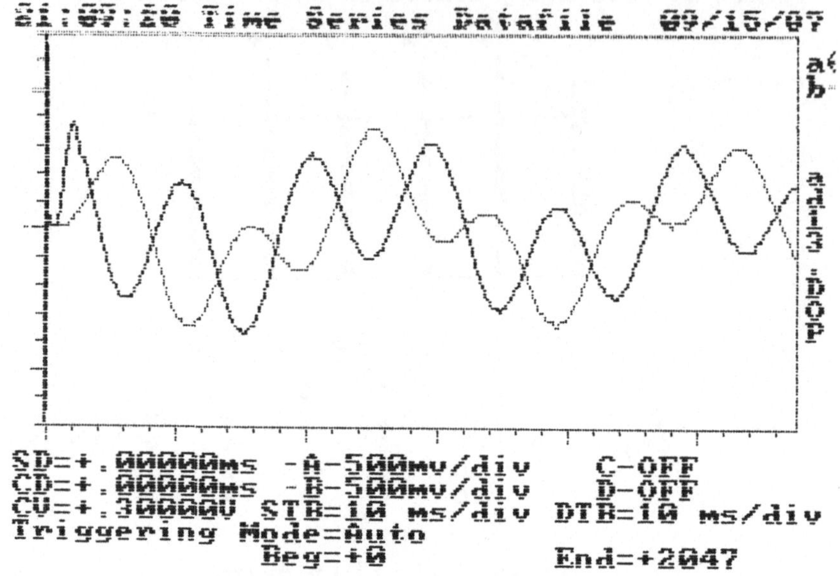

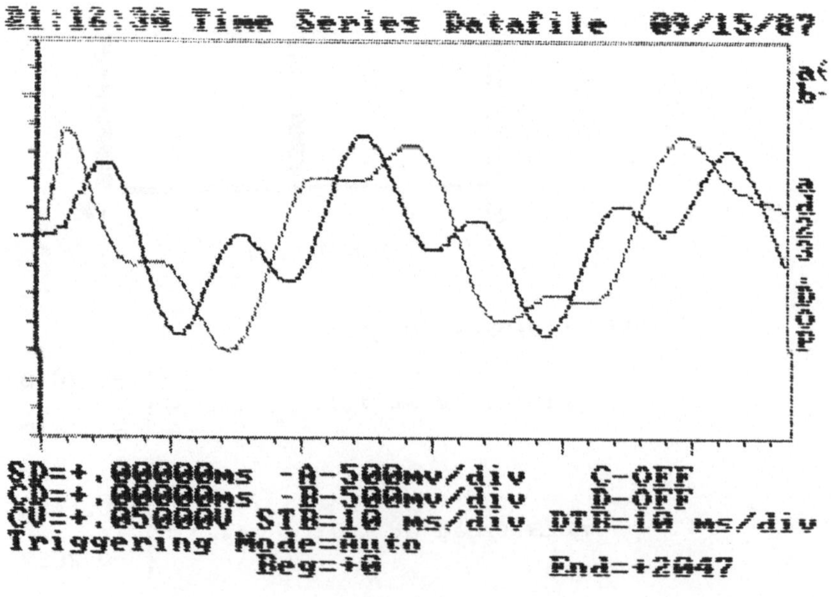

Figure 4. Measured and Simulated Response for 300 Steps for Test AD13

FILE: AD13, DOF 1

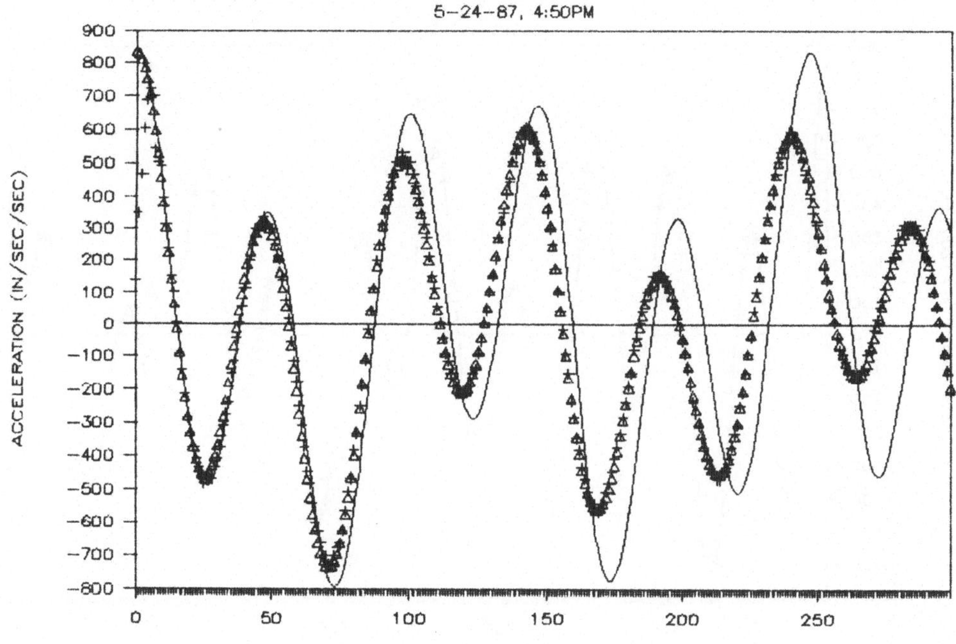

FILE: AD13, DOF 2

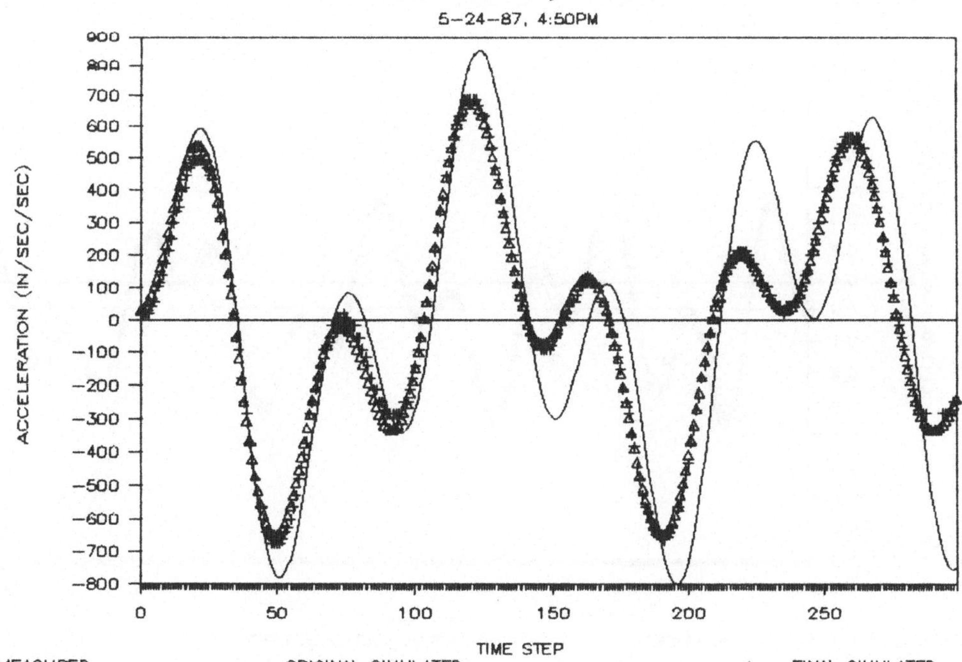

Figure 5a. Measured and Final Simulated Response for 2000 Steps of DOF 1 of Test AD13

FILE: AD13, DOF 1

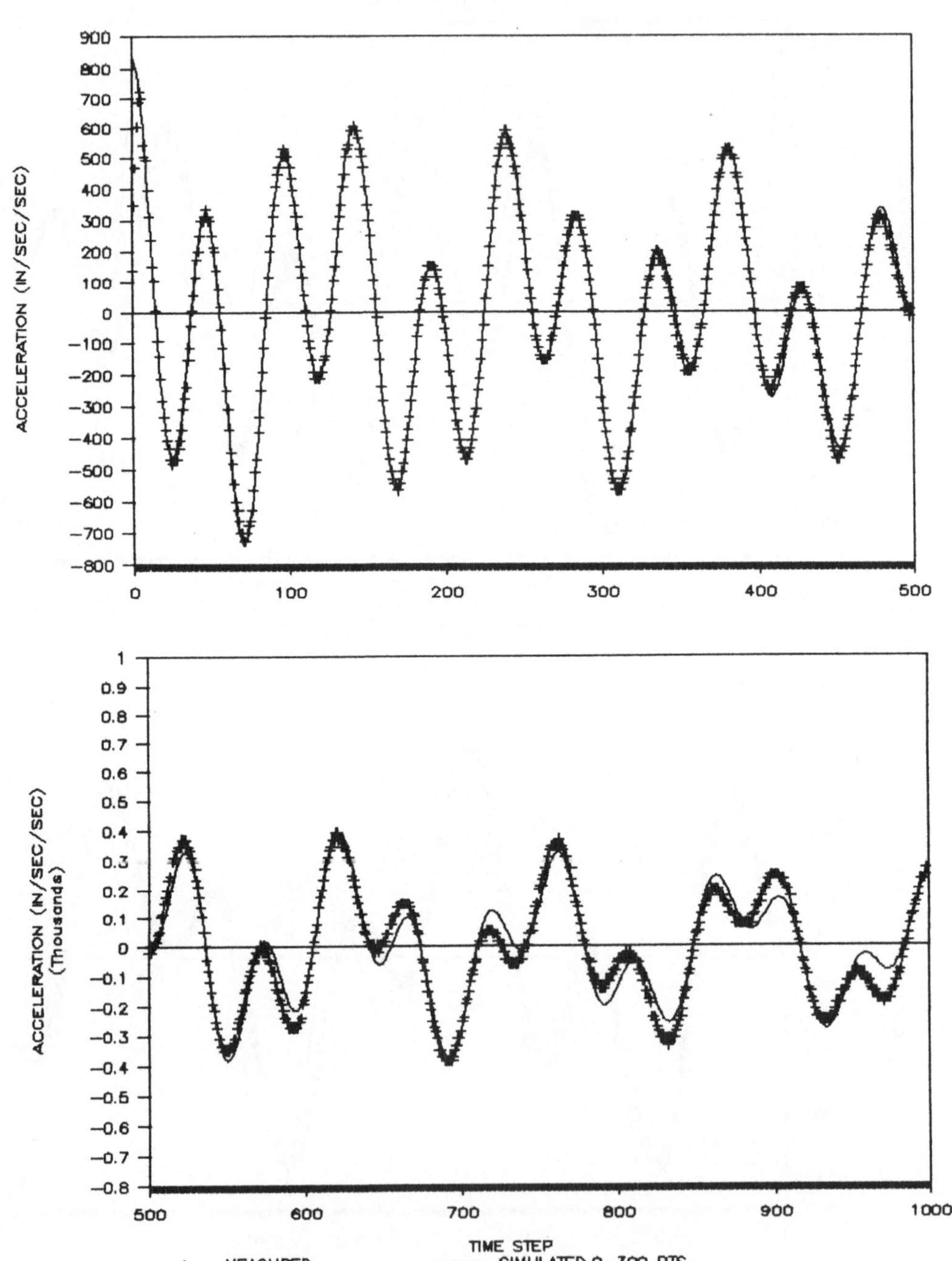

Figure 5b. Measured and Final Simulated Response for 2000 Steps of DOF 1 of Test AD13, Continued

FILE: AD13, DOF 1

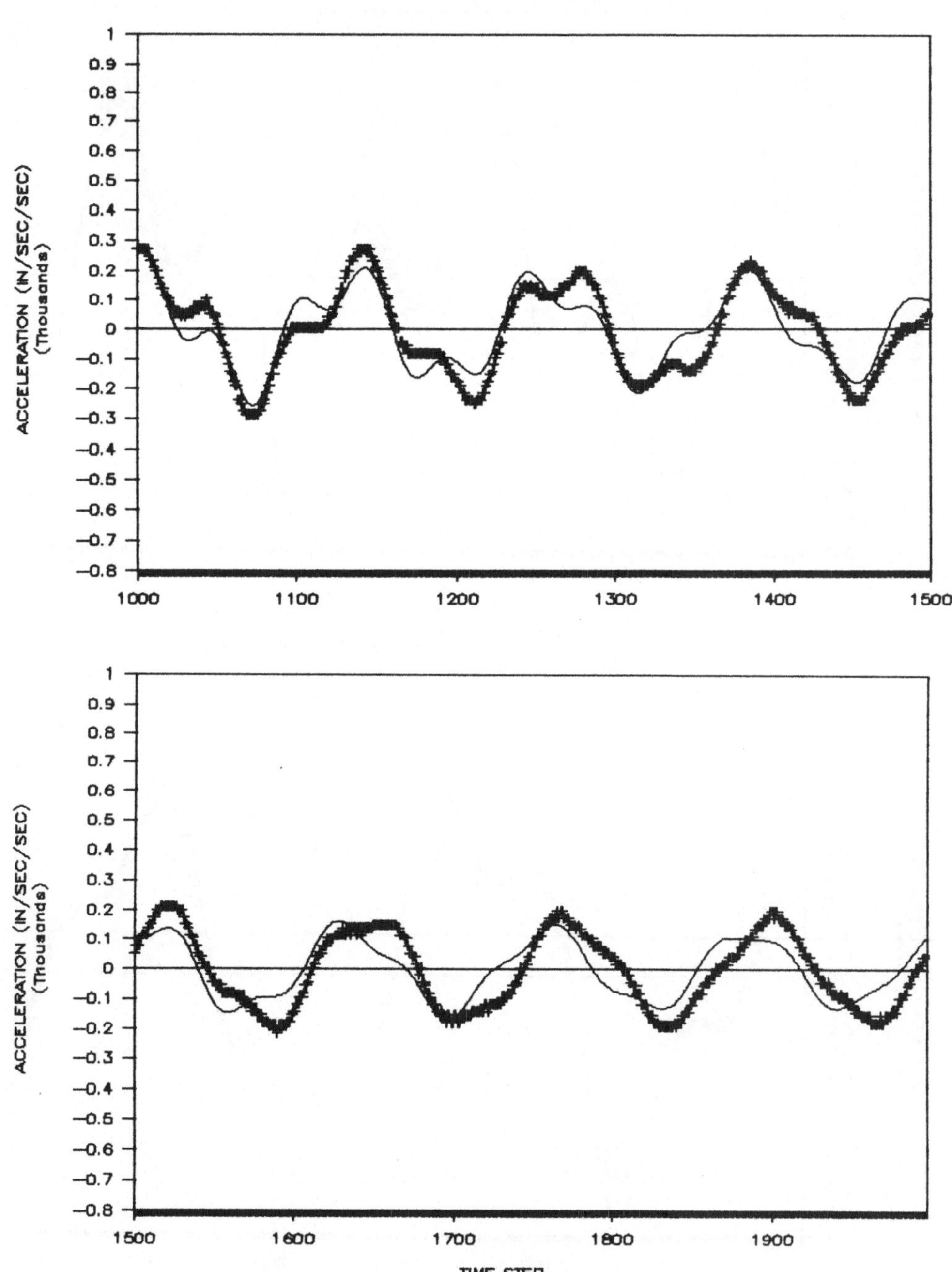

Figure 6. Measured and Original Simulated Response for
Test AD23

FILE: AD23, DOF 1

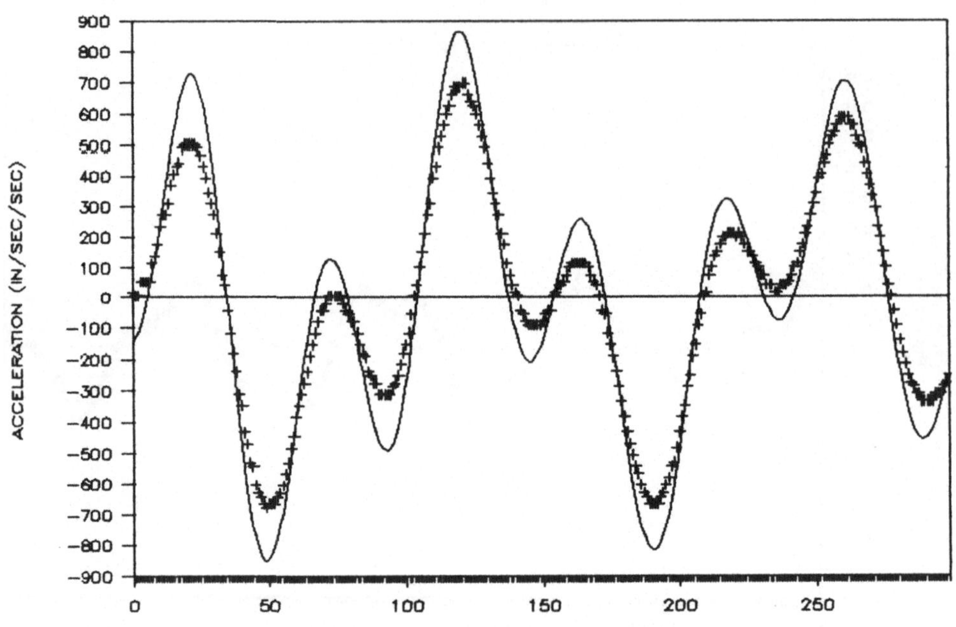

FILE: AD23, DOF 2

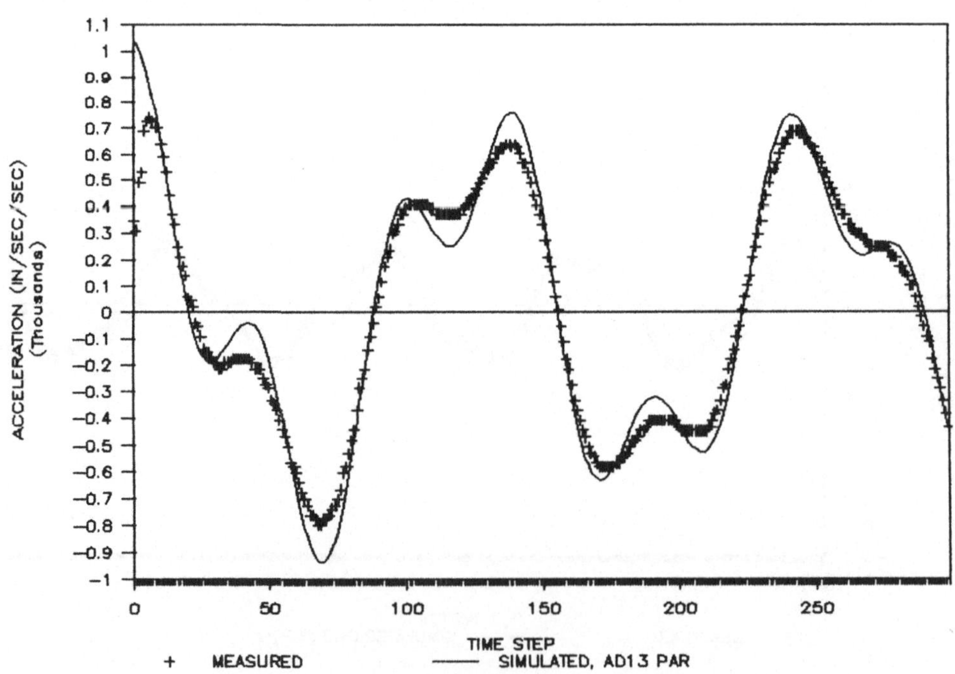

Figure 7. Measured and Final Simulated response for Test AD23

FILE: AD23, DOF 1

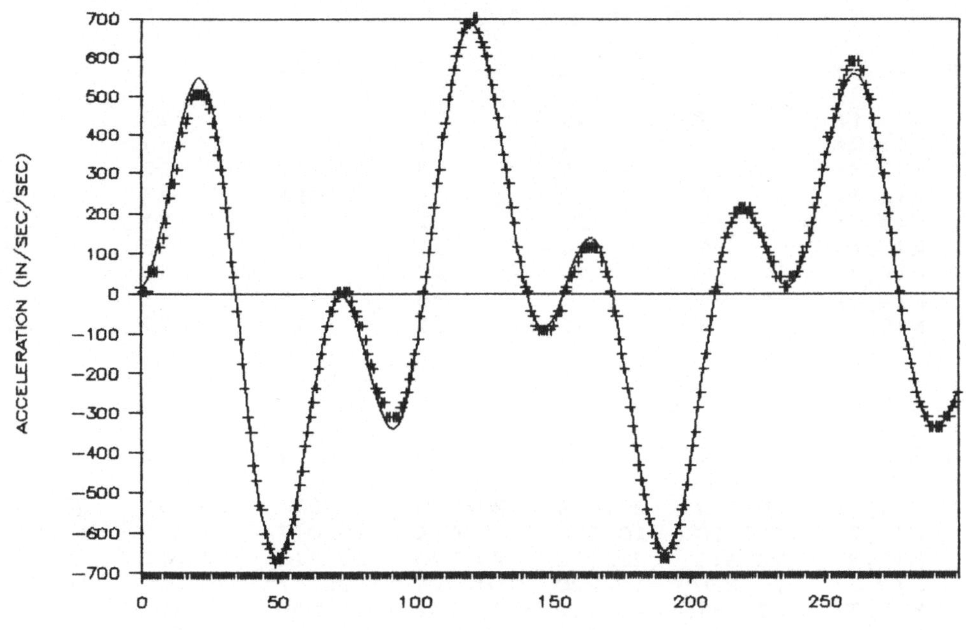

FILE: AD23, DOF 2

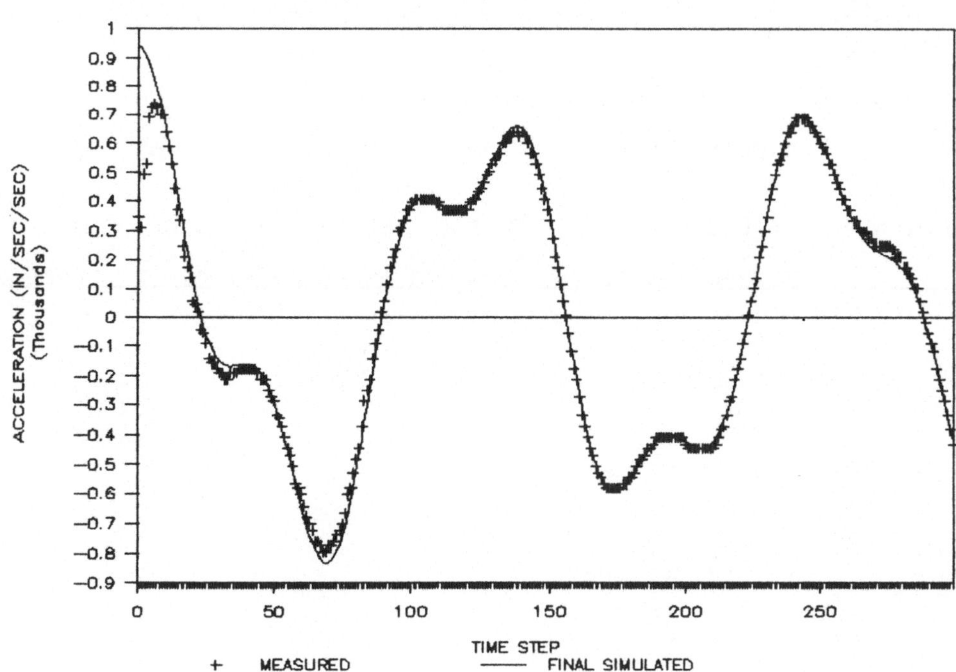

Table 1. Initial and Final Parameters for 5 DOF Structure using First Mode Excitation

I N I T I A L   P A R A M E T E R S

TRIAL -->	1	2	3	4	5	6
K(1,1)	19	21	23	21	18	16
K(1,2)	-9	-11	-12	-9	-9	-8
K(2,2)	19	22	18	19	22	22
K(2,3)	-10	-12	-8	-11	-12	-12
K(3,3)	18	21	21	22	21	24
K(3,4)	-9	-11	-12	-11	-12	-14
K(4,4)	19	22	22	18	19	16
K(4,5)	-8	-11	-8	-8	-8	-8
K(5,5)	9	12	11	11	12	12

F I N A L   P A R A M E T E R S

TRIAL -->	1	2	3	4	5	6
K(1,1)	20.025	20.034	20.028	20.033	20.019	20.033
K(1,2)	-10.010	-10.020	-10.014	-10.017	-10.010	-10.017
K(2,2)	20.316	20.316	20.320	20.313	20.322	20.305
K(2,3)	-10.220	-10.220	-10.224	-10.217	-10.227	-10.212
K(3,3)	20.161	20.167	20.162	20.160	20.152	20.160
K(3,4)	-10.000	-10.010	-10.002	-10.004	-9.992	-10.007
K(4,4)	20.196	20.195	20.194	20.193	20.200	20.194
K(4,5)	-10.180	-10.170	-10.177	-10.174	-10.190	-10.173
K(5,5)	10.164	10.159	10.163	10.160	10.175	10.159

F I N A L   S Q U A R E D   E R R O R S

0.00165	0.00165	0.00165	0.00165	0.00165	0.00165

F I N A L   S L O P E   O F   T H E   E R R O R   S U R F A C E

-9.7E-09	-4.3E-08	-1.0E-08	-5.8E-08	-1.4E-07	-2.9E-08

Table 2. Parameters for 3 DOF Reduced Model

GUYAN REDUCTION: INITIAL PARAMETERS   ERROR = 1077.0

$$[M] = \begin{bmatrix} 0.0125 & 0.0025 & 0.0000 \\ 0.0025 & 0.0150 & 0.0025 \\ 0.0000 & 0.0025 & 0.0125 \end{bmatrix} \quad [K] = \begin{bmatrix} 15.0000 & -5.0000 & 0.0000 \\ -5.0000 & 10.0000 & -5.0000 \\ 0.0000 & -5.0000 & 5.0000 \end{bmatrix}$$

GUYAN REDUCTION: FINAL PARAMETERS   ERROR = 148.3

$$[M] = \begin{bmatrix} 0.0161 & 0.0033 & -0.0021 \\ 0.0033 & 0.0018 & 0.0025 \\ -0.0021 & 0.0025 & 0.0125 \end{bmatrix} \quad [K] = \begin{bmatrix} 16.9260 & -4.7840 & -0.6407 \\ -4.7840 & 10.2700 & -5.1270 \\ -0.6407 & -5.1270 & 5.2040 \end{bmatrix}$$

INCOMPLETE MODES: INITIAL PARAMETERS   ERROR = 158.0

$$[M] = \begin{bmatrix} 0.0177 & 0.0024 & -0.0021 \\ 0.0024 & 0.0174 & 0.0028 \\ -0.0021 & 0.0028 & 0.0124 \end{bmatrix} \quad [K] = \begin{bmatrix} 18.8100 & -5.8990 & -0.4630 \\ -5.8990 & 10.6750 & -5.1140 \\ -0.4630 & -5.1140 & 5.1640 \end{bmatrix}$$

INCOMPLETE MODES: FINAL PARAMETERS   ERROR = 148.3

$$[M] = \begin{bmatrix} 0.0160 & 0.0033 & -0.0022 \\ 0.0033 & 0.0176 & 0.0025 \\ -0.0022 & 0.0025 & 0.0125 \end{bmatrix} \quad [K] = \begin{bmatrix} 16.8560 & -4.7360 & -0.6550 \\ -4.7360 & 10.2440 & -5.1190 \\ -0.6550 & -5.1190 & 5.2010 \end{bmatrix}$$

NUMBER OF DOFS	FREQUENCIES (HZ) 1	2	3	4	5
5	1.4325	4.1815	6.5917	8.4679	9.6581
3	1.4354	4.1617	6.6130		

STORY	MODE SHAPES 1	2	3	4	5
5 DOF STRUCTURE					
1	1.70	-4.56	-5.97	5.49	-3.26
2	3.26	-5.97	-1.70	-4.56	5.49
3	4.56	-3.26	5.49	-1.70	-5.97
4	5.49	1.70	3.26	5.97	4.56
5	5.97	5.49	-4.56	-3.26	-1.70
3 DOF STRUCTURE					
1	1.65	-4.55	-6.64		
3	4.55	-3.22	5.55		
5	5.98	5.46	-4.45		

Table 3. Parameters and Percent Error from Simulated Damage on 3 DOF Structures

PARAMETERS FROM UNMODIFIED EQUIVALENT 3 DOF STRUCTURE

M[1,1]	M[1,2]	M[1,3]	M[2,2]	M[2,3]	M[3,3]
0.01599	0.00329	-0.00215	0.01757	0.00254	0.01252

K[1,1]	K[1,2]	K[1,3]	K[2,2]	K[2,3]	K[3,3]
16.856	-4.736	-0.655	10.244	-5.119	5.201

PARAMETERS FROM STIFFNESS MODIFIED EQUIVALENT 3 DOF STRUCTURE

STORY CHANGED	M[1,1]	M[1,2]	M[1,3]	M[2,2]	M[2,3]	M[3,3]
1	0.01791	0.00314	-0.00243	0.01745	0.00266	0.01244
	12.0%	-4.7%	13.0%	-0.7%	4.7%	-0.6%
2	0.01771	0.00226	-0.00195	0.01684	0.00293	0.01228
	10.8%	-31.3%	-9.3%	-4.2%	15.4%	-1.9%
3	0.01451	0.00297	-0.00211	0.01808	0.00260	0.01256
	-9.3%	-9.9%	-2.0%	2.9%	2.4%	0.3%
4	0.01678	0.00294	-0.00197	0.01830	0.00242	0.01193
	4.9%	-10.7%	-8.4%	4.2%	-4.9%	-4.7%
5	0.01535	0.00384	-0.00209	0.01695	0.00223	0.01329
	-4.0%	16.7%	-2.8%	-3.5%	-12.1%	6.2%

STORY CHANGED	K[1,1]	K[1,2]	K[1,3]	K[2,2]	K[2,3]	K[3,3]
1	20.069	-5.096	-0.756	10.314	-5.082	5.181
	19.1%	7.6%	15.4%	0.7%	-0.7%	-0.4%
2	18.560	-6.192	-0.365	11.057	-5.097	5.122
	10.1%	30.7%	-44.3%	7.9%	-0.4%	-1.5%
3	16.793	-5.070	-0.653	10.053	-5.054	5.176
	-0.4%	7.1%	-0.3%	-1.9%	-1.3%	-0.5%
4	17.778	-5.314	-0.525	11.057	-5.684	5.705
	5.5%	12.2%	-19.8%	7.9%	11.0%	9.7%
5	16.228	-4.415	-0.768	10.734	-5.686	5.783
	-3.7%	-6.8%	17.3%	4.8%	11.1%	11.2%

PARAMETERS FROM MASS MODIFIED EQUIVALENT 3 DOF STRUCTURE

STORY CHANGED	M[1,1]	M[1,2]	M[1,3]	M[2,2]	M[2,3]	M[3,3]
1	0.01599	0.00352	-0.00206	0.01776	0.00237	0.01263
	0.0%	7.0%	-4.2%	1.1%	-6.7%	0.9%
2	0.01713	0.00376	-0.00238	0.01818	0.00245	0.01252
	7.1%	14.3%	10.7%	3.5%	-3.5%	0.0%

STORY CHANGED	K[1,1]	K[1,2]	K[1,3]	K[2,2]	K[2,3]	K[3,3]
1	15.990	-4.437	-0.649	10.242	-5.185	5.243
	-5.1%	-6.3%	-0.9%	-0.0%	1.3%	0.8%
2	17.354	-4.838	-0.698	10.379	-5.169	5.235
	3.0%	2.2%	6.6%	1.3%	1.0%	0.7%

Table 4. Parameters from Laboratory Experiments on a 3 DOF Structure

PRELIMINARY   PARAMETERS

MASS *   DAMPING   STIFFNESS

$$\begin{bmatrix} 0.003575 & 0.000000 \\ 0.000000 & 0.003575 \end{bmatrix} \quad \begin{bmatrix} 0.00000 & 0.00000 \\ 0.00000 & 0.00000 \end{bmatrix} \quad \begin{bmatrix} 42.37 & -21.14 \\ -21.14 & 21.14 \end{bmatrix}$$

FINAL   AD13   PARAMETERS

MASS   DAMPING   STIFFNESS

$$\begin{bmatrix} 0.003575 & 0.000000 \\ 0.000000 & 0.003550 \end{bmatrix} \quad \begin{bmatrix} 0.00006 & -0.00794 \\ -0.00794 & 0.01481 \end{bmatrix} \quad \begin{bmatrix} 45.73 & -24.59 \\ -24.59 & 25.29 \end{bmatrix}$$

MASS   DAMPING   STIFFNESS

$$\begin{bmatrix} 0.003575 & 0.000000 \\ 0.000000 & 0.003150 \end{bmatrix} \quad \begin{bmatrix} 0.01568 & -0.00128 \\ -0.00128 & 0.00033 \end{bmatrix} \quad \begin{bmatrix} 46.32 & -22.88 \\ -22.88 & 21.87 \end{bmatrix}$$

MODAL   PARAMETERS

	FREQUENCY (HZ)		MODE SHAPES	
	1	2	1	2
FINAL AD13	7.95301	20.97283	1	9.316	13.89
			2	13.95	-9.36
FINAL AD23	7.95147	20.95893	1	9.132	14.01
			2	14.93	-9.73

* NOTE: M[1,1] AND M[1,2] WERE NOT ACTIVE PARAMETERS.

List of References

1. Ewins, D. J., _Modal Testing: Theory and Practice_, Research Studies Press Ltd., Letchworth, Hertfordshire, England, 1984.

2. Hart, G. C., and Yao, J. T. P., "System Identification in Structural Dynamics," Journal of the Engineering Mechanics Division, ASCE, Vol 103, No. EM6, Dec. 1977.

3. Natke, H. G., "Multi-Degree-of-Freedom Systems -- A Review," _Identification of Vibrating Structures_, Courses and Lectures - No. 272, International Centre for Mechanical Sciences, Udine, Italy, Springer-Verlag, New York, 1982.

4. Bekey, George A., "System Identification - An Introduction and a Survey," _Simulation_, Vol 5, No. 4, October 1970, pp 151-166.

5. Matzen, V. C. and Hardee, Jr. J.E., "Mathematical Modeling of Indeterminate Trusses," Proceedings of the Second ASCE/EMD Specialty Conference on the Dynamic Response of Structures, January 1981, Atlanta, Georgia.

6. Luenberger, D. G., _Introduction to Linear and Nonlinear Programming_, Addison-Wesley Publishing Company, Reading, Massachusetts, 1973.

7. Matzen, V. C., "Time Domain Identification of Reduced Parameter Models," Proceedings of the 1987 Spring Conference on Experimental Mechanics, Houston, TX, June 14-17, 1987..

8. _CAL - 80, Computer Assisted Learning of Structural Analysis_, Edward L. Wilson, September 25, 1985. (Available from NISEE, University of California, Berkeley).

9. Bathe, K.-J., _Numerical Methods in Finite Element Analysis_, Prentice-Hall, Englewood Cliffs, NJ, 1982.

10. Matzen, V. C. and Murphy, C. E., "On Obtaining Mass Participation Factors Using 'Equivalent' Structures," The International Journal of Analytical and Experimental Modal Analysis, Vol. 1, No. 1, January 1986.

11. Wright, Brian, Report for Independent Study, CE 598, North Carolina State University, Raleigh, North Carolina, June 1987.

12. Rapid systems, Inc, 433 N. 34th St., Seattle, WA 98103.

Fuzzy Data Processing in Damage Assessment

Hitoshi Furuta and Naruhito Shiraishi
Department of Civil Engineering
Kyoto University, Kyoto 606, Japan

Abstract

In the safety and damage evaluation of structures, it is important to establish a rational method for handling or interpreting the available data obtained from inspection or experiment. Due to the vagueness or scarcity of available data, probabilistic methods may fail to provide a reliable model for the evaluation of damage state.

In this paper, attempts are made to apply the concept of fuzzy sets to data processing in the damage assessment of bridge structure. By using the fuzzy quantification theory, which was proposed for the discrimination of qualitative data, the inspection results obtained from the fact-finding inquiries are analyzed to classify the damage state of existing bridges. For the fatigue analysis of reinforced concrete bridge deck, a new interpretation of S-N curve is presented based on the concept of possibility distribution. Several numerical examples are presented to illustrate the procedure of data processing presented herein.

Keywords: Bridge Structure, Damage Assessment, Data Processing, Fatigue, Fuzzy Sets, Inspection, Quantification Theory, Regression Analysis

1 Introduction

The technology of designing and constructing bridge structures has been so remarkably developed in recent years that there arise some problems concerning methods for estimating structural safety from a new aspect. As many bridges are becoming superannuated, this issue is becoming more and more important in terms of maintenance (Shiraishi, Furuta and Sugimoto, 1985).

Since a considerable number of bridges have suffered from damage resulting from crack, corrosion and other types of deterioration, maintenance work is important to ensure their safety or serviceability.

To establish an appropriate maintenance program, it is necessary to evaluate their damage states in a quantitative manner. However, damage assessment of existing structures is not an easy task because of the scarcity of available data and the complex mechanism (Yao, 1979).

There are several kinds of information which are available in the damage assessment. For example, we can consider such information as 1) the examination of design documents, 2) the visual examination, 3) the field testing, 4) the laboratory testing, 5) the structural analysis. However, such information may not be available for all bridges, mainly due to the financial and technical constraints. While the second information (i. e. item 2)) is the easiest to collect among them, it is usually obtained in a qualitative manner which is meaningful but not precisely defined. In other words, the results of observation or inspection are generally reported in such forms as being ranked by A (severely damaged), B (damaged), C (slightly damaged), etc. The boundaries of A and B or B and C are not sharp even according to the existing inspection manuals.

In some cases, we can use the information obtained from the field testing or laboratory testing, whose quality and amount may be insufficient due to the financial and technical limitation. Especially, it is difficult to conduct the sufficient number of fatigue experiment for reinforced concrete bridge deck, because it requires a special loading condition and a large size of testing piece (Matsui, 1984).

In order to overcome the above problems concerning data available in the damage assessment of bridge structure, we attempt in this paper to apply the concept of fuzzy sets (Zadeh, 1965) to its data processing. By using the fuzzy quantification theory (Watada, 1983), which was proposed for the discrimination of qualitative data, the inspection results obtained from the fact-finding inquiries are analyzed to classify the damage state of existing bridges. For the fatigue analysis of reinforced concrete bridge decks, a new interpretation of S-N curve is presented based on the concept of possibility distribution (Zadeh, 1978). Several numerical examples are presented to illustrate the procedure of data processing presented herein.

2 Application of Fuzzy Quantification Theory

2.1 Fuzzy Quantification Theory

Hayashi's quantification analysis (Hayashi, 1984) is a kind of multivariate analysis. In the quantification theory there are three types of theory, among which type II theory is used in this paper. Type II theory was developed for the discrimination analysis, whereas type I and III theories were proposed for the regression and clustering analyses. The fuzzy quantification theory was proposed to deal with fuzzy data in the quantification theory. In many problems such as those of social and economic systems, it is difficult to obtain a quantitatively clear-cut expression of the information, because qualitative attributes derived from questionnaires are given in linguistic terminology which is vague and indistinct. Fuzzy sets are suitable for treating vagueness and ambiguity involved in the linguistic expression.

Type II fuzzy quantification theory aims to determine the linear discriminant function of fuzzy groups, which provides the maximum separation of these fuzzy groups in a real space. The discriminant function $y(x_\alpha)$ is defined as

$$y(x_\alpha) = \sum_{i=1}^{K} \sum_{k=1}^{1_i} a_{ik}\, \mu_{Aik}(x_\alpha), \quad \alpha = 1, 2, \cdots\cdots, n \qquad (1)$$

where μ_{Aik} is the membership function of fuzzy category A_{ik}, a_{ik} is a category weight, K is the total number of items, 1_i is the item number of the i-th category, and n is the sample number. The goodness of the discrimination is appraised by the fuzzy variance ratio η^2 of the between-groups variance to the total variance concerning the fuzzy groups. Then, our criterion is to maximize the fuzzy variance ratio:

$$\eta^2 = \frac{\sigma_G^2}{\sigma^2} \qquad (2)$$

where σ^2 and σ_G^2 denote the fuzzy total variance and the fuzzy variance between groups, respectively.

Supposing that n samples $x_\alpha\ (\alpha=1, 2, \cdots, n)$ are obtained from the sample space Ω, the number of fuzzy external criterion B_r is defined as follows, using the membership function $\mu_{Br}(x_\alpha)$.

$$N_{Br} = \sum_{\alpha=1}^{n} \mu_{Br}(x_\alpha) \quad (r=1, \cdots\cdots, M) \qquad (3)$$

Also, the total number included in the sample space is defined as

$$N = \sum_{r=1}^{M} N_{Br} \qquad (4)$$

Then, The fuzzy mean $m(B_r)$ and fuzzy variance $\sigma^2(B_r)$ are expressed as

$$m(B_r) = \frac{1}{N_{Br}} \sum_{\alpha=1}^{n} x_\alpha \mu_{Br}(x_\alpha) \tag{5}$$

$$\sigma_{Br}^2 = \frac{1}{N_{Br}} \sum_{\alpha=1}^{n} (x_\alpha - m(B_r))^2 \mu_{Br}(x_\alpha) \tag{6}$$

Using Eqs. 5 and 6, σ^2 and σ_G^2 are calculated as

$$\sigma^2 = \frac{1}{N} \sum_{r=1}^{M} \sum_{\alpha=1}^{n} (x_\alpha - m)^2 \mu_{Br}(x) \tag{7}$$

$$\sigma_G^2 = \frac{1}{N} \sum_{r=1}^{M} N_{Br} (m(Br) - m)^2 \tag{8}$$

where

$$m = \frac{1}{N} \sum_{r=1}^{M} \sum_{\alpha=1}^{n} x_\alpha \mu_{Br}(x_\alpha) \tag{9}$$

Eqs. 7 and 8 are rewritten as

$$\sigma^2 = \frac{1}{N} \sum_{r=1}^{M} \sum_{\alpha=1}^{n} \{ \sum_{i=1}^{K} \sum_{k=1}^{l_i} (\mu_{Aik}(x_\alpha) - \bar{\mu}_{Aik}) a_{ik} \}^2 \mu_{Br}(x_\alpha) \tag{10}$$

$$\sigma_G^2 = \frac{1}{N} \sum_{r=1}^{M} \sum_{\alpha=1}^{n} \{ \sum_{i=1}^{K} \sum_{k=1}^{l_i} (\bar{\mu}_{Aik}^r - \bar{\mu}_{Aik}) a_{ik} \}^2 \mu_{Br}(x_\alpha) \tag{11}$$

where $\bar{\mu}_{Aik}^r$ is the mean value of A_{ik} with respect to each external criterion B_r, and $\bar{\mu}_{Aik}$ is the mean value of A_{ik} with respect to the total external criterion.

The maximum value of η^2 is found out when its partial derivative with respect to a_{j1} is equal to zero.

385

$$\sum_{r=1}^{M} \sum_{\alpha=1}^{n} \sum_{i=1}^{K} \sum_{k=1}^{1_i} \mu_{Br}(x_\alpha)(\overline{\mu}_{Aik}^{r} - \overline{\mu}_{Aik})(\overline{\mu}_{Aj1}^{r} - \overline{\mu}_{Aj1}) a_{ik}$$

$$= \eta^2 \sum_{r=1}^{M} \sum_{\alpha=1}^{n} \sum_{i=1}^{K} \sum_{k=1}^{1_i} \mu_{Br}(x_\alpha)(\mu_{Aik}(x_\alpha) - \overline{\mu}_{Aik})$$

$$\cdot(\mu_{Aj1}(x_\alpha) - \overline{\mu}_{Aj1}) a_{j1} \qquad (j=1,\cdots\cdots,K;1=1,\cdots\cdots,1_j)$$

$$(12)$$

Eq. 12 implies that the underlying problem may result in an eigenvalue problem. The eigenvector corresponding to the maximum eigenvalue provides the category weights which are sought.

2.2 Numerical Example

To demonstrate the efficiency of type II fuzzy quantification theory in the damage assessment of bridges, consider an illustrative example. Table 1 presents a result of fact-finding inquiry which was conducted on ten existing bridges. Usually the following items are considered in the inspection of the damage state of bridges; 1) name, 2) road and river, 3) site, 4) length and width, 5) bridge type, 6) construction year, 7) design specification, 8) visual examination, 9)

Table 1 Linguistic Evaluation for Damage State

bridge \ items	1	2	3	4	5	6	7	8	9	damage rank
1	M	Vs	M	S	M	L	S	M	M	Sl
2	L	L	L	S	C	L	L	L	S	D
3	M	Vs	M	S	L	M	L	L	S	Sl
4	L	L	L	S	C	M	L	L	S	D
5	M	S	S	S	C	M	S	L	S	Sl
6	M	L	L	L	C	S	L	M	M	Se
7	S	M	L	S	C	M	L	M	M	M
8	M	S	L	S	C	S	S	L	S	Sl
9	M	S	S	S	M	S	S	M	M	Se
10	L	Vs	L	S	C	S	L	S	M	Se

item 1;superstructure, item 2;shoe, item 3;substructure, item 4; joint, item 5;N value, item6;liquefaction, item 7;sink, item 8; construction year, item 9;design specification design document, 10) ground condition, 11) design earthquake acceleration, 12) liquefaction, etc. (Yao and Furuta, 1986). Here, items from 6 through 10 are employed as investigating items. They are evaluated in terms of linguistic variables such as "Large (L)", "Medium (M)" and "Small (S)", paying attention to superstructure, shoe, joint and substructure. These linguistic variables are defined by the membership functions shown in Fig. 1. The damage state of superstructure is evaluated in the forms of L, M and S, which denote the damage ranks of "severely damaged", "moderately damaged" and "slightly damaged", respectively. Similar to the superstructure, the substructure is evaluated by L, M and S. For the shoe four linguistic variables as L, M, S and Vs are employed, where Vs denotes "very slightly damaged". In the evaluation of joint, only two variables L and S are used; L means that the joint has such defects as crack or corrosion, and conversely S means no defect.

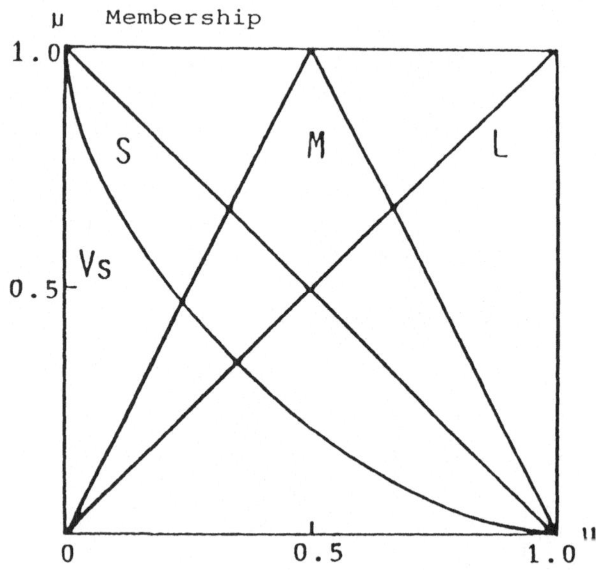

Fig. 1 Membership Functions for Linguistic Variables

Ground condition is classified using the N value. For example, L is given if the depth of soil with an N value more than 25 is greater than 20 m. M is assigned if that depth is from 10 m to 19.9 m, whereas S is assigned if the depth is between 0 and 9.9 m. The symbol C denotes that data are not available. The liquefaction is classified into three categories. The sink of ground is evaluated whether it exists (L) or not (S). The construction year is classified into three categories and the design specification is also classified into three categories. It is assumed that the total assessment of their damage states is given by a well-experienced inspector as shown in Table 1, where the damage state is evaluated in such a verbal form that the damage is either of "No (N)", "Slight (Sl)", "Moderate (M)", "Severe (Se)" and "Destructive (D)".

Using these fuzzy data, the category weight of each damage factor is calculated as shown in Table 2, through the type II fuzzy quantification theory. In Table 2 the symbols A, B and C mean three categories corresponding to the linguistic variables employed. The difference between the maximum category weight and the minimum category weight shows the effect of the item on the damage assessment of bridges. The results indicate that the items 2 and 4 play important roles in the damage assessment.

Table 2 Numerical Results of Category Weight

item	cat.	value	item	cat.	value
1	A	0.854	6	A	0.885
	B	0.501		B	-0.010
	C	0.066		C	0.274
2	A	-1.093	7	A	-0.181
	B	-0.317		B	-0.431
	C	-0.070			
	D	-0.921	8	A	0.176
				B	-0.416
3	A	0.846		C	-0.128
	B	0.448			
	C	0.235	9	A	0.469
				B	0.082
4	A	-2.084		C	0.238
	B	-3.294			
5	A	0.055			
	B	-0.071			
	C	0.107			
	D	-0.054			

3 Application of Fuzzy Regression Analysis

3.1 Fuzzy Regression Analysis

In general, the fatigue characteristic of structures under a constant repeated load is expressed in terms of the amplitude of load S and the repeated cycle to fatigue failure N (Yao et al., 1986).

$$\log N = C - m \log S \tag{13}$$

where C and m are constants to be determined from experiments. Eq. 13 is called an S-N curve. In this paper the S-N curve is modeled as a possibility model using the fuzzy regression analysis (Heshmaty and Kandel, 1985; Tanaka, 1984). To illustrate the fuzzy regression analysis, its outline is summarized in the following.

Suppose that the following data are given:

$$
\begin{aligned}
&(y_1, \; x_{11}, \cdots\cdots\cdots, \; x_{1n}) \\
&\quad \cdots\cdots\cdots\cdots\cdots \\
&\quad \cdots\cdots\cdots\cdots\cdots \\
&\quad \cdots\cdots\cdots\cdots\cdots \\
&(y_m, \; x_{m1}, \cdots\cdots\cdots, x_{mn})
\end{aligned}
\tag{14}
$$

where x_{ij} and y_j are input data and output data, and m and n are the numbers of data and variables, respectively. Here, Y_i, the estimate value of y_i, is defined as

$$\widetilde{Y}_i = \widetilde{A}_1 x_{i1} + \widetilde{A}_2 x_{i2} + \cdots + \widetilde{A}_n x_{in} \tag{15}$$

The above fuzzy parameters $\widetilde{A}_j$ are determined so that y_i are included in $\widetilde{Y}_i$ with some membership degree. The symbol $\sim$ denotes a fuzzy quantity. Using L and R functions developed by Dubois and Prade (Dubois and Prade, 1979), $\widetilde{A}_j$ is specified by the parameters α_j and c_j as shown in Fig. 2. All computations on fuzzy quantities are performed based on the extension principle (Dubois and Prade, 1980).

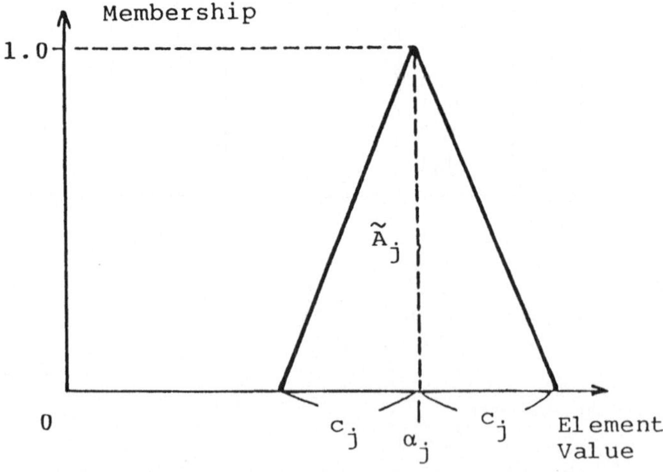

Fig. 2 Membership Function for $\widetilde{A}_j$

As a measure for the consistency between $\tilde{Y}_i$ and y_i , we introduce a parameter h which is defined on [0,1]. By using this parameter, a constraint is derived as follows:

$$\min (\mu_{Y_i}(y_i)) \geq h \qquad (16)$$

The above inequality means that the fuzzy number $\tilde{Y}_i$ must include y_i with a membership grade more than or equal to h. Then, the goal is to find $\tilde{A}_j$ such that the total ambiguity regarding $\tilde{A}_j$ is minimized:

$$S = c_1 + c_2 + \cdots\cdots + c_n \quad \to \quad \min \qquad (17)$$

The parameter h corresponds to the width of the regression curve. By changing the value of h, it is possible to take the engineering judgment of the decision maker into consideration. Usually, the value of h is determined by experienced engineers through the examination of available experimental data.

Fuzzy regression analysis refers to the following linear programming problem (Tanaka, 1984):

$$\text{Objective function : } S = \Sigma \ c_j \to \min \qquad (18)$$

$$\text{subject to} \quad (1-h) \Sigma \ c_j \left| x_{ij} \right| + \Sigma \ \alpha_j x_{ij} \geq y_i$$

$$(1-h) \Sigma \ c_j \left| x_{ij} \right| - \Sigma \ \alpha_j x_{ij} \geq -y_i \qquad (19)$$

$$c_j \geq 0$$

The above formulation is applied to the modeling of the S-N curve:

$$\log \tilde{S} = \tilde{A}_0 + \tilde{A}_1 \log N \qquad (20)$$

$$\log \tilde{N} = \tilde{A}'_0 + \tilde{A}'_1 \log S \qquad (21)$$

In Eqs. 20 and 21, either the load level $\tilde{S}$ or the repeated cycle to fatigue failure $\tilde{N}$ is expressed by a fuzzy set. Since the above formulations of regression analysis are different from the usual formulation, central values of $\tilde{A}_0(\tilde{A}'_0)$ and $\tilde{A}_1(\tilde{A}'_1)$, $\alpha_0(\alpha'_0)$ and $\alpha_1(\alpha'_1)$, may be different from those obtained from the usual analysis.

3.2 Illustrative Example

Based on the experimental data given in (Kameda and Morita, 1984), an S-N curve is obtained through the fuzzy regression analysis. In Fig. 3, several circle points represent the experimental results. It is assumed that the S-N curve can be represented by a straight line with the ordinate of log P/P_d and the abscissa of log N. P is the axial load of vehicles and P_d is the axial load yielding the reinforcing steel, respectively. In Fig. 3, P/P_d is defined as a fuzzy number, whereas N is considered as an ordinal number. This figure corresponds to the possibility model defined by Eq. 20. The solid line provides the central value of the S-N curve, the alternate long and short dash lines show the bounds with h=0.7, and the dash lines show the bounds with h=0.5. As the value of h increases, the width of the S-N curve becomes broader. This figure also indicates that $\tilde{A}_1$ is a fuzzy number, but $\tilde{A}_0$ is an ordinal number without scatter. This result implies that the width of the curve becomes larger as N becomes larger. The slope of this line (corresponding to the solid line) is considerably different from that obtained by the usual regression analysis. This is due to the fact that the fuzzy regression analysis has such a different criterion that the width of the line should be minimized.

Fig. 4 presents the result where the abscissa and ordinate are exchanged. In this case N and P/P_d are considered to be a fuzzy number and an ordinal number, respectively. This case corresponds to the possibility model given by Eq. 21. Alternatively, in this case $\tilde{A}'_1$ and $\tilde{A}'_0$ are obtained as an ordinal number and a fuzzy number. Therefore, the S-N curve obtained has the same width all over the values of P/P_d. From Figs. 3 and 4, it can be seen that what is a fuzzy number greatly influences on the numerical results.

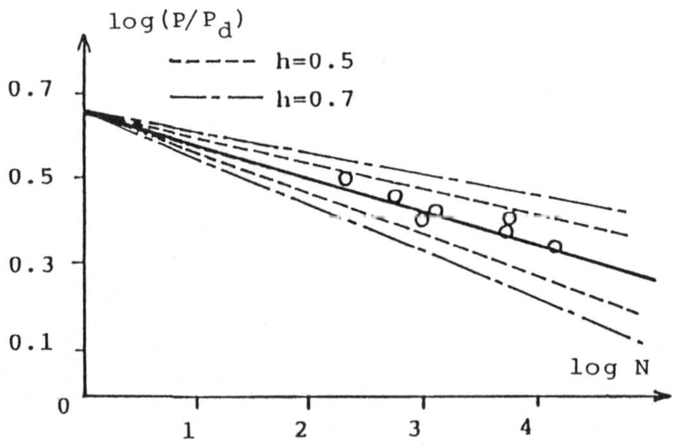

Fig. 3 S-N Curve Obtained from Eq. 20

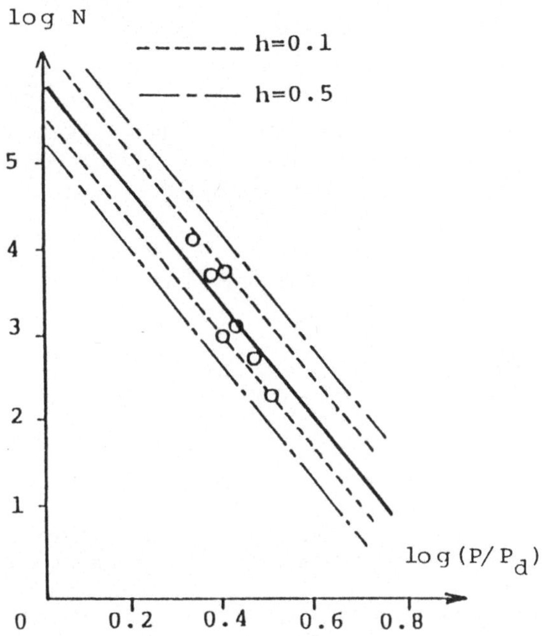

Fig. 4 S-N Curve Obtained from Eq. 21

4 Conclusions

In this paper we gave an insight for the treatment of data available in the damage assessment of bridge structure. The problem of damage assessment of bridge is included in twilight category (Hayashi, 1984; Shiraishi and Furuta, 1985); its contexture may not be clearly defined, there are a lot of factors to be considered, and it is difficult to collect sufficient data in amount and/or quality.

In order to account for the ambiguities and uncertainties involved in the inspection results which are reported in vague and indistinct linguistic terminology, the type II fuzzy quantification theory was applied for rating the damage state of existing bridges. The fuzzy quantification theory is a generalization of the usual quantification theory by a fuzzification process. The discriminant function obtained by this method can assess the degree to which each attribute affects the differentiation of fuzzy groups. Furthermore, based on the discriminant function, it is possible to classify new bridges into one of several damage states which are specified in terms of fuzzy sets.

Paying attention to the scarcity of experimental data available for the fatigue analysis of concrete bridge decks, the concept of possibility distribution is introduced into the modeling of the S-N curve. The possibility distribution is a fundamental basis of the fuzzy sets theory which is a counterpart of the probability theory. Since the possibility model of the S-N curve is not based on the concept of frequency, it is useful for the case without sufficient amount of data. The use of this model enables us to evaluate the fatigue life of the concrete decks in a more informative and realistic form. The final results are obtained in terms of fuzzy sets which provide us with all the potential possibilities with corresponding grades.

References

Dubois D. and H. Prade (1979). "Fuzzy Real Algebra : Some Results", Fuzzy Sets and Systems, 2, 327-348.

Dubois, D. and H. Prade (1980). Fuzzy Sets and Systems : Theory and Applications, Academic Press.

Heshmaty, B and A. Kandel (1985). "Fuzzy Linear Regression and Its Applications to Forecasting in Uncertain Environment", Fuzzy Sets and Systems, 15, 159-191.

Kameda, H. and S. Morita (1984). "Analysis of Damage Factors and Influence of Live Loads on Fatigue of RC Decks in Urban Expressways", In Report of Hanshin Expressway Corporation, 108-146. (in Japanese)

Matsui, S (1984). "Study on Design and Analysis of Fatigue of Concrete Decks of Road Bridges", presented in Partial Fulfillment of Doctor of Engineering, Osaka University. (in Japanese)

Shiraishi, N., H. Furuta and M. Sugimoto (1985). "Integrity Assessment of Bridge Structures Based on Extended Multi-Criteria Analysis", Proc. of ICOSSAR-4, 1, 505-509.

Shiraishi, N. and H. Furuta (1985). "Assessment of Structural Durability with Fuzzy Sets", Proc. of NSF Workshop on Civil Engineering Applications of Fuzzy Sets, IN, USA, 193-218.

Tanaka, H. (1984). "Possibility Model and Its Applications", Systems and Control, Japan, 28, 447-451. (in Japanese)

Yao, J. T. P. (1978). "Damage Assessment and Reliability Evaluation of Existing Structures", J. of Eng. Struc., 1, 245-251.

Yao, J. T. P. and H. Furuta (1986). "Probabilistic Treatment of Fuzzy Events in Civil Engineering", J. of Probabilistic Mechanics, 1, 1, 58-64.

Yao, J. T. P., F. Kozin, Y.-K. Wen, J.-N. Yang., G. I. Schueller and O. Ditlevsen (1986). "Stochastic Fatigue, Fracture and Damage Analysis", Structural Safety, 3, 231-267.

Watada, J. (1983). "Theory of Fuzzy Multivariate Analysis and Its Applications", presented in Partial Fulfillment of Doctor of Engineering, Univ. of Osaka Prefecture.

Zadeh, L. A. (1965). "Fuzzy Sets", Information and Control, 8, 338-353.

Zadeh, L. A. (1978). "Fuzzy Sets as a Basis for a Theory of Possibility", Fuzzy Sets and Systems, 1, 3-28.

Concepts

A Systems Approach to Fire Safety Engineering
J. A. Purkiss

Experimental Vulnerability Detection in Civil Structures
P. Ibanez

The Machinery Vibration and Wear Advancement Identification and Forecasting
C. Cempel

Bridge Inspection by Dynamic Tests and Calculations Dynamic Investigations of Lavant Bridge
R. G. Flesch, K. Kernbichler

System Identification Approaches in Structural Safety Evaluation
H. G. Natke, J. T. P. Yao

Structural Damage Assessment Using a System Identification Technique
J.-Ch. Chen, J. A. Garba

A Systems Approach for Fire Safety Engineering

J A Purkiss

Department of Civil Engineering, Aston University
Birmingham, U.K.

Abstract

Upto recently the main approach to Fire Safety Engineering has been on an adhoc basis in that little planning has been done to minimise the required fire endurance period by considering, say, compartmentation, or fire load or indeed the detail method of construction at the incipient planning stage. This has been partly due to the use of a Regulatory Approach and not to the use of calculation methods.

The paper considers two examples, that of an isostatic structural element within a small fire compartment before considering any constraints acting on the system. These constraints may take the form of either acceptable failure risks or be of an economic or practical nature. The paper concludes by considering a general system for structural fire safety engineering.

Introduction

Before the application of a systems approach to Fire Safety Engineering can be assessed it is necessary first to consider briefly what the objectives are behind Fire Safety Engineering. Witteveen (1) has summarized them as follows:-

i)   Reduction in the risk of injury and death of persons, and

ii)   Reduction in the risk of damage to, or loss of, the building, contents and environment.

From these objectives, three main elements in any rational analysis of the Fire Safety Engineering concept can be identified:-

i)   Agreed levels of life and property safety

ii)   Quantitative methods of assessing potential hazard, and

iii) Quantiative methods of assessing the effectiveness of protective measures to meet such agreed levels of safety under any potential hazards.

These three elements cannot be considered in isolation, but form sections of a complex whole which may be identified as a system. For most structures, the effect of fire is usually only considered by the Engineer as pertaining directly to structural safety, whereas the whole system is far more complex in that requirements of say fire fighting , whether by internal sprinkler systems or fire brigades, fire escapes and smoke control must be considered. This paper will however simplify the situation in only considering the structural aspects of the problem.

A systems approach has already been formulated (2) for fire spread through buildings in the form of the NPFA decision tree and its development into a Building Fire Safety Model for use in domestic building structures. The NFPA approach only sets out the decisions required to reduce the effects of fire and does not consider the structural safety aspect. The Building Fire Safety model does allow a probabalistic assessment to be made fire spread based on the fire load position and structure geometry.

Justification for a systems approach

In order to be able to evaluate the need for a systems approach to structural Fire Safety Engineering, it is first necessary to consider the traditional approach. This is based on a classification system. defined in Fig. 1 whereby the required fire duration period, t_{fd}, is obtained from Statutory Regulations and is compared with the fire resistance period t_{fr}, obtained from standard test data.

This method is inadequate for a series of reasons (1, 3):-

i)   The temperature-time response of a standard fire test to BS 476 Part 8 (4) in the UK (which is similar to ISO 834 (5)) bears little or no relation to a "real" fire, in that no account is directly taken of the effects of the type of fuel, compartment characteristics or ventilation that occur in a real fire. It should also be noted that the fire duration period specified by Regulations can often be abitrary, in that they are based on no rational argument, or determined for emotive reasons.

ii)   Structural behaviour in a fire test can bear little relationship to the performance

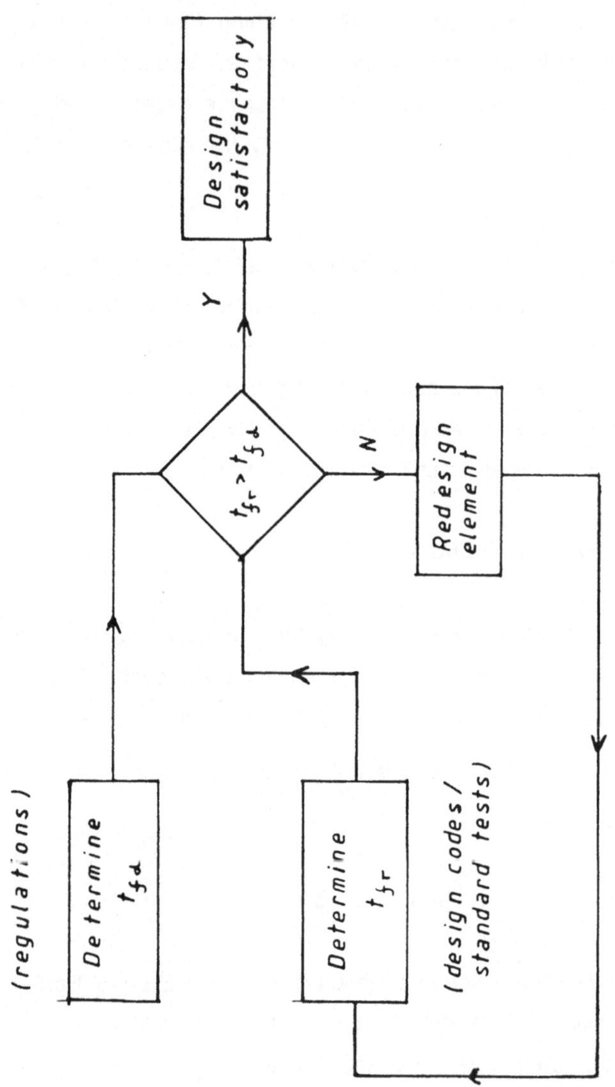

FIG 1. Historical approach based on classification

in a "real" fire, since fire tests are either carried out on isolated elements lacking the continuity or restraint found in actual structures, often on elements of restricted dimensions (usually length or height) caused by furnace availability or carrying imposed loading or heating conditions which do not produce the most detrimental effects.

It is thus clear that the classification system used for determining fire performance has no place when compared to the sophisicated approaches, in the form of limit state analysis, which are used for ambient conditions, and that a calculation approach should be made available for both the structural performance and the required endurance period.

Rather than to attempt to assemble the whole system in one, it has been decided to consider structural performance using two examples, before imposing the required endurance period as a constraint. It should be noted that other constraints may also exist - these are essentially of an economic or practical nature and will be considered later.

Example 1: A simply supported beam in a small single fire compartment

The system under investigation is illustrated in Fig. 2(a), together with the required steps in the analysis of the system in Fig. 2(b).

i)   Calculation of temperature-time response for the compartment

The method developed by Petterson (6) may be used. Essentially it involves an interactive solution to the heat flow equation within the compartment sketched in Fig 3.

The following parameters are required for the analysis:

a)   the fire load, q, which is a measure of the amount of combustible material mass of wood cribs/unit floor area.

b)   the available ventilation, $(A_w\sqrt{H_w})/A_t$, where A_w is the window area, H_w the window height and A_t the total compartment surface area.

c)   the weighted mean thermal inertia of the walls which is defined as $(k\rho c)^{1/2}$, where k is the thermal conductivity, ρ the density and c the specific heat.

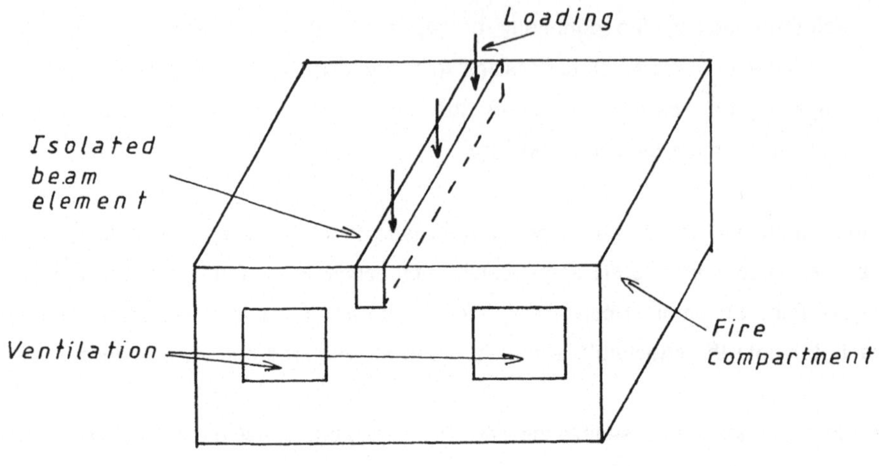

(a)

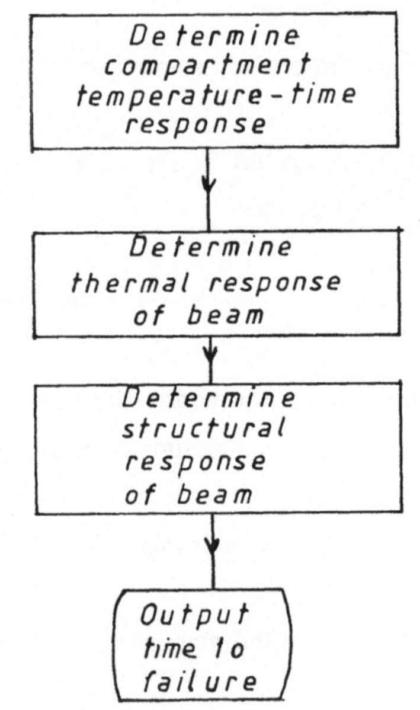

(b)

FIG 2. Single element in fire compartment

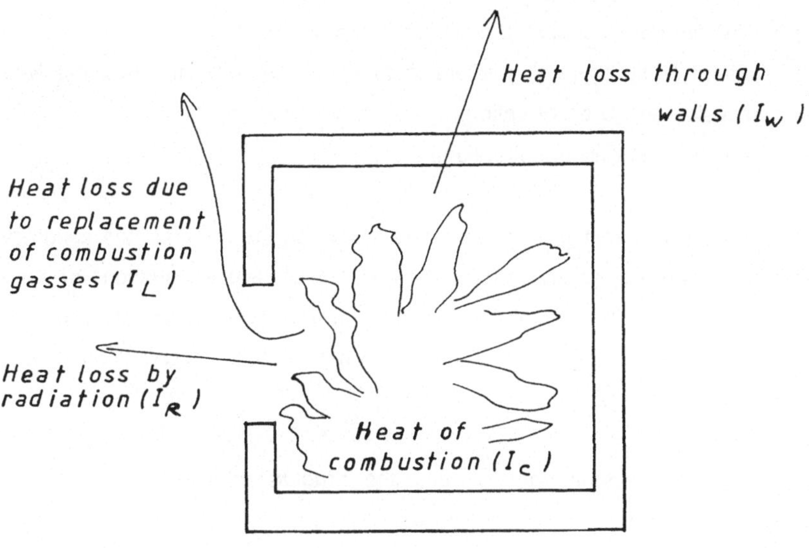

$$I_C = I_L + I_W + I_C$$

FIG 3. Heat balance equation for a fire compartment

The main concern here is with the assumptions contained in the method which can be summarized as follows:-`

- a)   All the fuel ignites simultaneously,
- b)   All the ventilation is present at flashover,
- c)   The fire does not spread outside the compartment,
- d)   There is no thermal gradient in the compartment, and that the walls are considered to be of uniform mean thermal inertia,
- e)   All heat flow is uni-directional.

Clearly given these assumptions and the need to assess the fire load, the accuracy of the result must be assessed, as also should the effect of statisitical variations of say the fire load and any other possible parametric variation. It is of some consolation that tests and calcuations have shown a good correlation, even though that correlation is with mean compartment temperatures.

ii)   Calculation of the temperatures within the structural element.

The basic problem is the solution of the two (or three) dimensional Fourrier heat transfer equation with temperature and space dependant thermal properties. The only way to solve such a problem is to use specially written computer programs such as FIRES-T2 (7), FIRES-T3(8) or TASEF-2 (9).

All these programs require as input data values of thermal properties of the constituent materials of the structural element. Clearly it is neither practicable nor possible to obtain values by test from the actual materials used, and recourse must be had to reperesentative values obtained from standard references (10,11). The remaining problem is with estimating values of surface heat transfer coefficients (3). Some calibration has been done against calculated temperature response in a standard furnace test and suitable values are dependant on the heat flux which in general is not monitored in a test.

Similarly here an assessment should be made of the accuracy of prediction.

iii)   Calculation of Structural Response

Again the problem is complex and ideally requires the use of computer programs such as FIRES-RC (12), CONFIRE (13), or FASBUS II (14, 15, 16).

The problem again arises with the assessment of materials behaviour, not only in obtaining requisite values (or a mathematical expression) but in ensuring that basic actual, rather than design, strength parameters are known. Recourse again for materials behaviour will be to representative values (10,11).

Suppose the aim in these calcuations was to determine the time to which the element could no longer support its load, and that the determined value was t_{calc}.

Given the assumptions and uncertainties within the three calculation stages, the assumption that this value of t_{calc} is correct cannot be made, and that a more likely value t'_{calc}, should be given by

$$t'_{calc} = t_{calc}/(\gamma_f.\gamma_h.\gamma_s) \qquad (1)$$

where γ_f is a factor allowing from the uncertainties in the fire temperature calculations, γ_h in the heat flow calculations and γ_s in the structural calcuations.

To date, it appears that no systematic attempt has been made to assess the effect of such γ values or to assess the accuracy of such calculations.

It should be noted, in the standard fire test some allowance is made in the fire endurance periods are rounded down to the nearest 30 mins.

Some calibration has been carried out for both thermal and structural response (3) indicating that analysis will give a similar response to that found from tests, it should be noted that any discrepancies are partly due to simplifications made in the models and partly due to representative values of material parameters required for the analysis. It is not clear whether if a minimum error calculation were made to obtain a best fit on the temperature response using modified thermal parameters would give a better structural response. It is likely that the materials models required at elevated temperatures are too complex to allow any best fit procedure on the structural response, although such a procedure may indicate the most important segments of the materials model that require more experimental data.

Example 2:   A single floor of a multi-storey continuous structure.

Although the basic procedure of calculation of compartment temperatures, thermal response of the structure elements, calculation of structural response remains, each is now subject to greater variation of unknowns. Some indication of this effect is given in Fig. 4, where although the basic flow chart of Fig. 2 (b) will be recognised, there are a large number of qualifications to be made.

It must be noted that at present there is no systems model capable of analyzing such an example. It should also be apparent that such a model, albeit currently hypothetical, would enable the Engineer to make rational judgements as to say the effects of greater compartmentation on the resultant costs of implementing fire safety.

Before continuing by laying out the basis of a full systems approach to fire safety engineering, it is necessary to consider any constraints to such a system.

System Constraints

Upto this point it has been considered sufficient to determine (or assess) the effects of a natural fire on a structure, or part structure, and to determine the end point of that structure when its level of safety falls below an acceptable level. It is necessary to establish what level of safety is acceptable. It is usual to relate the level of safety to the period in which the structure becomes incapable of carrying the applied load either through collapse or excessive deflection. It is also necessary to condsider other restraints of either a practical or economic nature.

i)   Level of safety

The current method whereby Statutory Authorities impose a fire duration based solely on usage, floor space and compartment volume is clearly incompatible with a calculation approach to the overall structure response.

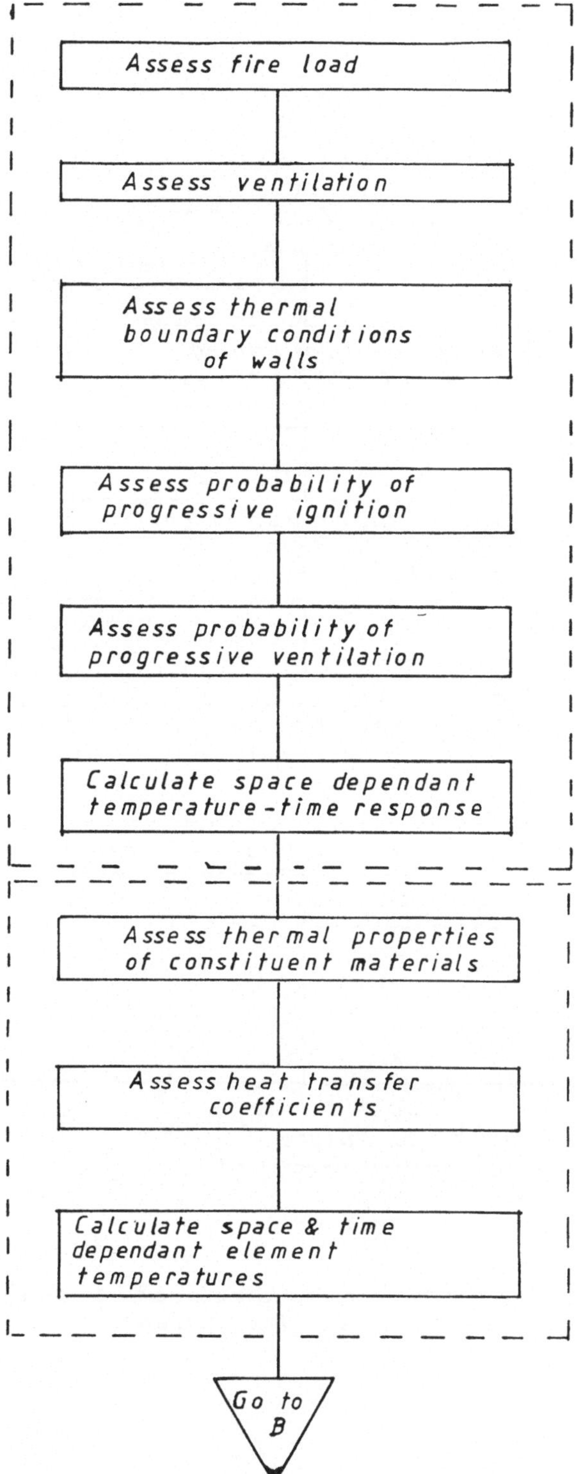

Assess fire load
Assess ventilation
Assess thermal boundary conditions of walls
Assess probability of progressive ignition
Assess probability of progressive ventilation
Calculate space dependant temperature-time response
Assess thermal properties of constituent materials
Assess heat transfer coefficients
Calculate space & time dependant element temperatures
Go to B

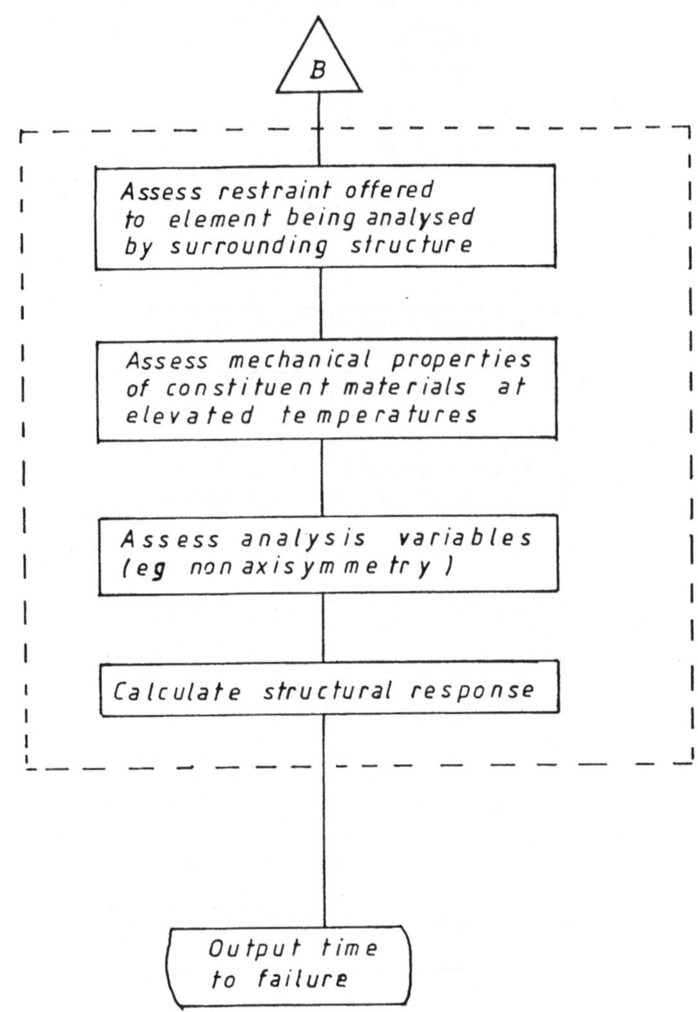

FIG 4.   System flow chart for a single storey fire compartment

Rasbash (18) has considered this question both for high risk industries and buildings, and concludes that the target failure probabilities should be dependant on the structure usage to a great extent and the number of stories (one or multi) to a lesser extent. The suggested probabilities of failure lie within 10^{-3} to 10^{-7} for the lifetime (of 50 years) of a single storey structure. For ambient design a typical value is 10^{-6}.

Kersken-Bradley (18) gives various examples of where Statutory Bodies in some countries are allowing a partial probalistic calculation of the required fire duration. In this paper, however, focus will be made on the method proposed in North America. Only a summary will be attempted here, full details will be found in Harmathy and Mehaffy (19).

It should be noted that the calculation procedure is related to test performance in a standard fire test and thus is more properly related to the concept of equivalent fire duration (20, 21) and implies that any structural calculations should be performed with the standard ISO 834 temperature-time curve imposed as the boundary condition in the thermal calculations.

The three variables, fire load/unit area, ventialtion and fire test period are treated as random variables with the design fire resistance requirement subject, therefore, to some uncertainty.

a)   Fire Load

The design value of fire load q_d, is given by

$$q_d = \bar{q} + \beta \sigma_q \qquad (2)$$

where $\bar{q}$ is the mean fire load and σ_q the standard deviation of fire load for a given occupancy. β will depend on the probability of a given fire load being exceeded (ie   for a 5% probability, $\beta = 1.64$).

b)   Ventilation

The normalized heat flow in a single compartment H' is given by

$$H' = (qA_f \times 10^6)(11\delta + 1.6)/(A_t\sqrt{kpc} + 935\sqrt{\emptyset}qA_f) \qquad (3)$$

where A_f is the floor area, $\sqrt{kpc}$ the thermal intertia of the compartment, A_t the total compartment area and the ventilation factor $\emptyset$, δ being defined by equation 4.

$$\delta = 0.79 \sqrt{H^3{}_w}/\emptyset \quad >1.0 \tag{4}$$

where H_w is the height of the opening.

The ventilation factor $\emptyset$, characterizes the air flow into the compartment and it is suggested, as it will produce a more severe case, that the minimum value, $\emptyset$ min, be used,

$$\emptyset_{min} = \rho_a A_w \sqrt{gH_w} \tag{5}$$

where ρ_a is the density of air and H_w the area of openings.

c)   Fire Test Period

From standard tests the fire test period, τ, can be related to the heat input H" by the equation,

$$\tau = 0.11 + 0.16 \times 10^{-4} H'' + 0.13 \times 10^{-9}(H'')^2 \tag{6}$$

It remains to relate H' and H", this can be done by using a "second moment' analysis giving

$$H'' = H' \exp \left(\beta\left((\sigma_{H''}/\bar{H}'')^2 + (\sigma_{H'}/\bar{H}'')^2\right)^{1/2}\right) \tag{7}$$

where $\sigma_{H''}/\bar{H}''$ and $\sigma_{H'}/\bar{H}'$ are the variances in the heat flows.

The calculated value of τ is then set equal (after rounding to the nearest half hour) to the required fire resistance.

ii)   Practical or Economic Constraints

It is impossible to delineate all these so a few examples will be given as illustrations:-

a)   The additional cost of extra fire walls, possibly only to last part of the required total fire duration, in order to reduce the temperatures reached in the fire and hence reduce the cost of fire protection. In this case an additional cost will derive from the possible loss of flexibility in the usage of the compartment.

b)   The cost of increasing a section size in order to reduce the stresses under service conditions (which will normally give the imposed loading to be taken during fire conditions) and hence increase the period the element will last during a fire. (This is really only applicable to steel work).

c)   For concrete, the magnitude of the cover to the main reinforcing is important and this should not be reduced, for the fire situation, below that required for normal durability requirements.

Overall Systems Approach to Fire Safety Engineering

An embryonic flow chart for this is given in Fig. 5. Although the broad outline is correct, it may well need ammending in the light of any developments in any of the basic models used for calculation. It is also to be noted that some of the calculation procedures needed may not currently be available. Although the system indicates an exit point when the requirements are satisfied, should there be a case of gross over-design then the procedure should be repeated. For clarity this is omitted from Fig. 5.

Conclusions

In view of the complex interaction of separate elements within the Fire Safety Engineering of Structures and the interation with ambient design criteria, a systems approach should give a method whereby an economic and practical solution can be arrived at. This paper has attempted only to lay down some of the ground rules, and it should also stimulate development in calculation procedures needed in order to make a complete systems approach viable in what will, hopefully , be the near future.

Acknowledgement

Whereas the views expressed in the paper, especially with regard to Statutory Regulation, are the author's own, I must thank my colleagues in the field who have stimulated my interest especially Bill Malhotra and Bob Anchor.

References

1.   Witteveen J <u>Trends in Design Methods for Structural Fire Safety.</u> Three Decades of Structural Fire Safety, BRE, Garston, 1983. pp 21 - 27.

2.   Roux H J, Berlin G N, <u>Toward a Knowledge -based Fire Safety System.</u> Design of Buildings for Fire Safety. Smith E E and Harmathy T Z (Editors), ASTM, STP685, Philadelphia, 1979, pp 3 - 13.

3.   Purkiss J A, Weeks N J, <u>A Computer Study of Reinforced Concrete Columns in</u>

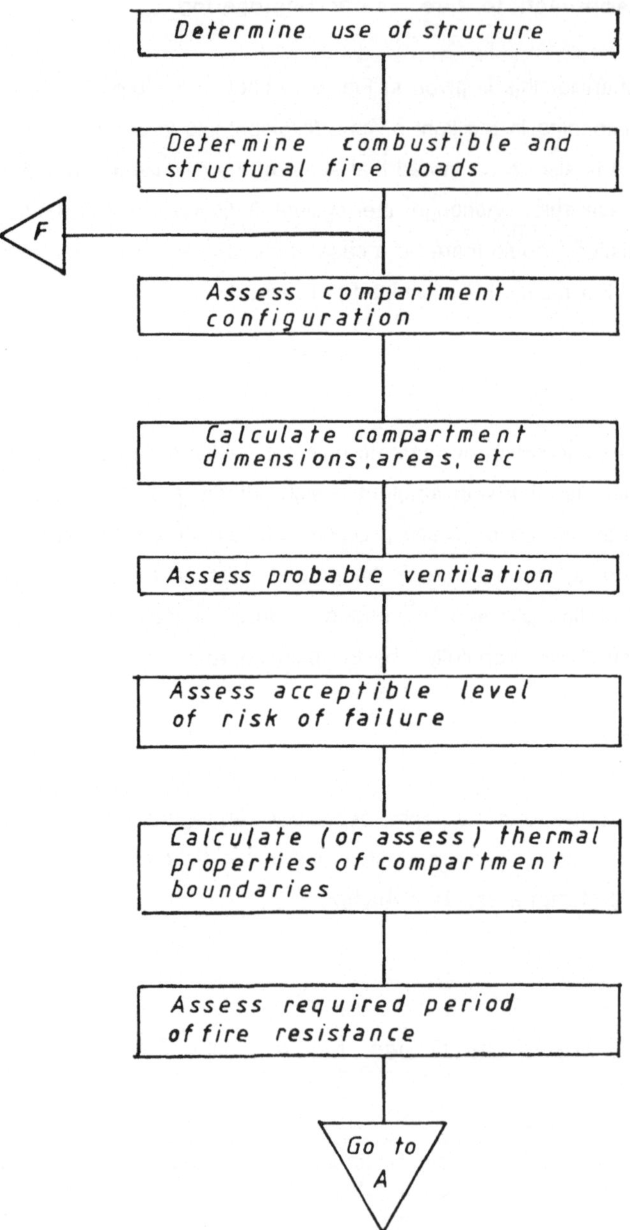

Determine use of structure
Determine combustible and structural fire loads
F
Assess compartment configuration
Calculate compartment dimensions, areas, etc
Assess probable ventilation
Assess acceptible level of risk of failure
Calculate (or assess) thermal properties of compartment boundaries
Assess required period of fire resistance
Go to A

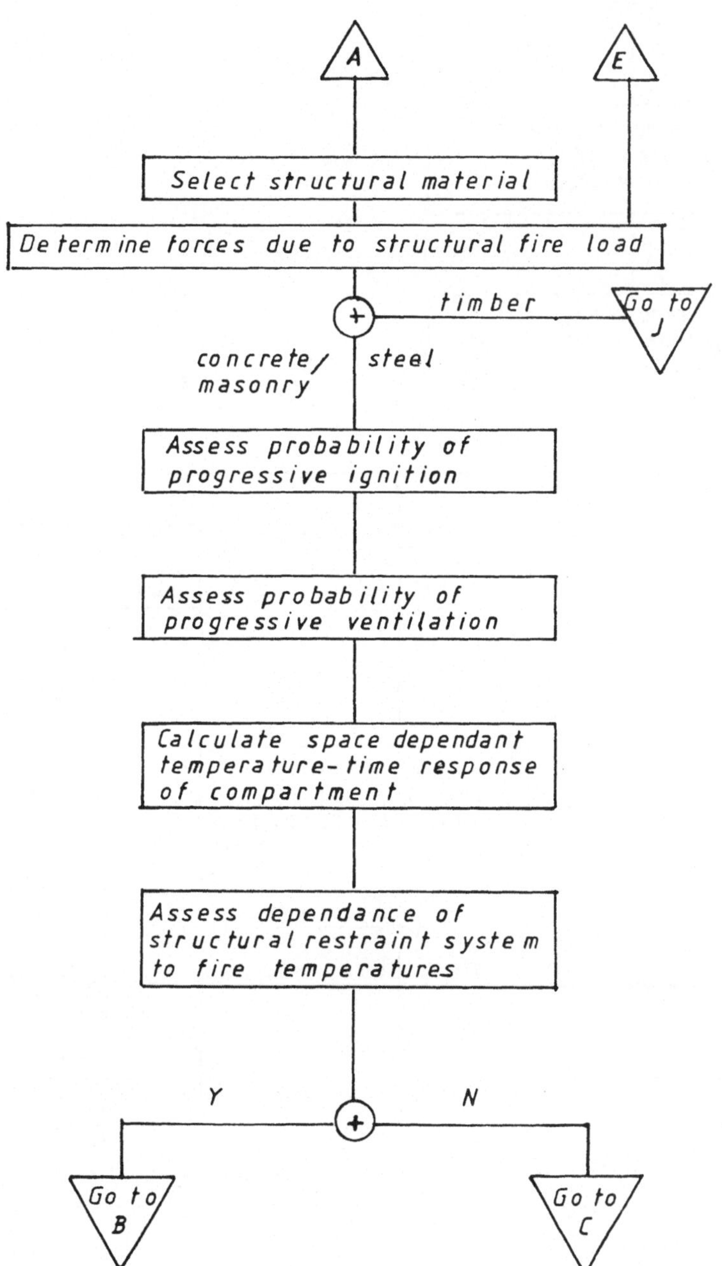

A
E
Select structural material
Determine forces due to structural fire load
timber
Go to
J
concrete/
masonry
steel
Assess probability of
progressive ignition
Assess probability of
progressive ventilation
Calculate space dependant
temperature-time response
of compartment
Assess dependance of
structural restraint system
to fire temperatures
Y
N
Go to
B
Go to
C

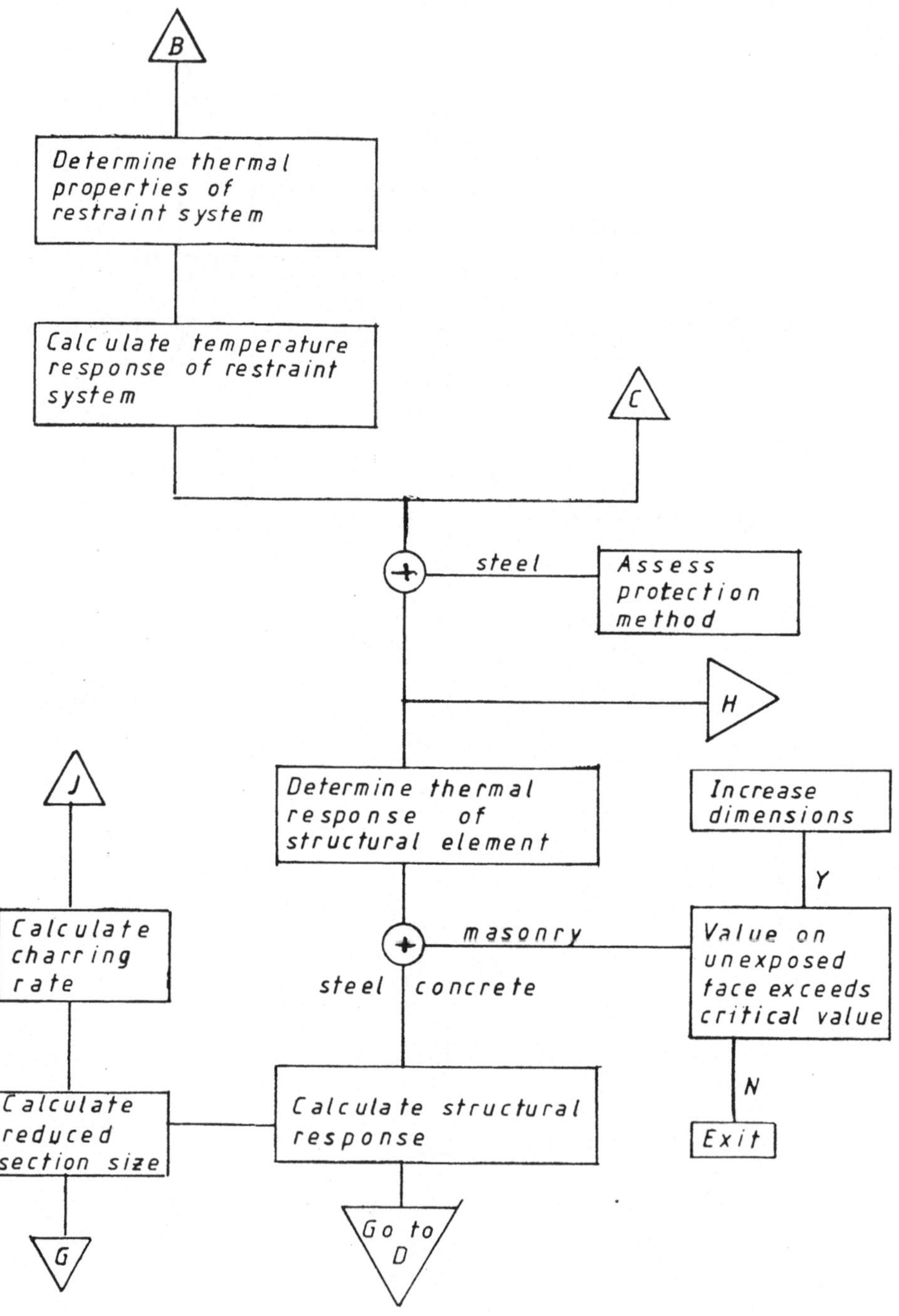

B
Determine thermal
properties of
restraint system
Calculate temperature
response of restraint
system
C
steel
Assess
protection
method
H
J
Determine thermal
response of
structural element
Increase
dimensions
Y
Calculate
charring
rate
masonry
Value on
unexposed
face exceeds
critical value
steel   concrete
Calculate
reduced
section size
Calculate structural
response
N
Exit
G
Go to
D

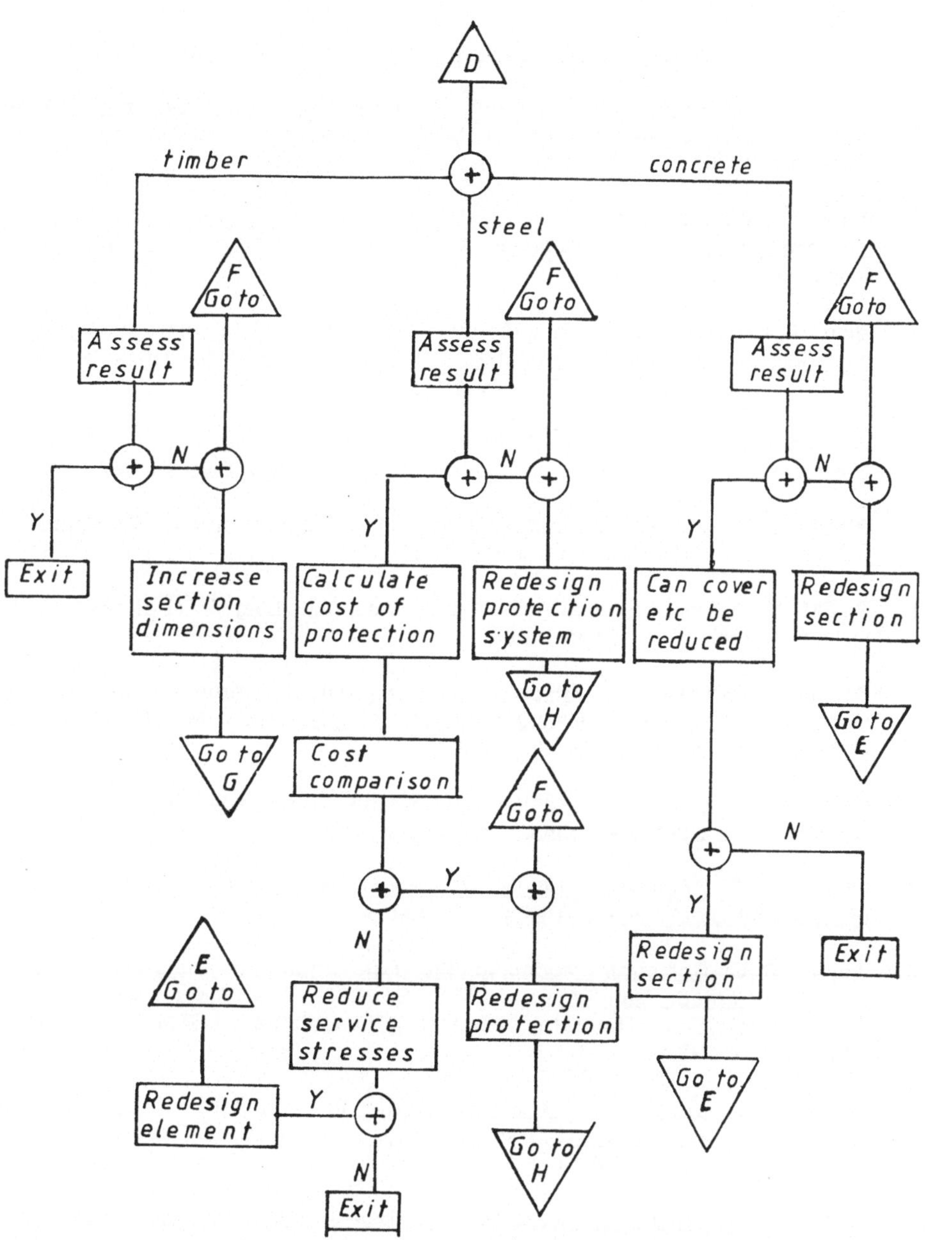

FIG 5. Generalized systems approach for Fire Safety Engineering.

a Fire. The Structural Engineer, _65B,_ (1987), 22 - 28.

4.   BS 476, Part 8, Fire Tests on Building Materials and Structures.
London British Standards Institution 1972.   (under revision and due to
be published as BS 476 Parts 20, 21, 22).

5.   ISO 834 _Fire Resistance Tests - Elements of Building Construction,_1985.

6.   Petterson O, Magnusson S-E, Thor J   _Fire Engineering Design of Steel
Structures_ Swedish Institute of Steel Construction, Publication No 50,
Stockholm 1976.

7.   Becker J, Bizri H, Bressler B, _FIRES-T A computer program for the fire
response of structures -_ Thermal, Report No UCB-FRG 74-1, University of
California, Berkeley, 1974.

8.   Iding R J, Bresler B, Nizamuddin Z, _FIRES-T3. A Computer Program for
the Fires Response of Structures - Thermal._ Report No UCB FRG 77 -
15, University of California, Berkley 1977.

9.   Wicksröm U, _TASEF-2. A Computer Program for Temperature Analysis
of Structures exposed to Fire._ Lund Institute of Technology, Sweden, 1979.

10.   RILEM - Properties of Materials at High Temperatures, Schneider U
(Editor), Kassel University, Germany, 1985.

11.   RILEM - Behaviour of Steel at High Temperatures, Anderberg Y   (Editor),   Lund
Institute of Technology, Sweden 1983.

12.   Becker J, Bressler B, _FIRES - RC. A computer program for the fire
response of structures - reinforced concrete frames._ Report No UCB-FRG 743,
University of California, Berkeley. 1974.

13.   Forsén N E , A Theoretical Study of the Fire Resistance of Concrete
Structure, FCB-SINTEFF, Trondheim, Norway. 1982.

14.   Jeanes D C, _Application of the Computer in Modelling Fire Endurance
of Structural Steel Floor Systems._ Fire Safety Journal. _9._ (1985) 119 - 135.

15.   Jeanes D C, _Computer Modelling the Fire Endurance of Floor Systems
in Steel-framed Buildings._ ASTM STP882, Fire Safety, Science and
Engineering, T Z Harmathy (Editor), Philadelphia, ASTM Publications,
1985, 223 - 238.

16.   Jeanes D C, _Developing Desing Concepts for Structural Fire Endurance
Using Computer Models._ Design of Structures Against Fire, Anchor R D
et al (Editors), London, Elsevier, 127 - 153.

17.   Rasbash D J, _Criteria for Acceptibility for Use with Quantitative Approaches to
Fire Safety._ Fire Safety Journal, _8._ (1984/5), 141 - 158.

18.   Kersken-Bradley M, _Probablistic Concepts in b Fire Engineering._ Design of
Structures Against Fire, Anchor R D et al (Editors), London,   Elsevier, 21 -39.

19.   Harmathy T Z, Mehaffy J R, _Design of Buildings for Prescribed Levels of Structural Fire Safety_, ASTM STP882 Fire Safety, Science and Engineering, T Z Harmathy (Editor), Philadelphia, ASTM Publications, 1985, 160 - 175.

20   Law M _A Basis for the Design of Fire Protection of Building Structures._ Structural Engineer, _61A_, (1983), 25 - 33.

21   Smith C Ian, _Structural Fire Engineering Design._ The Structural Engineer, _65A_ (1987), 51 - 55.

Experimental Vulnerability Detection in Civil Structures

P. IBANEZ

Paper not available

The Machinery Vibration and Wear Advancement Identification and Forecasting

Czesław Cempel

Poznań University of Technology, Institute of
Applied Mechanics
60-965 Poznań, Piotrowo 3

<u>key words</u>: machinery wear, vibration, life curve, condition
forecasting, break-down prediction.

Summary

The paper considers the interrelation between wear processes
and vibration of running machines. Starting with proposed
equality for energy and vibration amplitude differentials the
differential equation of machine vibration evolution was obtained. It enables analytical definition of time to machine
break-down and even its measurement during machine prototype
studies. This equation enables also vibration amplitude or
condition prediction as well as time to break-down assessment
when one posses the set of machine vibration observations
when running. This possibility was verified experimentally
in case of railroad diesel engine vibration diagnostics.

1. Introduction

The wear of machinery parts occurs in many ways but it always
is closely related to dynamic phenomena such as vibration,
acoustic noise and ultrasound. Taking into account the main
types of wear one should first consider the fatigue phenomena
which, if they occur throughout the machinery part, are the
cause of its integrity loss. If they occur on the surface
they can cause pitting of moving machinery parts /bearings
and gears/ or fretting/fatigue corrosion/ in inmovable joints
working in a corrosive atmosphere. The second important type
of wear is adhesive and/or abresive wear which proceeds in
every moving rotating or sliding joint. The intensity of this
type of wear depends on the lubrication quality, unit pressure as well as on relative vibration amplitude. The third type
of wear of machinery parts is creep, particularly in higher
temperatures. This distortional phenomenon is highly dependent on temperature, mean working stress, and high frequency
vibration /ultrasound/. The last type of wear i.e. erosion

is caused by such phenomena as cavitation, corrosion, impact
of liquid or gaseous stream with solid particles, and so on.
These particular ways of wear are less sensitive to vibration
but they are intensive sources of sound and vibration themselves.

In general all the important wear phenomena are vibration
sensitive beeing the cause and/or by - product of wear. This
is possible because machine, as an energy system, transforms
the input energy into both usable energy and residual wear
energy. This energy fraction is dissipated in the wear processes which include vibration as both the cause and the effect.
This way of reasoning gave lately so called tribovibroacoustical /TVA/ machine model [1], /published in Wear [1] by the
present author/. But it may betreated as the starting point
for further considerations of the wear-vibrational relation
in machines, their generalizations and applications. It is
shown in the paper that from machine prototype measurements
one can estimate the break-down time. And observing the succesive vibration readings taken from machine body one can assess its future vibration and condition as well. This was possible due to two assumptions only: 1° Wear advancement is
equivalent to dissipated energy, 2° Differential increment of
vibration is equivalent to the same dissipated energy increment.

2. Dissipated energy as the wear advancement measure.

In the appendix to this paper it is shown that each form of
wearing occurs when energy is dissipated in that particular
way. In other words total work done to each particular form
of wearing or dissipated energy may be quantitative measure
of that form of wear advancement. As the wear phenomena proceed in life time of the machine designated here to O   one
may write

$$Z_i \, O \; = E_{ti} \, O \; , D, \text{ design parameters} \qquad /1/$$

where $Z_i \, O$ = wear advancement for i-th form of wear, E_{ti} ...
= energy dissipated to that form of wearing which depends on
life time O , vibration amplitude D and machine design parameters /see appendix/. For the given machine type, where

all design parameters are set up, we may write simply

$$Z_i(\theta) = E_{ti}(\theta, D)$$

where D means vibration amplitude in general but for different form of wear it may have different meaning. For example volumetric fatigue is governed by dynamic stress amplitude which is simple proportional to the vibration velocity amplitude [2,3] . From the other side surface fatigue-like fretting is governed simply by vibration displacement amplitude /see appendix A3/.

It is abvious that each form of wear, in application to given machinery element, has its own limit value of dissipated energy /work done/ E_{til} . When this value of work done to the element is reached the element looses its utility features giving for example abnormal bearing clearance or even the lost of integrity of the element. This way of reasoning gives us the tool for element as well as the machine condition $X(\theta)$ difinition

$$X_i(\theta) = E_{til} - E_{ti} = \begin{cases} > 0 \; , & \text{good condition} \\ \leq 0 \; , & \text{faulty condition} \end{cases}$$

Now we have to transfer our consideration from the element level of the machinery structure to the level of machine system where input power $-N_i$ is distributed to usable power $-N_u$ and power dissipated for tribological /wear/ processes $-N_t$. From the mathematical point of view one can simply sum up all wearing forms: i=1,...5, (1) /see also appendix/ and machine elements: j=1,...n, obtaining wear advancement measure $Z(\theta)$ as total dissipated energy

$$Z(\theta) = E_t(\theta, D) = \sum_{j=1}^{n} \sum_{i=1}^{5} E_{tij}(\theta, D) \qquad /2/$$

In reality a machine has arranged in rows reliability structure which is governed by the rule of the weakest element. But there are many so called weak elements such as bearings, gear tooths, seals and so on, and it is impossible to predict apriori for a given time θ which critical element will fail first. This is mainly due to unknown distribution of dissipated power for different form of wearing and different element

during the machine life time. For example for some machines
at idle speed the volumetric fatigue may be prevaling mode of
wear, while at full load pitting and abrasive wear may domi-
nate.

Knowing that we will consider the power flow and the dissipated energy accumulation as for whole machine system but we
will expect failures of some most unreliable parts like bearings, seals and other. The decisive factor, which determines
the wear intensity of a given machine type or even machine
itself, is its mechanical efficiency $-\eta$. For machines without thermodynamic conversions this may be simple defined in
terms of power as on Figure 1 and formula (3)

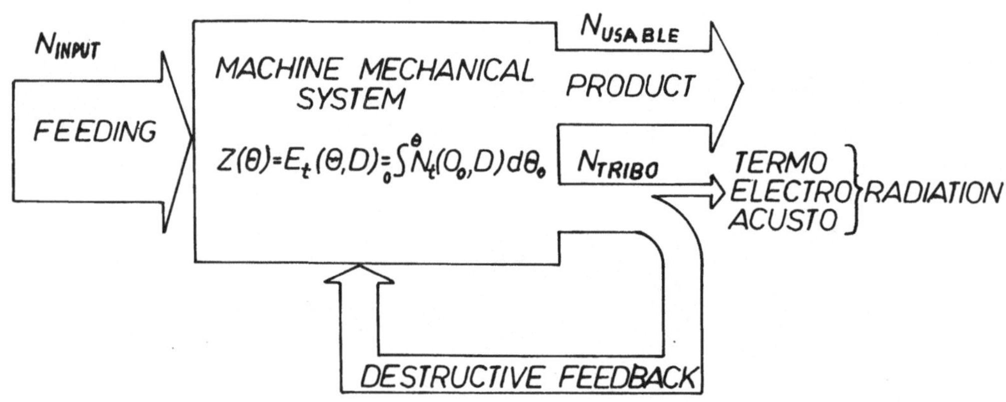

Fig.1. Machine as an acting energy system with
destructive feedback and wear advancement
as accumulated dissipative energy E_t .

$$\eta = \frac{N_u}{N_i} = \frac{N_i - N_t}{N_i} < 1 \qquad\qquad /3/$$

Hence: $N_t = (1 - \eta) N_i$.

That part of input power N_i is dissipated for tribological
processes and accumulated into wear advancement because

$$Z(\theta) = E_t(\theta,D) = \int_o^\theta N_t(\theta,D) d\theta_o \qquad\qquad /4/$$

The less the efficiency value η for a given machine type –
the much energy is transformed into wear and vibration. The
higher repair and lubrication quality the smaller fraction of

input power will be accumulated in wear processes advancement
$Z(\theta)$ hence the less will be the vibration amplitude.
This is known from machine condition monitoring practice of a
large group of machinery where the advancement of the wear is
covariable with vibration amplitude - D observed on some
criticall parts [4,5]. So on the basis of /4/ and the above
one can write

$$Z(\theta) = E_t(\theta,D) \sim D(\theta) \qquad /5/$$

This is the foundation of vast amount of practical knowledge
and technology of vibration condition monitoring, and also
this is the basis for new branch of science - vibroacoustical
diagnostics - emerging from the latest monographs /see [6 , 7]
for example/. But for a pity there is no explicit formulation
of that problem /5/ yet from the analitycal point of view.

3. Differential model of wear and vibration in a machine.

Expression /5/ being the paradigm of vibroacoustical diagnostics shows the global relation of covariability between wear
advancement and vibration amplitude. But this may be the result of some equality relationship on some local level such
as differentials of vibration amplitude and the loss energy.
In the earlier authors paper [1] the tribovibroacoustical
loss factor γ was introduced as this part of dissipated
energy which is going to be transformed into vibration. Let
us generalize this factor to the shape of life time θ dependent TVA loss function $\gamma(\theta)$ which is defined as previously
but on differential level. Going from above and the basic paradigm of vibroacoustical diagnostic /5/ we can write

$$dD(\theta) = \gamma(\theta) \cdot dE_t(\theta,D) \qquad /6/$$

This means that for arbitrary machine life time moment θ
some fraction $\gamma(\theta)$ of dissipated energy increment of dE_t is
converted into differential increment dD of machine's vibration amplitude.
This relationship can serve also as the definition and the way
of measurement of tribovibroacoustical loss function, namely

$$\gamma(\Theta) \equiv \frac{dD}{dE_t} \approx \frac{\Delta D(\Theta)}{\Delta E_t(\Theta,D)} \qquad /7/$$

It is seen here that this quantity has deep experimental meaning, it can be measured and interpreted during the machine
prototype design and after each repair procedure. The less
$\gamma(\Theta)$ value the better design or repair quality.
The differential of dissipated energy /6/ can be calculated
as a sum of two components

$$dE_t(\Theta,D) = \frac{\partial E_t}{\partial D}\, dD + \frac{\partial E_t}{\partial \Theta}\, d\Theta = E_t'(\Theta,D)dD + N_t(\Theta,D)dD. \quad /8/$$

The first term of the above expansion can be calculated if we
assume that derivative of dissipated power is constant being
explicit independent of time Θ

$$\frac{\partial N_t}{\partial D} = N'(\Theta,D) = \text{const} \qquad /8a/$$

Then we have

$$E_t'(\Theta,D) = \frac{\partial}{\partial D} \int_0^\Theta N_t(\Theta_o,D)\, d\Theta_o = \int_0^\Theta N_t'(\Theta_o,D)\, d\Theta_o =$$

$$= \int_0^\Theta N_t'(D)\, dO_o = N_t'(D) \cdot \Theta \qquad /9/$$

Putting the results of /9/ and /8/ into equality /6/ one can
obtain

$$dD(\Theta) = \gamma(\Theta)\left[\Theta \cdot N_t'(D)\, dD + N_t(D)d\Theta\right] \qquad /10/$$

And finally we obtain the differential equation governing the
transformation of a wear into vibration which may be called
the tribovibroacoustical /TVA/ differential model of the machine

$$\frac{dD}{d\Theta} = \frac{\gamma(\Theta)\, N_t(D)}{1 - \gamma(\Theta) N_t'\, D \cdot \Theta} \cdot \qquad /11/$$

Please note here that the wear advancement is equal to the
amount of dissipated energy which is covariable to the vibration amplitude /see(5)/. But the evolution of vibration ampl-

itude is governed by dissipated power N_t and TVA loss function $\gamma\,\Theta$. Please note also that for the new machine start-up: we have $\Theta \sim 0$, $D \sim D_o$ and the increment of vibration amplitude with respect to time is linear

$$dD \;=\; \gamma(0)\; N_t(\,D_o)\cdot d\Theta$$

what is in a good agreement with diagnostic practice $\lceil 5$ chapt. $1.3\rfloor$. In turn for the high value of the life time Θ , when denominator of /11/ is approaching zero in particular, the small time increment $\Delta\Theta$ gives very high vibration amplitude value $\Delta D \to \infty$, like during the machine break-down. Hence we can write

$$\text{If}\quad 1 - \gamma(\Theta)\; N'(\,D)\,\cdot\,\Theta \;\Rightarrow\; 0 \quad \text{then}\quad \Theta \Rightarrow \Theta_b$$

An the machine break-down time Θ_b is defined as

$$\Theta_b \;\equiv\; \Big[\gamma(\Theta)\; N'_t(\,D)\Big]^{-1} \tag{/12/}$$

Let us notice that we can use here the loss function definition $\gamma(\Theta)$, /7/ and the expression of power derivative. Doing that one can find

$$\Theta_b \;=\; \Big|\,\gamma(\Theta)\,N'_t(D)\,\Big|^{-1} \;=\; \frac{dE_t}{dN_t} \tag{/13/}$$

$$\underset{\sim}{} \;\; \frac{\Delta E_t}{\Delta N_t}$$

Thus measuring the dissipated energy and the power increment of a running – in machine one can assess its time to break--down Θ_b . It seems that this finding is of a great value, if valid, and can estimate machine time-to-break-down during its design stage and prototype study.

Now, going back to the differential equation /11/, one can put there the break-down time obtaining

$$\frac{dD}{d\Theta} \;=\; \frac{\gamma(\Theta)\,N_t(D)\cdot\Theta_b}{\Theta_b - \Theta} \tag{/14/}$$

But according to deffinition of Θ_b /13/ and $\gamma(\Theta)$ /7/ we can also write

$$\Gamma(\Theta) \cdot \Theta_b = \frac{dD}{dN_t} \qquad \text{or} \qquad dD = \Gamma(\Theta)\,\Theta_b \cdot dN_t \qquad\qquad /15/$$

So combining above two differential equation one can finally obtain

$$\frac{dN_t}{N_t} = \frac{d\Theta}{\Theta_b - \Theta} \qquad\qquad /16/$$

This is the differential equation which according to our assumption governs machine's dissipated power N_t during its life time Θ . We will analyze it in detail in next paragraph.

4. The vibration - life curve of a machine.

It is obvious here that our goal is to use the whole diagnostic potential of the differential equation /11/. Going to its transformed form /14/ we obtained differential equation for dissipated power

$$\frac{dN_t}{N_t} = \frac{d\Theta}{\Theta_b - \Theta} \qquad\qquad /17/$$

Solution of the above gives formula /18/ and Fig.2 ,

$$N_t(\Theta) = \frac{N_{to} \cdot \Theta_b}{\Theta_b - \Theta} \ , \qquad N_{to} = N_t(0) \qquad\qquad /18/$$

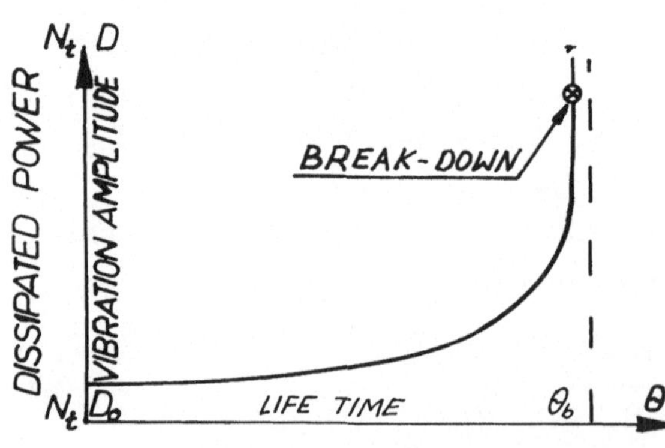

Fig.2. The time behaviour of a machine dissipated power; vibration amplitude, as a solution of differential equations /17/ /20/.

As we can see from Fig.2 the power of the wear processes $N_t(\Theta)$ increases slowly at the begining and very rapidly near the break-down time value - Θ_b . It seems to be true as complete self destruction of the machine during its break-down in not a seldom case. If we use just aquired knowledge on $N_t(D) = N_t\big[D(\Theta)\big] = N_t(\Theta)$ behaviour puting it to left hand side equation /14/ we may find

$$\frac{dD}{d\Theta} = \frac{\Gamma(\Theta)N_{to} \cdot \Theta_b^2}{(\Theta_b - \Theta)^2} \qquad /19/$$

Above differential equation presents the evolution of vibration amplitude $D(\Theta)$ during the life time of the machine. None of this equations /11,14,19/ was known yet and it is hard to assess their meaning already. But we may call it /19/ as another form of TVA machine model of which solution should give vibration-life curve of a machine.

Just to solve equation /19/ let us assume for a while the constancy of TVA loss function $\Gamma(\Theta) = \Gamma = \text{const}$. Hence almost imediately one can find

$$D = D0 = \frac{N_{to} \cdot O_b^2}{O_b - O} = \frac{D_o \cdot O_b}{O_b - O} \qquad /20/$$

$$D_o = N_{to} \, O_b$$

where of course $D_o = D(O)$ is initial vibration amplitude value of a machine.

As one can notice the behaviour of a machine vibration amplitude is colinear with dissipated power what is shown in common in Fig.2. In the broader sense of dependence between vibration D and dissipated power N_t coveriability is always valid. This may be easy shown when one remembers differential equation /15/ that is

$$dD = \Gamma(\Theta) \; \Theta_b \; dN_t$$

which implies solution

$$D = \Gamma(\Theta) \, \Theta_b \, N_t \quad \text{with} \quad D_o = D(\Theta = 0) = \Gamma(O) \, \Theta_b \, N_{to} \qquad /21/$$

So recalling differential equation /19/ we can write in simpler form

$$\frac{dD}{d\Theta} = \frac{D_o \cdot \Theta_b}{(\Theta_b - \Theta)^2} \qquad\qquad /22/$$

But recalling earlier form of this equation namely /14/ and combining with acquired just covariability /21/ one can find the simplest form of TVA differential model

$$\frac{dD}{d\Theta} = \frac{D\,\Theta_b}{\Theta_b - \Theta} \qquad\qquad /23/$$

It has to be taken into account that both above differential equations have identical solutions of the form of /20/ as it is illustrated on Figure 2. But the different forms of TVA machine model have been shown here because there may be preference of some form in some specific applications. One thing is worth here for discussion namely just found covariability relation /21/ between vibration amplitude D and the dissipated power N_t . Confronting it with our initial assumption /8a/ namely $\frac{dN_t}{dD}$ = const, so it follows

$$\frac{dD}{dN_t} = \text{const.} = \gamma(\Theta)\,\Theta_b \quad . \qquad\qquad /24/$$

Hence it follows from the above that increase in tribovibro-acoustical loss function $\gamma(\Theta)$ decreases the break-down time and vice versa. This conclusion seems to be in good agreement with observed machines behaviour, it confirms also the validity of our initial assumtion /8a/.

5. Applications of TVA machine model.

The theory presented here is quite new and it seems that all its implication and possible application are yet not fully known. Despite of this the three results of this paper have strong applicational meaning: analytical definition of a time to break-down /13/, vibration-life curve /20/ and TVA machine model /23/ in its difference approximation form in particular. This possibilities have been discussed extensively in earlier authors publication $[16]$, hence here we will take into account condition monitoring applications only. This area of our results application is similar to $[1]$ - the prediction of future vibration amplitude i.e. condition and time to break-down assessment from series of observation $D(\Theta_t) = D_t$, $t = 1,2,\ldots$ of a running machine. It may be done on two mays; by using the vibration-life curve /20/ and curve fitting procedure segmentally, and secondly by using differential ewuation /22/ in its finite difference form.

Starting to vibration amplitude and the same the machine condition forecasting from /20/ we will be looking for D_{t+1} , while having previous readings D_t , $t=1,\ldots$

$$D_{t+1}^{P} = \frac{D_o \cdot \Theta_b}{\Theta_b - \Theta_{t+1}} \qquad /25/$$

where Θ_b is not known and D_o is known only under the assumtion of $\gamma(\Theta) = \gamma = $ const. /then $D_o = D_1$ /.

But forming the expression for the average square error

$$B = \frac{1}{N} \sum_{n=1}^{N} \left(D_n^{P} - D_n \right)^2 = \frac{1}{N} \sum_{n=1}^{N} \left(\frac{D_o \, \Theta_b}{\Theta_b - \Theta_n} - D_n \right)^2 \quad /26/$$

and taking the exstremality condition with respect to unknowns $\frac{\partial B}{\partial D_o} = 0$, $\frac{\partial B}{\partial \Theta_b} = 0$ one can find

$$D_o = \frac{\displaystyle\sum_{n=1}^{N} \frac{D_n}{\Theta_b - \Theta_n}}{\displaystyle\sum_{n=1}^{N} \frac{\Theta_b}{(\Theta_b - \Theta_n)^2}} \qquad /27/$$

where D_n^{P} is predicted amplitude, D_n - measured amplitude for a life time Θ_n , and the Θ_b calculation may be done only numerically from the second extremality condition.

Another way to prediction of D_{t+1} and Θ_b comes from differential equation /22/. Writing it in finite difference form we have

$$\frac{\triangle D}{\triangle \Theta} = \frac{D_o \cdot \Theta_b}{(\Theta_b - \Theta)^2} \qquad /28/$$

what for starting point $D_t = D(\Theta_t)$ gives prediction formula

$$D^P_{t+1} = D_t + \frac{D_o \cdot \Theta_b (\Theta_{t+1} - \Theta_t)}{(\Theta_b - \Theta_{t+1})^2} \qquad /29/$$

The way calculation D_o , Θ_b leads as previously through average square error minimization like in /26/. Here after some calculation one can obtain

$$\tilde{D}_o = \frac{\sum_{n=1}^{N} \left[D^P_n - D_{n-1} \frac{\Theta_b (\Theta_n - \Theta_{n-1})}{(\Theta_b - \Theta_n)^2} \right]}{\sum_{n=1}^{N} \left[\frac{\Theta_b (\Theta_n - \Theta_{n-1})}{(\Theta_b - \Theta_n)^2} \right]^2} \qquad /30/$$

The time to break-down Θ_b estimation may be done only by numeric way as previously.

Summing up all possible application of our results we may say that first way of Θ_b assessment /13/ may be done by experiment only. But the second way together with vibration amplitude prediction is possible when we have a series of vibration observations taken from running machine. We will deal with that in the next point.

6. Examples of forecasting by means of TVA machine model.

For verification of prognostic potential of our TVA machine model let us apply it to the series of vibration observation taken from the body of 12-cylinder railroad diesel engine [8]. The succesive prognosis D^P_{t+1} were calculated by means of two formulae /29/ and /33/ resulting from TVA model. These have been compared with the simplest exponential prognostic technique [9,10] as below

$$D^P_{t+1} = D_t \exp \lambda_t \cdot (\Theta_{t+1} - \Theta_t) \simeq D_o \exp \lambda_t \cdot \Theta_{t+1} \qquad /31/$$

where

$$\lambda_t = \frac{1}{t} \sum_{n=1}^{t} \frac{\ln \dfrac{D_n}{D_{N-1}}}{\Theta_n - \Theta_{n-1}} \qquad /32/$$

enables to assess the time to break-down because: $\Theta_b \simeq \lambda_t^{-1}$.
Such prognostic calculations have been carried out for nine
vibration measuring points of engine body [16] and typical
results are shown in Fig.3 . Here the engine life time is measured in thousend of kilometers and locations of vibration
acceleration measuring point is shown in upper left hand side
of figure. In the middle we have thick curve of vibration observations and thin forecastings curves according to TVA model
/25/ and exponential model - E /31/. In the lower right hand
side the assessment of the time to break-down Θ_b by the TVA
model is also shown.
The average prognosis error of a last point for each technique
is done in parenthesis e.g TVA, /2.3%/ and E, /4,0%/. In general /including prognostic cases not shown here/[16], one may
say that model TVA /curve fitting/ gives more averaged prognosis results with slightly higher error than finite difference model. What concers time to break-down Θ_b estimation
the results are less stable as concerned arrows go autside the
range of the Figure. The ultimate results of Θ_b calculations
for techniques TVA and E oscilate between $400 \div 900 \cdot 10^3$ km, and
finite difference model is much more unreliable and unstable.

Summing up calculated and presented results one can say both
TVA models are prognostic oriented although the fitting of
vibration-life curve technique gives more stable results for
vibration amplitude and time to break-down prediction.

7. Conclusions

The paper concerns interrelation between wear processes and
vibration of a machine; interrelation being the paradigm of
vibrational diagnostics. At first the measure of wear advancement as the energy amount dissipated into tribological processes is postulated. This is colinear with measured vibration
amplitude, hence differential equivalence of vibration and

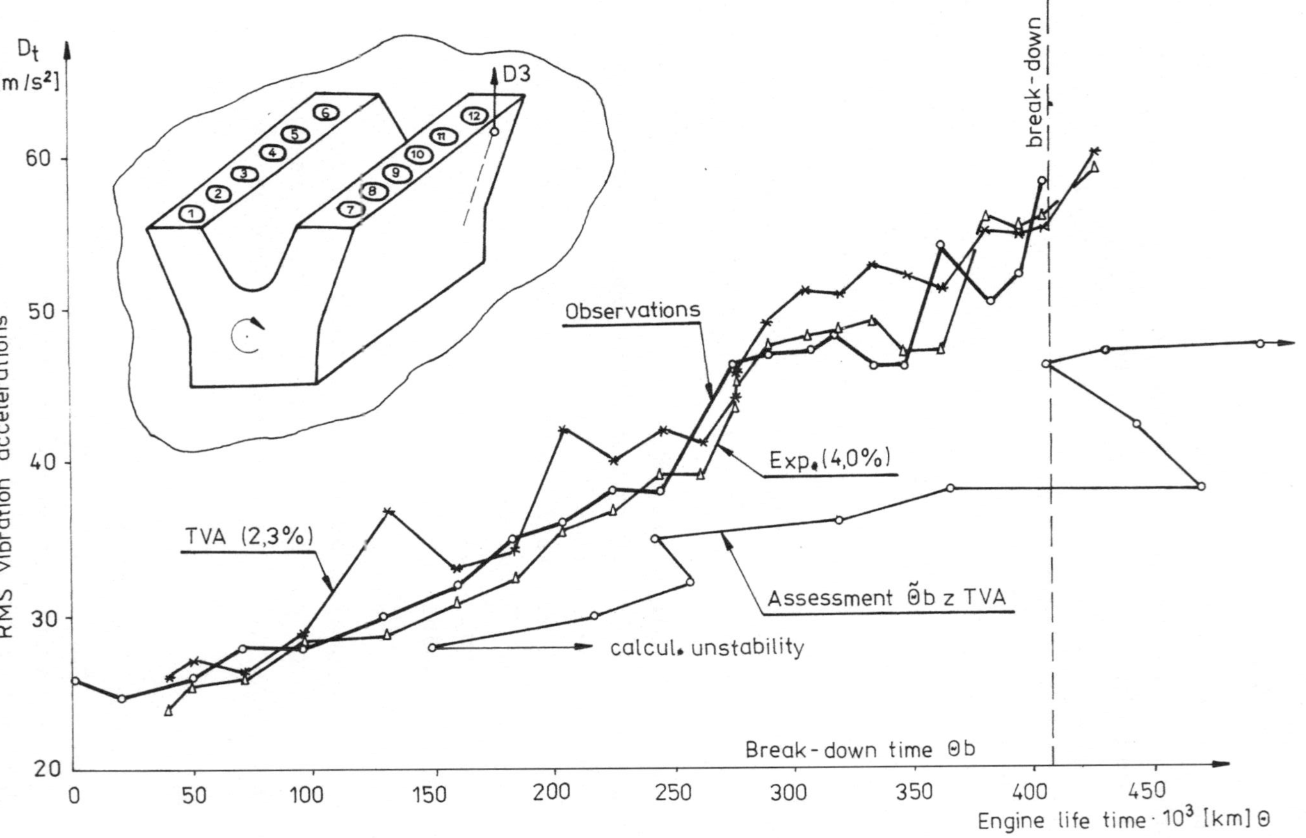

Fig 3. Vibration accelerations of railroad diesel engine (point D3) in comparison with the results of different prognostic techniques according to TVA and exponential (E) models (values in parenthesis = last prognosis error).

dissipated energy increments was proposed. As a result the
differential equation /TVA model/ governing the evolution of
machine vibration amplitude was obtained. Analysis and solution of this equation gives:

1. Analitycal definition of time to break-down of a machine
   and proposal for its experimental verification.
2. Vibration life-curve of the machine which can be used with
   succes to machine vibration prediction and time to break-
   down assessment in the running phase of machine life time.
3. Finite difference TVA machine model which can be use for
   the some purposes as above.

The prognostic abilities for wear advancement identification
of both version of TVA model were experimentally verified
giving good results. But it seems that much more theoretical
and experimental work is needed in order to assess the diagnostic potential of the presented theory.

APPENDIX   <u>Energy realion of a wear and vibration in machinery
elements.</u>

<u>A. Fatigue</u>

A.1. <u>Volumetric fatigue</u>

According to Morrow hipothesis Kliman [1] gives the following
way of fatigue energy calculation

$$E_{t1} = E_f = 3 \left(\frac{\sigma_a}{k}\right)^{1+n} \cdot N_f = 3\left(\frac{\sigma_a}{k}\right)^{1+n} \cdot f \cdot \Theta$$

where: k, n – material constants, σ_a – alternative stress
amplitude, N_f – number of cycles to failure, f – frequency of
stress pulsation, Θ – life time of element.

A.2. <u>Surface fatigue – pitting</u>

According to Palmgren relationship [12] the energy dissipated
by pitting may be assessed as below

$$E_{t2} = E_p = c_1 p^3 N_f = c_1 p^3 f \Theta$$

where additionally to previous notation: c_1 – proportionality
coefficient, p – unit pressure in kinematic pair.

A.3. Corrosive fatigue - fretting

The wearing energy will be here proportional to the volume - U of removed material [12]

$$E_{t3} = E_{fr} = c_2 U = c_2 \cdot \left[\frac{k_o \cdot \sqrt{p} - k_1 \, p}{f} - k_2 \, A \, p \right] \cdot f \cdot \Theta$$

where additionally: k_o, k_1, k_2 - material constants, A - vibration displacement amplitude, c_2 - proportionality coefficient.

B. Friction wear

The wearing energy will be proportional here, as previously, to the volume of abrasively removed material - U, which is governed by Archard's law [12,13]

$$E_{t4} = E_a = c_3 U = \frac{c_3 \cdot k_3 \cdot p \cdot V \cdot \Theta}{Re}$$

where in addition: c_3 - proportionality coefficient, k_3 - material constants, V - sliding velocity in kinematic pair, Re - yield stress of a material.

C. Distortional wear - creep

It may be acknowledged here that dissipated energy is proportional to creep strain - ε_{cr} and this is higly dependent from [14]: stress - σ , temperature - T , and amplitude and frequency of ultrasound [15]

$$E_{t5} = E_{cr} = c_4 \cdot \varepsilon_{cr} = c_4 \left[\frac{\sigma}{E} + e \, (\, T, f) \, \sigma^d \cdot \Theta \right]$$

where additionally: c_4 - proportionality coefficient, e,T,f - creep function of temperature and frequency, E - Young's modul, d - material constant.

D. Erosion

With the exclusion of chemical corrosion the mechanical vibration are cause and/or results of erosion. The most pronounced it is for erosion by cacitation, whery the volume of material removed by cavitation, and the same dissipated energy is

$$E_{t6} = E_{cav} = c_5 U = c_5\ B \bullet \Theta \bullet V^{n(\Theta)}$$

where in addition to previous notation: c_5 - proportionality constant, V - material constant, V - velocity of liquid flow, $n(\Theta)$ - cavitation exponent.

E. The vibration and the wear energy

Now we have to acknowledge that each load parameter in relationship E_{ti} has its own dynamic component of: stress $\sigma = \sigma_m + \sigma_a$, unit presure $p = p_m + p_a$, velocity $V = V_m + V_a$, and these dynamic components are proportional to vibration amplitude: σ_a , p_a , $V_a \sim D$. In dependenceof a wear form it will to the vibration velocity amplitude /like in fatique/ the vibration displacement amplitude /like in fretting/ and so on. Suming up now the all forms of wear we can write the folowing quantitative relation for wearing energy

$$E_t = \sum_1^6 E_{ti} = E_t\left(\Theta, D, k_j, R_e, V_m, p_m, \sigma_m, T, \text{design par.}\right)$$

For given machine type, i.e established set of design parameters, and for established working and load condition this energy will depend only on life time - Θ and vibration amplitude

$$E_t = E_t(\Theta, D)$$

The accumulation process of a wear dissipated energy will depend of course on instant dissipated power - $N_t(\Theta, D)$ as

$$E_t(\Theta, D) = \int_0^\Theta N_t\left(\Theta_o, D\right) d\Theta_o .$$

Finally for continuous running machinery with constant load and uniform wearing the dissipated power will explicit depend on vibration only.

$$N_t(\Theta, D) = N_t\left(D(\Theta)\right)$$

Hence

$$E_t(\Theta, D) = \int_0^\Theta N_t\left(D(\Theta_o)\right) d\Theta_o .$$

References

1   Cempel C., The tribovibroacoustical model of machines, Wear, 105 /1985/, 297-305.

2   Lyon R.H., Statistical energy analysis of dynamical systems, MIT Cambridge Press, 1975, chapt.1.

3   Cempel C., The fatique limit for vibration of machines and structural elements, Zagadnienia Eksploatacji Maszyn, No 4 /56/, 1983, 551-561.

4   Collacott E.A., Mechanical fault diagnostics and condition monitoring, Chpman and Hall, London, 1977.

5   Cempel C., The foundamentals of vibroacoustical diagnostics, WNT Press, Warsaw, 1982 /in Polish/.

6   Yavlenskyi K.N., Yavlenskyi A.K., Vibrodiagnostics and quality forecastings of mechanical systems, Mashinostroeniye, Leningrad 1983, /in Russian/.

7   Cempel C., Vibroacoustical diagnostics of machinery, T.U. Poznań Press, 1985 /in Polish/.

8   Tomashevsky F., Vibration diagnostics of railroad diesel engine, PhD T.U. Poznań, 1987, /in Polish/.

9. Cempel C., Simple forecasting techniques in vibration condition monitoring, VII Diagnostic School DIAGNOSTICS-85, Rydzyna, September 1985, 141-150 /in Polish/.

10 Cempel C., Simple condition forecasting techniques in vibroacoustical diagnostics, Mechanical Systems and Signal Processing Vol.1, No 1, 75-82.

11 Kliman V., Fatigue life estimation under random loading using the energy criterion. International Journal of Fatigue, No 1, 1985, 30-44.

12 Łuczak A., Mazur T., Physical wear of machinery elements, WNT Press, Warsaw 1981, chapt.2 /in Polish/.

13 Engel P.A., Impact wear of materials, Elsevier, New York, 1976, chapt.1.

14 Juvinall R.C., Stress, strain, and strength, Mc Graw Hill Book Comp. New York 1967, chapt.19.

15 Severdenko C.P., and others, Ultrasound and plasticity, Minsk, 1976, /in Russian/.

16 Cempel C., The wear process and machinery vibration - a tribovibroacoustical model, Bull. Pol. Ac. Techn. Vol 35, No ..., pp ... /just in print/.

Bridge Inspection by Dynamic Tests and Calculations Dynamic Investigations of Lavant Bridge

R. G. FLESCH [1], K. KERNBICHLER [2]

1. OBJECT

A dynamic method for the damage evaluation of large bridges is developed by BFVA and Technical University Graz. The method is a combination of dynamic in-situ tests and dynamic calculations and is also useful for the general improvement of mathematical modelling for static and dynamic calculations. Experiences with prestressed bridges were obtained. In general damages decrease the stiffness and increase damping. The structures are very large and the influences of cracks on dynamic parameters are very small. The basic concept of the method is to elaborate a dynamic model in the virgin state of the structure. Damages of the structure will lead to a certain pattern of deviations of dynamic parameters from the virgin state and can be used for localisation and quantification of damages in a global manner. The following steps are used:

* elaboration of a mathematical dynamic model for vertical direction
* carrying out of dynamic in-situ tests on the bridge
* fitting of mathematical model to the test results →optimum virgin model
* carrying out of dynamic in-situ tests after a certain time of use
* if changes of modal frequencies, modeshapes and damping ratios are detected the mathematical model must be fitted again to the test results
* comparing the old and the new mathematical model e. g. changes of stiffness in certain regions could be detected and quantified

Authorities are very interested also in methods for existing structures where no virgin model is available. Test phases carried out with different

---

1) Head of departement for Structural Dynamics/BVFA-Vienna
2) Institute for R/C Constructions, Technical University Graz

levels of excitation force could result in different nonlinear reactions in areas of damage. If the bridge behaves mainly linear in non damaged condition maybe only some dynamic parameters, especially damping, could change. Hence, an identification could be possible.

The method can be divided into the following parts (see fig. 1):
* mathematical modelling
* in-situ test - technique
* analysis of test results (1. step of identification: elaboration of modal test model)
* systematic system identification (fitting of mathematical model to modal test model)

Especially for the last problem a cooperation with Curt-Risch-Institute was started.

Hence, in the "classical" approach parametric identification is used. Natke and Yao /1/ suggest the use of non-parametric identification for routine inspection. If the variations are within a defined range, no significant damage of the structure is indicated and no parametric model is necessary for localisation.

In this paper a new approach will be presented, using a structural dynamics modification (SDM) software. The test results of Lavant bridge were analysed by a modal analysis software. The sensitivity of the modal parameters to the decrease of stiffness and the increase of damping in certain locations was investigated.

2. DEVELOPMENT OF THE METHOD

2.1. Projects of the past

From the investigations of bridge Raach /2/ it was found that the method works in principle.

During the investigation of bridge Obernberg /3/ the test technique could be improved significantly. Considering the modes 1 - 7, a maximum difference Δf = 0,006 Hz between measured and calculated modal frequencies was obtained. Two of the higher modes could also be fitted well to the test results via a more detailed modelling of the open caisson foundation of pier 2. The best results were obtained from the investigations of Gänstorbrücke /4,5,6/ in Ulm/FRG. The tests were carried out before and after repair work. Three phases were investigated. Changes of modal bending frequencies and modal torsional frequencies were detected. The increase of stiffness resulting from the repair work was elaborated from the mathematical models.

2.2. Test technique

The developement of the test technique was reported in /5,7/. It is important for practical use that at least great parts of the tests can be carried out under traffic. As frequency changes of 0,01 Hz must be detected a very precise technique is necessary. Different methods can be used for excitation. In our opinion the use of moveable sweep sine exciters is preferable. All other methods (impulse, ambient) have limitations e. g. energy problems, too short duration of excitation, limited frequency band of excitation which can be controlled hardly, fixed points of excitation, no possibility for excitation of higher modes, etc. In the past an eccentric mass exciter was used. By a static frequency changer the frequency can be controlled with an accuracy of 0,003 Hz. The exciter has a total mass of 1500 kg and need not to be fastened to the structure. The disadvantage of the eccentric mass exciter is the quadratic force and the very low force amplitude in the low frequency range.

From the experience, a minimum force of 1 kN should be available in the range from 0.2 - 10 Hz (or more). At the moment BVFA develops a reaction mass exciter driven by an hydraulic actuator. The excitation force can be kept constant during the sweep and much more force is available in the low frequency range.

The response is measured by velocity transducers Hottinger SMU 30 A. As there are sometimes distances up to 1 km between transducer and magnetic tape recorder, the amplifiers are mounted in boxes together with the transducers. The transfer function of every transducer was elaborated on the BVFA - shaking table. The function is approximated analytically and used for the correction of the measured transfer functions via a complex division. Hence, the lowest frequency which can be measured accurately is 0,5 Hz. BVFA plans to develope also a transducer for measurements down to 0.1 - 0.2 Hz in the future.

Earlier the tests were divided into a short - sweep and a precision measurement /5/. Only a small part of the information available is obtained by the precision measurement. Hence, the tendency is towards an improved (short) sweep. More investigations are necessary to find the optimum total sweep time and sweep rate. The sweep rate could be e. g. decreased when transducers show resonances etc. Further the range could be swept more then one time. Taking into account frequent damping ratios it should be checked if the resonances can really develop.

With the hydraulic exciter also white noise can be used as control signal. This kind of excitation would be good to eliminate disturbances e. g. by traffic vibrations, but maybe the energy is not sufficient for the broad band excitation.

It is assumed that 2 - 3 positions of the exciter will be necessary to obtain 40 - 60 modes, depending on the dynamic system. It was shown that also higher modes, which are often sensitive to damages, can be excited well using resonance methods. For the bridges with open cross sections also torsional modes are helpful to evaluate damages. Additional points of excitation could be necessary.

Depending on the location of the excitation some modes behave more ore less complex. The phase angle depends on the excitation point.

Investigations will be carried out. It is assumed that information about the condition of the structure can be obtained also from these informations.

Tests at different force levels will be carried out to increase information.

Additional masses (e. g. 40 ton lorry) placed systematically during the tests could be used to increase the opening of cracks and hence to decrease the stiffness further for a higher probability of detection.

If the tests must be carried out under traffic, disturbances must be minimized further. In the project Lavant mean values and variances of the coherence function were calculated for every mode.

2.3. Analysis of measurements

From records of the respose and of the excitation force transfer functions are obtained by FFT - Analysis. In the earlier projects only short sweep responses /5,7/ were recorded on tape. The results of precision measurement were calculated on line by a computer. Peak picking was used to get the modeshapes from the transfer functions. Circle fit algorithm were used to analyse the data from precision measurement.

In the future digital zooming will be used to increase frequency resolution.

For the analysis of Lavant bridge the SMS modal analysis software MODAL 3.0 could be used the first time. The software is very modular and provides good fitting algorithm and very comfortable handling possibilities.

Especially the parameter estimation using rational fraction polynomials /8/ works very well. The total concept seems to be trimmed toward mechanical engineering problems. Some special problems of our tests on very large civil engineering structures had to be solved by additional software, e. g.:

* complex division of transfer function by transducer transfer function and division by ω^2 because of quadratic force, both before scaling to internal data format of MODAL 3.0 due to accuracy problems
* calculation of coherence and of a fit quality criterion
* the MODAL 3.0 strategy to estimate frequencies and damping from one or few selected transfer functions and to obtain the residues by autofitting the transfer functions does not work well. Too much information is lost. Further, the global fit option which should elaborate frequencies and damping from all transfer functions gave very bad result (e. g. negative damping or very low damping ratios).
* with a prefit - program the problem was solved. The program detects automatically every peak of the imaginary part of the transfer function. A printout shows all measuring points were a special peak occurs. In addition the mean coherence of the "potential" mode is listed. With this information the operator selects band wides and number of DOF's for parameter estimation. Then all transfer functions are fitted in that way. Before averaging to obtain global frequencies and damping ratios bad data are eliminated taking into account the following criteria:
  - deviation from mean value
  - negative damping
  - damping less than 0,8 %
  - phase angle
  - fit criterion

2.4. Dynamic calculations

The dynamic calculations were carried out by Technical University Graz. SAP IV and FLASH were used for mathematical modelling. During every project attempts were made to find the model with the adequate level of accuracy. Only beam elements and additional spring elements were used. In our opinion the inclusion of plate elements e. g. will not improve the modelling, because the material parameters of civil engineering structures are often very uncertain. Further, a real structure has deviations from plan - cross section, variances of mass density etc. All this must be taken into account via correction factors.

About 1200 DOF's were used for the models. The haunched beams caused problems at the beginning. Now the moments of inertia are approximated well using harmonic averaging /9/.

In general, the bending stiffness can be modelled well but problems can arise for the modelling of torsional stiffness and cross section deformation for open cross sections.

To get precise results important details must be modelled well. In some cases the transversal coupling of adjacent bridges via the carriageway slab had to be modelled. The problem was solved by eccentric connected beam elements /3,5/. The influence of the foundation was sometimes modelled well by additional springs. In the case of Obernberg bridge with one predominant open caisson foundation two higher modes were very sensitive on the modelling of that foundation /3/.

Modes with vectors in longitudinal direction are often sensitive to elongations due to temperature changes. This had to be taken into account when fitting the model to test results /4,5,6/.

2.5. System identification

The state of the art is known well by the participants.

In the projects of the past the system identification was carried out in two relatively simple steps:
  * the origin model was fitted to the test results in a global manner
    via the modulus of elasticity. Hence, the modulus was the central fit
    parameter, representing also uncertainties and deviations of cross
    section, mass densities and boundary conditious etc.
  * by trial and error in the areas of repair work, especially in the
    project Gänstorbrücke /4,5,6/
For routine inspection systematic methods are necessary. Methods were developed by Natke but they are limited to models with 30 DOF's at the moment.

3. DYNAMIC INVESTIGATION OF LAVANT BRIDGE

3.1. Dimensions and program

Lavant bridge is one of the greatest prestressed framed bridges in Europe and consists of a large main bridge and a smaller hill side bridge. The main bridge has 6 spans. The maximum span is 160 m. The maximum shaft height is 130,35 m. The dimensions and cross sections are shown in fig.2.

Dynamic in-situ tests were carried out in November 1985 in the virgin state. The bridge was opened for traffic in June 1986. The bridge was investigated in horizontal and in vertical direction. The aim of horizontal modelling was to improve modelling for earthquake calculations. The results are given in /9/ and will be presented at the 9. WCEE in Tokyo.

The results of the tests in vertical direction are used to elaborate the virgin modal model for the bridge.

The tests were carried out without traffic disturbances.

Although the test technique was at a high level during the tests some principle mistakes occured. From errors a lot can be learned for further projects, hence, the mistakes will be discussed:
* as it was easier to move the exciter than 8 transducers the "fixed" transducers - "roving" exciter concept was used.
  It was planned also to excite torsional modes, hence, the exciter roved first along the centerline and then eccentrically along the eastern guardrail. During eccentric excitation the short sweep was too quick with the given force level. Realistic shapes could be only elaborated along the centerline.
* no measurements were carried out at the top of the piers. For some modes vertical displacements could have been measured.
* no measurements and excitations were carried out in longitudinal direction of the bridge
* the old concept short sweep + precision measurement was used. It

would have been better to try to improve the short sweep because the new analysis concept needs complete transfer functions.

For the calculation SAP IV and a 2-D model were used. The model consists of 248 elements. The foundation was modelled by 24 springs.

3.2. Results

If the in-situ tests would be carried out again using the experience from the first test, the new hydraulic exciter and the improved sweep technique more and better data would be obtained. In the given situation no digital zooming was used. For the 10 Hz band the frequency resolution is

Δf = 0,025 Hz. Using the procedure described in 2.3 a systematic evaluation of global frequencies and damping was possible. Before the elaboration of modal frequencies and damping ratios bad data were eliminated in the following cases:
   * frequency values out of range mean value $\mp$ 0,025 Hz
   * negative damping or damping < 0,8 %
   * phase of residue between 45° and 135° or between 225° and 315°

Due to the problems reported in 3.1. no information about torsional modes is given.

At the moment the signals of transducer 5 and 7 are analyzed. Some further modes could be elaborated from the signals of transducer 1 and 3. Hence, disregarding the useless information of the transducers in eccentric position, 50 % of information is available at the moment.

The modal frequencies and damping ratios obtained from the test are given in table 1. The results of the calculations are given in table 2. The SMS user program "Modal Assurance Criteria" was used to identify statistical correlations between the modes. Some results of "Auto MAC" calculations available at the moment are given in Table 3. The correlation is not disscussed in the paper but will be the basis for the elimination of bad results. Maybe some results will be replaced by modes elaborated from the transducer signals 1 or 3.

Using 9   modes for the global fit a modulus of elasticity of $3,9.10^6$ N/cm^2 was obtained. The maximum difference between measurement and calculation is 0,056 Hz.

Some examples of calculated and measured modeshapes are given in fig. 3,4.

In chapter 3.3. a new approach is carried out to investigate the sensitivity of the dynamic properties to structural changes using the SDM - software. The modal model is the basis for the simulations. It is assumed from now on that modal frequencies and damping ratios are correct up to three decimals. Taking into account the improvements planned, this could be true in the future. For the moment, it is the aim to show the principle possibilities.

3.3. Sensitivity of dynamic properties to structural changes

3.3.1. Theory of SDM

The SDM software can be used together with Modal 3.0. The theory is described in /10/.

The unmodified modal model is used to calculate the effect of local increase or decrease of mass, stiffness and damping on the modal properties. Before using SDM the modes must be normalized to unit modal mass.

Mass changes can be modelled at one DOF. Changes of stiffness and damping can be modelled between two DOF's. For the modification measured DOF's or DOF's interpolated by constraint equations can be used.

Stiffness modifications are implemented using a technique called Eigenvalue Modification first developed by Kron /11/. The equation of motion of the modified system in modal coordinates is:

$$[\bar{M}]\{\ddot{z}(t)\} + [\bar{C}]\{\dot{z}(t)\} + [\bar{K}]\{z(t)\} = [\Phi]^T\{f(t)\} \tag{1}$$

with:

443

$$[\overline{M}] = [\diagdown\, \diagdown] + [\emptyset]^T [\triangle M][\emptyset] \tag{2}$$

$$[\overline{C}] = [\diagdown\, 2\xi\Omega] + [\emptyset]^T [\triangle C][\emptyset] \tag{3}$$

$$[\overline{K}] = [\diagdown\,\Omega^2] + [\emptyset]^T[\triangle K][\emptyset] \tag{4}$$

and:

$\qquad$ $[\triangle M]$... matrix of mass modifications

$\qquad$ $[\triangle C]$... matrix of damping modifications

$\qquad$ $[\triangle K]$... matrix of stiffness modifications

$\qquad$ $[\emptyset]$... matrix of unmodified eigenvectors

$\qquad$ $[\diagdown\, \diagdown]$... identity matrix (unit modal masses)

$\qquad$ $[\diagdown\, 2\xi\Omega\diagdown] = [\emptyset]^T[C][\emptyset]$ ... modal damping matix

$\qquad$ $[\diagdown\,\Omega^2\diagdown] = [\emptyset]^T[K][\emptyset]$ ... modal stiffness matrix

Solving the determinant equation (5)

$$\|[\overline{M}]s^2 + [\overline{C}]s + [\overline{K}]\| = 0 \tag{5}$$

the eigenvalues

$$\overline{p}_\kappa = \overline{\sigma}_\kappa + i\,\overline{\omega}_\kappa \qquad k = 1....m \tag{6}$$

of the modified structure are obtained.

Solving the homogeneous equation (7)

$$\left\{ [\overline{M}]\,\overline{p}_\kappa^2 + [\overline{C}]\overline{p}_\kappa + [\overline{K}] \right\}\{\overline{u}_\kappa\} = 0 \tag{7}$$

the eigenvectors of the modified structure are found to be

$$[\overline{\emptyset}] = [\emptyset][\,\{\overline{u}_1\}, \{\overline{u}_2\}, \dots\dots \{\overline{u}_m\}\,] \tag{8}$$

To investigate the influence of local cracks of bridges, the decrease of
bending stiffness must be modelled. In /12/ a solution is presented. Rib
stiffeners can be approximated well by a 3 - DOF beam element (see fig.5).
In our case a negative stiffener had to be modelled. The matrix of the
stiffness modification is found to be:

$$[\triangle K] = [\triangle K_1] + [\triangle K_2] + [\triangle K_3] \tag{9}$$

with

$$[\Delta K_1] = \frac{3EI}{a^2 b} \begin{bmatrix} -1 & 1 & 0 \\ 1 & -1 & 0 \\ 0 & 0 & 0 \end{bmatrix} \tag{10}$$

$$[\Delta K_2] = \frac{3EI}{ab^2} \begin{bmatrix} 0 & 0 & 0 \\ 0 & -1 & 1 \\ 0 & 1 & -1 \end{bmatrix} \tag{11}$$

$$[\Delta K_3] = \frac{3EI}{abL} \begin{bmatrix} 1 & 0 & -1 \\ 0 & 0 & 0 \\ -1 & 0 & 1 \end{bmatrix} \tag{12}$$

These stiffnesses are now in terms of the basic SDM scalar stiffness modifications and can be implemented easily.

3.3.2. Expected effects of different damages on dynamic properties

In general damages will decrease stiffness and increase damping. Concrete bridges are mainly damaged by bending cracks. Not only the crack itself, but also the adjacent regions of disturbed bond result in an stiffness decrease. Friction damping in the area of disturbed bond and at the crack edges increase damping depending on the magnitude of the relative displacement. Only seldom shear failure occurs. For other types e.g. composite constructions also shear could cause damages.

Changes of torsional modes are useful for damage evaluation of bridges with open cross sections. Due to the cross section of Lavant bridge they are of no interest in this project.

Assuming linearity, the prestressing force itself or changes of prestressing, settlements of foundation, changes of the system's centerline, do not directly change the dynamic parameters. Many of these changes produce constraint stresses. Often prestressed bridges react to unforeseen constraints by cracking. Hence, these changes could be detected indirectly via the cracks.

In principle static stresses could move the center of the small amplitude vibrations to another position of the nonlinear material working line. This would give a changed local tangential modulus of elasticity resulting in a changed local stiffness. Therefore changes of static stresses or additional constraint stresses could produce a stiffness change. Further in the opinion of some authors static stresses could have an influence on material damping. Influences of temperature on the test results must be studied very carefully, especially on modes with significant longitudinal movement. The influences of damages of bearings and carriageway expansions must be studied in the future.

Some expected effects are compiled in fig. 9.

3.3.3. Results of simulations

Simulations were carried out only for the main bridge.

Measurements are available in the points 41 - 64 exept of the pier points. For the points 220 - 235 shape vectors were calculated using the constraint equation. The situation is given in fig.6 together with a good approximation of the moment of inertia distribution which was proved by the good results of the calculations. Exactly, the distribution of $3EI/l^3$ is given as an equivalent spring, which is equivalent to equ. (10) and (11) for a = b = length of two adjacent elements.

Now the sensitivity of the bridge is investigated implementing a "roving" negative rib stiffener in three adjacent DOF's.

In this way influence lines could be elaborated, giving the decrease of the modal frequencies due to the reduction in two adjacent elements. To elaborate these influence lines an amount of 10 % of the equivalent springs was used. The differences can be recalculated linearly for any other amount.

Similar, as a first approach, a "roving" damping element using also three adjacent DOF's was modelled. In this case an unique equivalent damper with 10000 Ns/cm was used for calculating the influence lines.

The results show clearly that some modes are very sensitive on changes in certain areas which is good for localisation of damages. From our experience cracks would result in a stiffness reduction of few percent for such long elements, but the results show that the differences could be detected experimentally. Some examples of frequency changes in past projects are given in table 4. In the future the number of measuring points will be increased to increase the probability for damage localisation.

Finally the influence of a "roving" mass was studied to get information on the influence of test equipment and possible additional masses which could be used to open cracks.
Some results are given in fig. 7 - 8.

The following investigations will be carried out:
* replacing of bad modes by results from transducers 1 and 3
* eventually detection of additional modes
* interpolation of additional points in one bridge span to reduce element length for the sensitivity-investigations
* improvements in modelling of 3 DOF roving negative stiffener and damper
* study of changes of modeshapes due to simulated changes

Results will be given in forthcoming papers.

4.CONCLUSION

It is assumed that structural dynamic modification is a powerful mean to get information on the sensitivity of structures to changes of stiffness and damping. The model used is directly coming from the real structure. A lot of possibilities was discussed for a further improvement of the elaboration procedure.

It is assumed that also the use of non parametric information suggested by Natke and Yao /1/ will be helpful to establish limits of negligible changes for the use in routine inspection procedures.

If localisation and quantification is necessary a procedure using influence lines could be used:

$$\left\{ (\omega_i^V - \omega_i^M) \right\} + [A]\,\{d\} = 0 \tag{13}$$

and

$$\left\{ (\xi_i^V - \xi_i^M) \right\} + [B]\,\{d\} = 0 \tag{14}$$

with

$\omega_i^V \text{ and } \xi_i^V$ ... i-th modal frequency and damping
in virgin state

$\omega_i^M \text{ and } \xi_i^M$ ... measured values

$[A]$ ... stiffness change influence matrix with
elements $a_{i,j}$

$a_{i,j}$ ...difference of mode i due to 10 % stiffness
reduction in element-group j

d_j ... damage factor for the j-element-group, e. g.

$[B]$ ... damping-change-influence matrix $\quad 0 \leq d_j \leq 8$
organized similar to $[A]$

5. REFERENCES

/1/ Natke H.G., and J.T.P. Yao: System Identification Approach in
Structural Damage Evaluation. Bericht des Curt-Risch-Institutes
CRI-F-2/86, Hannover, 1986.

/2/ Kernbichler , K., und R. Flesch: Static and Dynamic Tests,
their Qualification for Bridge Inspection and Long-Term
Oberservations of Bridge Structures. RILEM Symposium,
Budapest, 1984.

/3/ Flesch, R., et.al.: Dynamic Testing and Modelling of Obern-
berg-Bridge. Int. Conf. on Num. Meth. for Transient and
Coupled Problems, Venice, 1984.

/4/ Flesch, R., et. al.: Brückeninspektion mittels dynamischer
Untersuchungen. 8. GESA-Symposium, Duisburg, 1984.

/5/ Kernbichler, K., et. al.: Dynamische Untersuchungen von Großbrücken (Massivbrücken), in-situ Versuche und Rechenmodelle.
Tagung Dynamische Probleme, Univ. Hannover, 1984.

/6/ Flesch, R., et al.: Dynamische in-situ Versuche und Rechenmodelle - Praktische Anwendung auf Großbrücken /Massiv-
brücken . ÖIAZ, 131. Jg., (1986), Heft 10.

/7/ Rauscher, G., und R. Flesch: Meßtechnik zur genauen Be-
stimmung der Modalparameter in Hinblick auf die dynamische
Bauwerksinspektion. VDI-Tagung in Bad Soden, VDI-Verlag,
1984.

/8/ Richardson, H., and D.L. Formenti: Parameter estimation
from frequency response measurements using rational fraction
polynomials. SMS-technical note 85-3.

/9/ Gritsch, J.: Erdbebenuntersuchung des Talüberganges Lavant.
Diplomarbeit, TU-Graz, 1985.

/10/ Formenti, D., and S. Welaratna: Structural Dynamic Modification - An Extension to Modal Analysis. Manual of SDM software, appendix C.

/11/ Kron, G.: Diakoptics. Macdonald, 1963.

/12/ Modelling Rib Stiffeners with the structural dynamics modification system. SMS-technical note 85-2.

Dynamic method for safety inspection

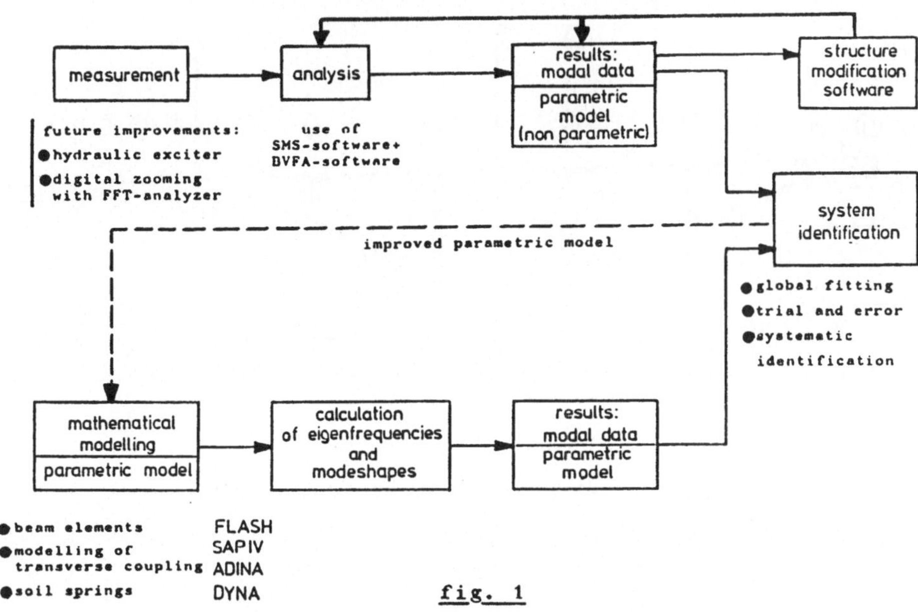

fig. 1

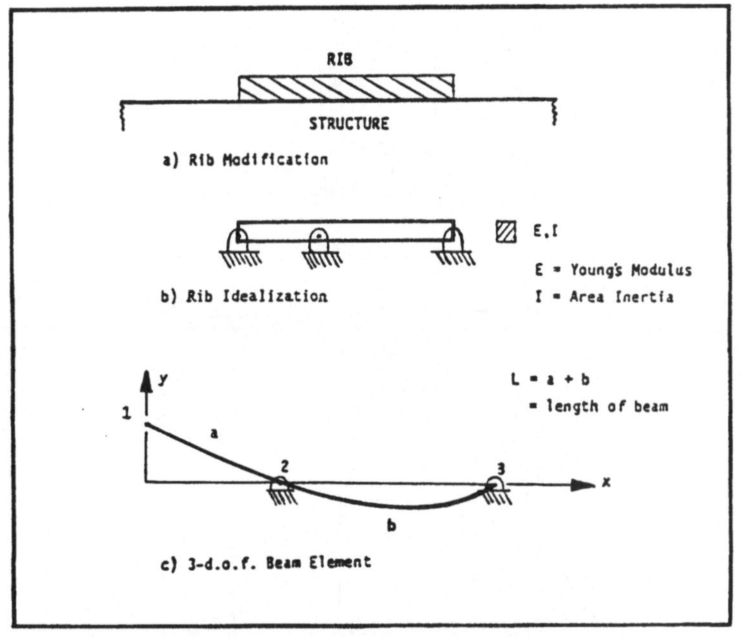

fig. 5: idealization of a Rib Stiffener

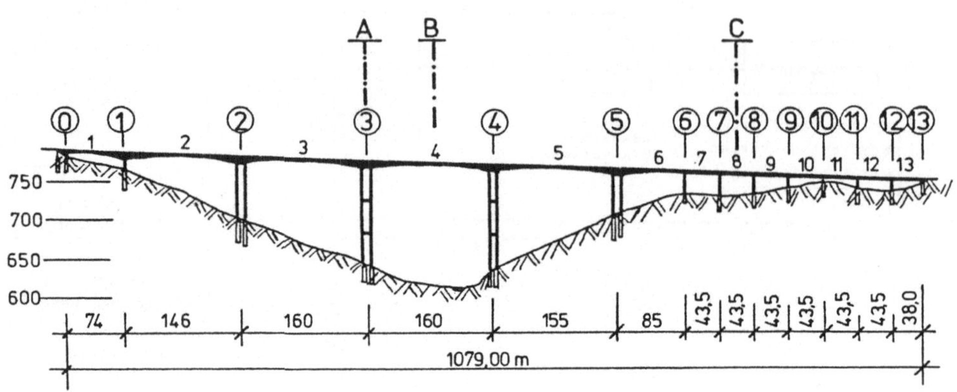

fig. 2: dimensions of Lavant bridge

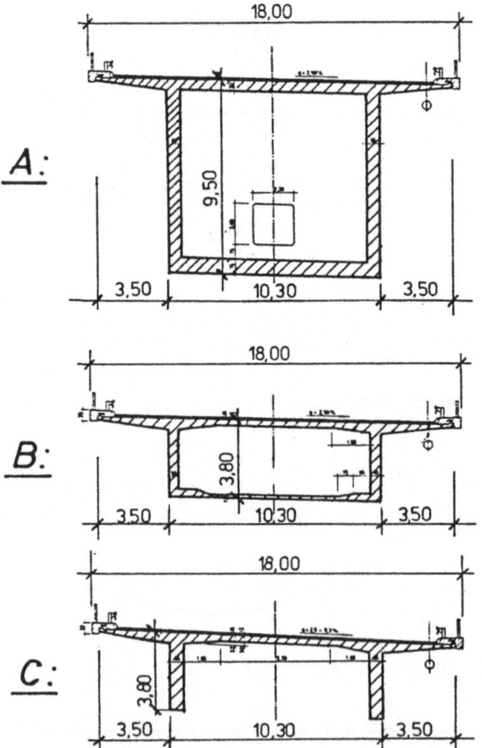

measured mode no.	modal frequency [Hz]	modal damping ratio [%]	identical with calculated mode no.
1	0.657	3.070	? ?
2	0.819	2.856	
3	0.957	2.870	3
4	1.023	1.219	
5	1.256	1.906	5
6	1.361	2.216	6
7	1.487	2.094	7
8	2.224	1.221	8
9	2.386	1.262	9
10	2.555	1.288	12
11	2.677	1.228	
12	2.895	1.145	15
13	3.060	1.205	16
14	3.327	1.078	17
15	3.685	0.959	
16	3.739	1.244	18
17	3.983	1.182	19
18	4.210	1.366	20
19	4.314	1.045	
20	4.601	1.232	
21	4.753	1.375	
22	5.066	1.190	
23	5.256	1.230	
24	5.299	0.873	
25	5.419	1.475	
26	5.595	1.424	
27	6.031	1.124	
28	6.332	0.958	
29	6.505	0.861	
30	6.713	1.130	
31	6.873	1.583	
32	7.360	1.340	
33	8.166	1.265	
34	8.553	0.951	
35	8.820	0.811	
36	9.041	1.080	
37	9.196	0.796	
38	9.561	1.156	

tab. 1

calc. mode no.	f_c [Hz]	measured mode no.	f_m [Hz]	Fit A (9 modes)		Fit B (13 modes)	
				$\bar{f_c}$ [Hz]	Δf [Hz]	$\bar{f_c}$ [Hz]	Δf [Hz]
1	0,806						
2	0,894						
3	0,969	3	0,957	0,987	-0,03	0,999	-0,042
4	1,126						
5	1,217	5	1,256			1,254	0,002
6	1,297	6	1,361	1,321	0,04	1,337	0,024
7	1,427	7	1,487	1,453	0,034	1,471	0,016
8	2,171	8	2,224	2,211	0,013	2,237	-0,013
9	2,342	9	2,386	2,385	0,001	2,413	-0,027
10	2,423						
11	2,519						
12	2,564	10	2,555	2,611	-0,056	2,642	-0,087
13	2,611						
14	2,661						
15	2,837	12	2,895	2,889	0,006	2,924	-0,029
16	3,013	13	3,060	3,068	-0,008	3,105	-0,045
17	3,316	14	3,327	3.377	-0,05	3,417	-0,09
18	3,486	16	3,739			3,592	0,147
19	3,793	17	3,983			3,909	0,074
20	3,927	18	4,210			4,047	0,163

tab. 2

f_m ..... measured modal frequency

f_c ..... calculated, non fitted modal frequency

$\bar{f_c}$ ..... fitted modal frequency

$$\Delta f = f_m - \bar{f_c}$$

$$\bar{f_c} = f_c \sqrt{1 + D_m}$$

$$D_M = \frac{1}{n} \sum_{i=1}^{n} D_i$$

$$D_i = 2 \frac{f_m - f_c}{f_c} + \left(\frac{f_m - f_c}{f_c} \right)^2$$

1. EIGENFORM
3. EIGENFORM
0,957 Hz
Coh: 0.078
0.338(max)
5. EIGENFORM
1,256 Hz
Coh: 0.083
0.315(max)
6. EIGENFORM
1,361 Hz
Coh: 0.198
0.431(max)
7. EIGENFORM
1,487 Hz
Coh: 0.172
0.43 (max)
measured shapes
calculated shapes
①
②
③
④
⑤
⑥

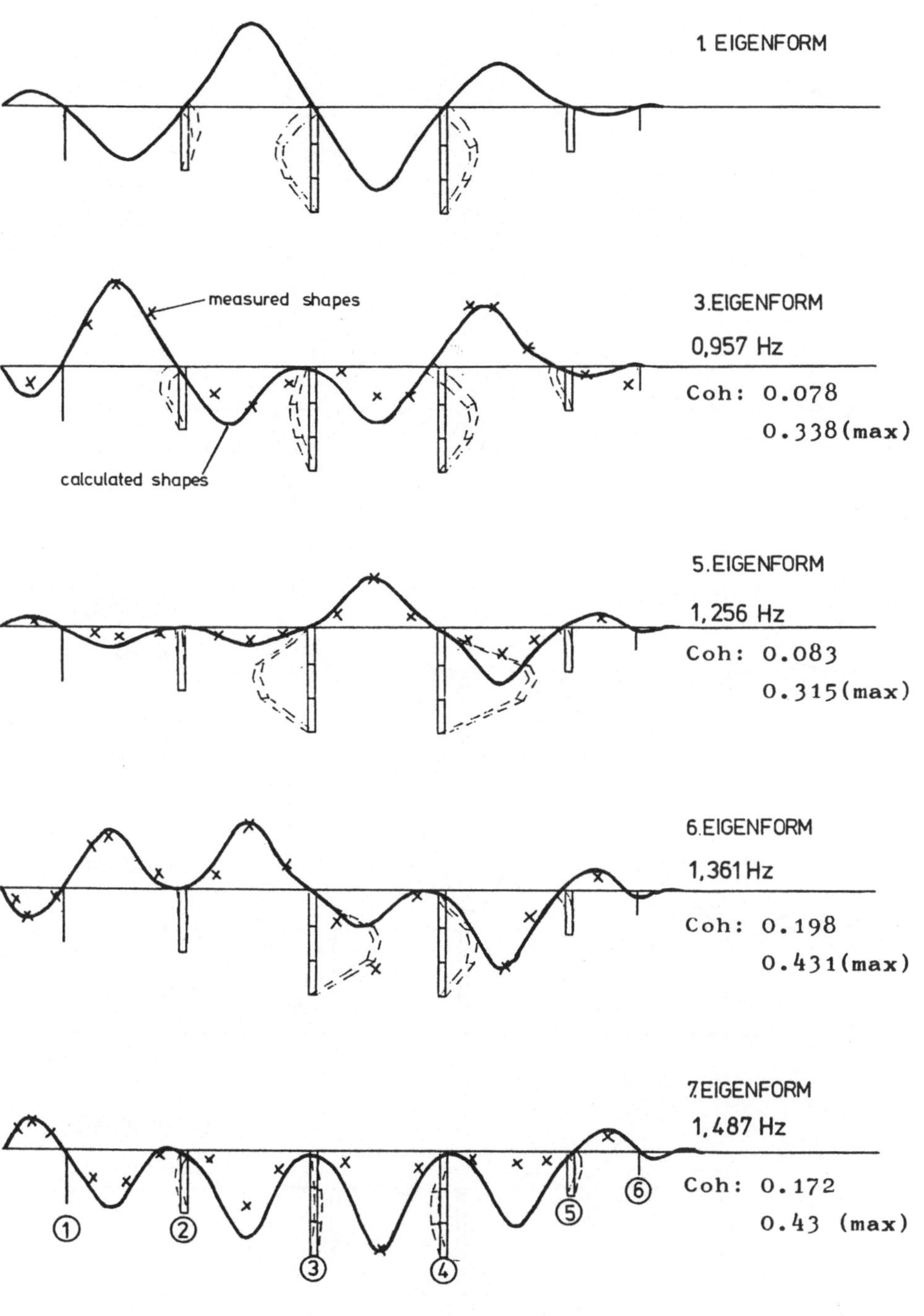

fig. 3

8.EIGENFORM
2,224 Hz
Coh:0.382
0.601(max)
9.EIGENFORM
2,386 Hz
measured shapes
calculated shapes
Coh:0.299
0.696(max)
15.EIGENFORM
2,895 Hz
Coh:0.608
0.909(max)
17.EIGENFORM
3,327 Hz
Coh:0.642
0.748(max)
19.EIGENFORM
3,983 Hz
Coh:0.471
0.937(max)

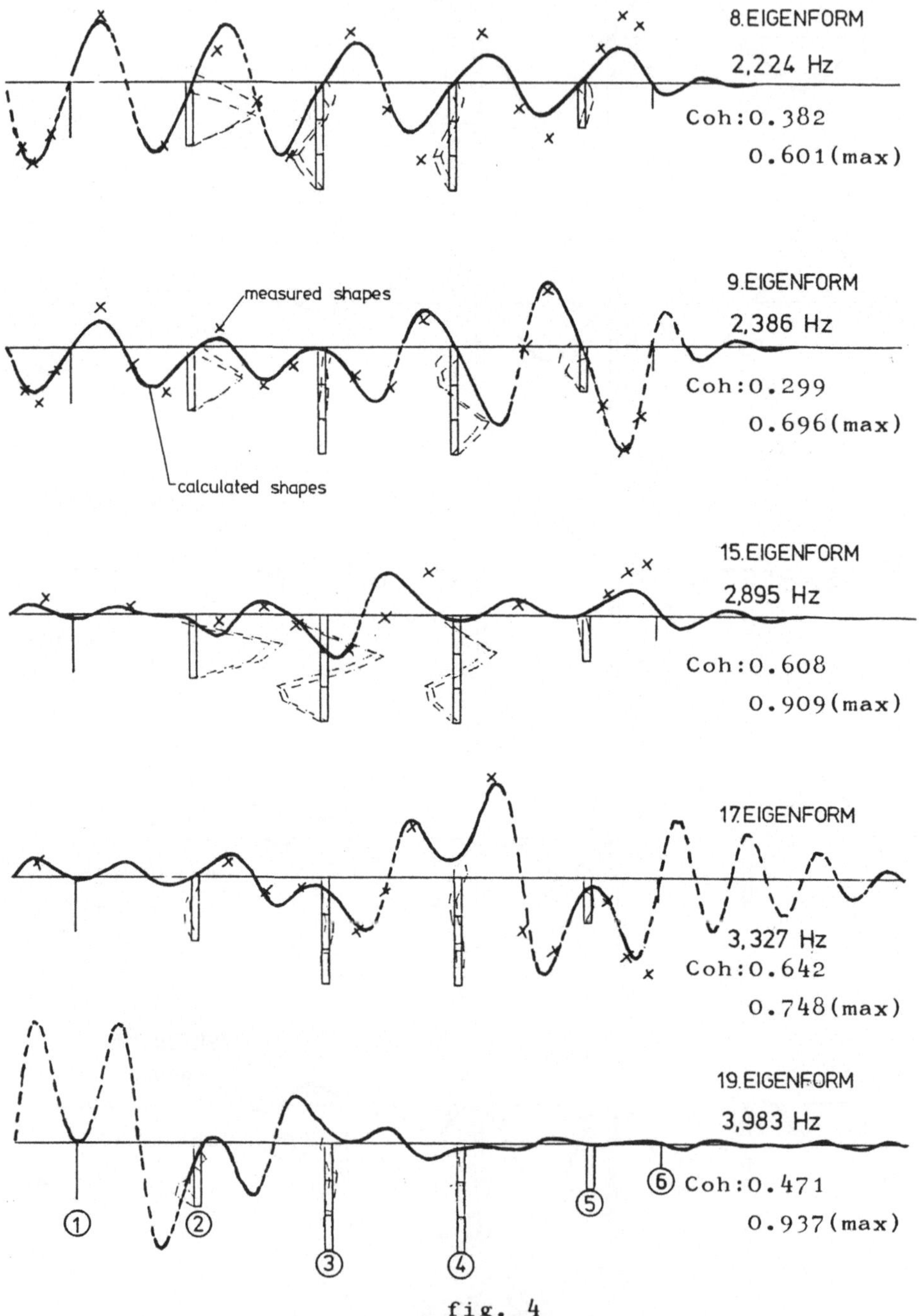

fig. 4

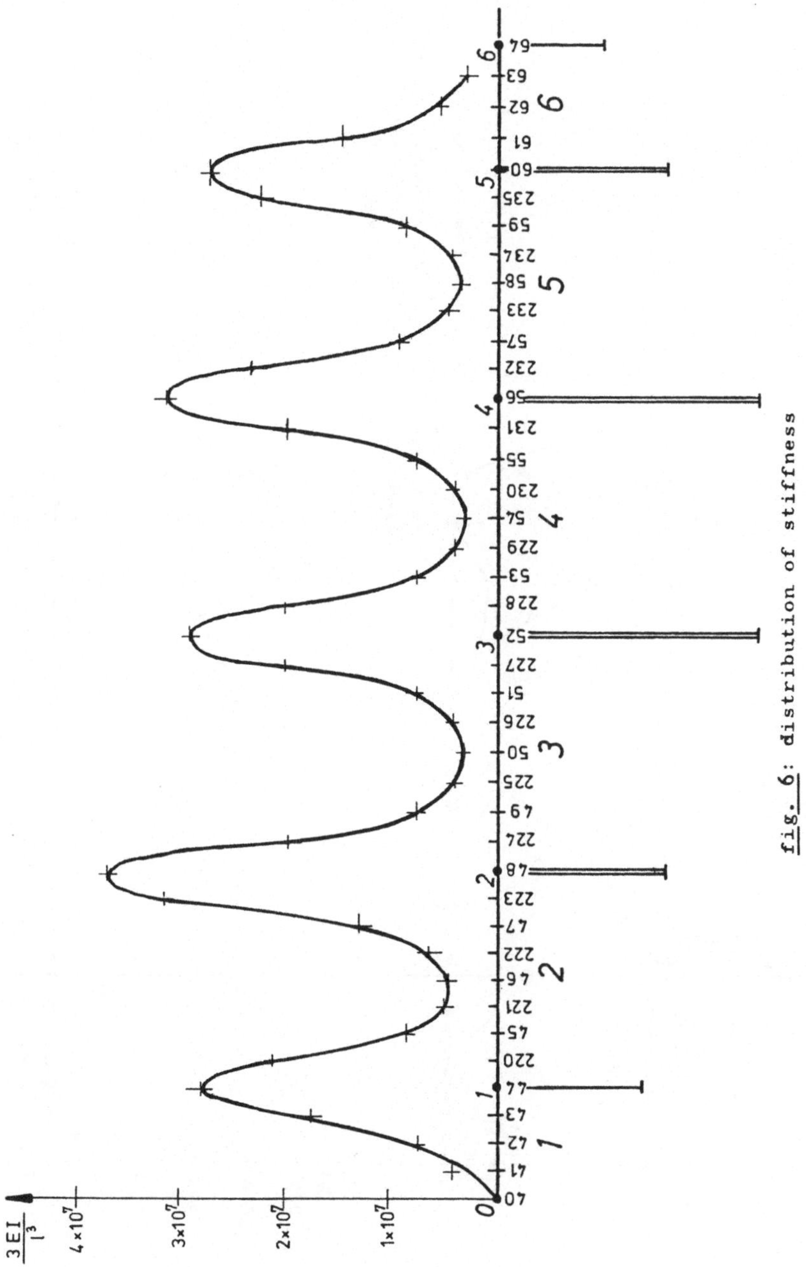

fig. 6: distribution of stiffness

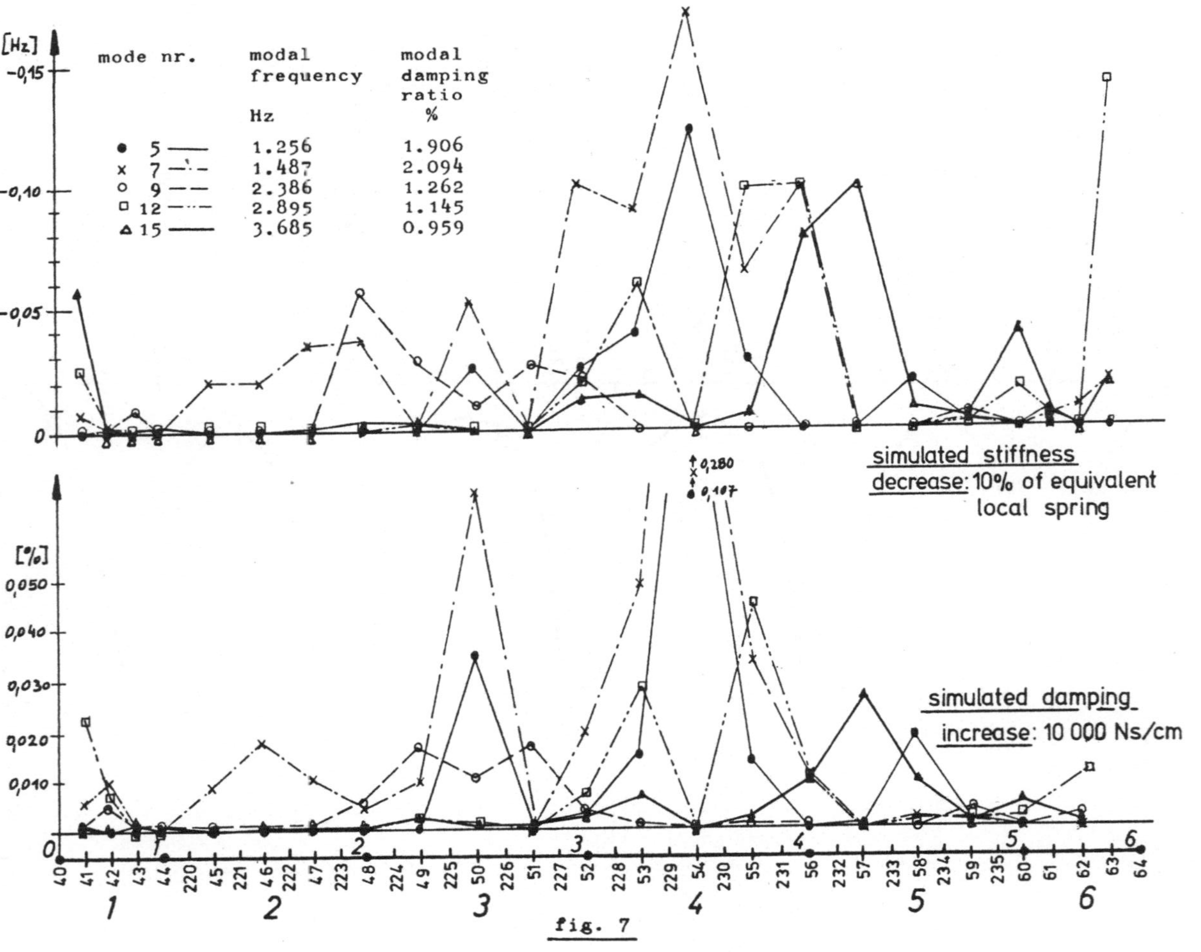

[Hz]
-0,15
-0,10
-0,05
0
mode nr.   modal frequency   modal damping ratio
                  Hz   %
5 ●   1.256   1.906
7 ×   1.487   2.094
9 ○   2.386   1.262
12 □   2.895   1.145
15 ▲   3.685   0.959
simulated stiffness
decrease: 10% of equivalent
local spring
0,280
0,107
[%]
0,050
0,040
0,030
0,020
0,010
0
simulated damping
increase: 10 000 Ns/cm
fig. 7

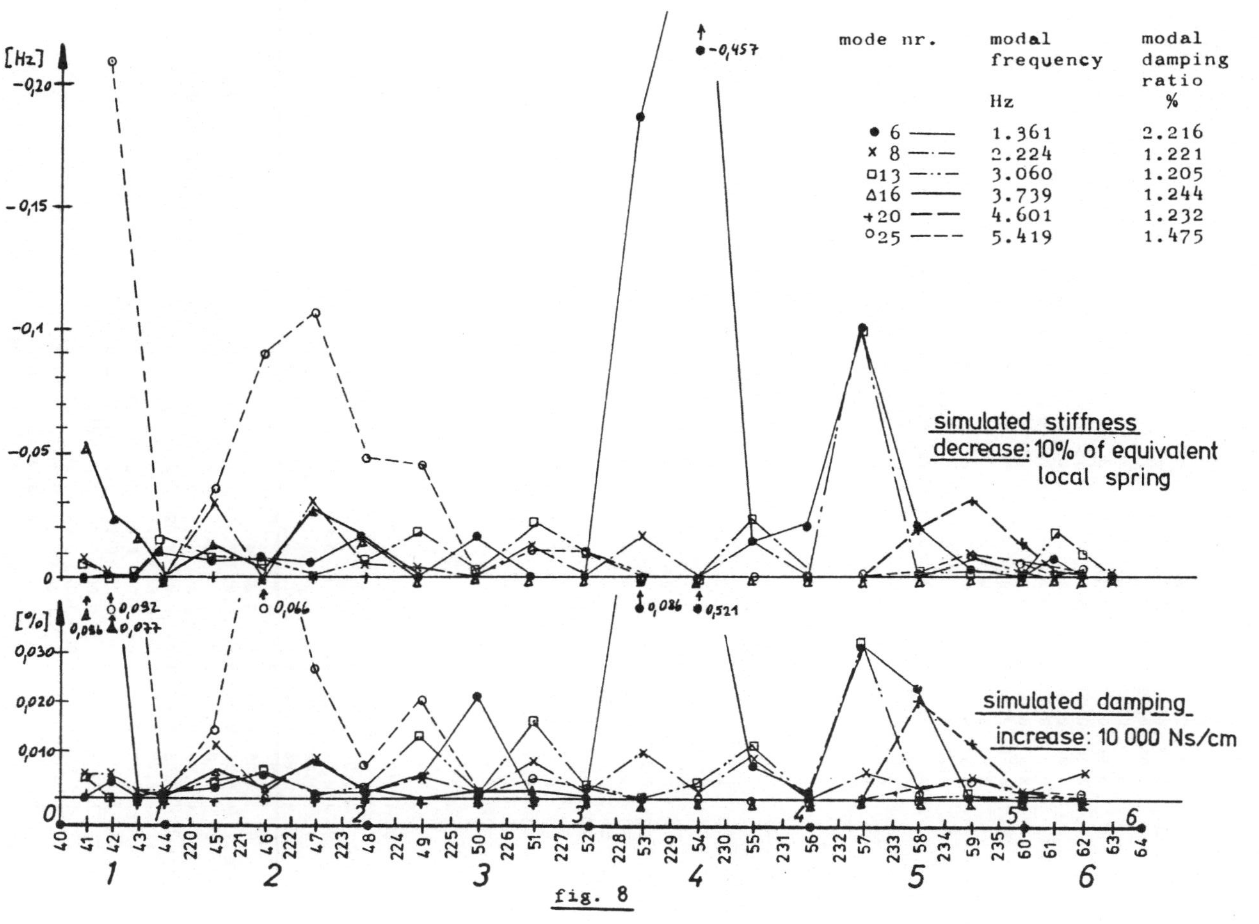

[Hz]
mode nr.   modal frequency   modal damping ratio
Hz   %
● 6   1.361   2.216
× 8   2.224   1.221
□ 13   3.060   1.205
△ 16   3.739   1.244
+ 20   4.601   1.232
○ 25   5.419   1.475
simulated stiffness
decrease: 10% of equivalent local spring
simulated damping
increase: 10 000 Ns/cm
[%]
fig. 8

```
                 *** AUTO M.A.C. MODE NO. ***
```

MODE	1	2	3	4	5	6	7	8	9	10	11	12
1	1.00											
2	0.07	1.00										
3	0.23	0.41	1.00									
4	0.08	0.09	0.01	1.00								
5	0.00	0.00	0.07	0.01	1.00							
6	0.03	0.05	0.00	0.05	0.06	1.00						
7	0.02	0.01	0.00	0.00	0.39	0.10	1.00					
8	0.11	0.00	0.01	0.03	0.00	0.01	0.01	1.00				
9	0.08	0.00	0.03	0.02	0.02	0.13	0.05	0.00	1.00			
10	0.00	0.04	0.00	0.00	0.00	0.00	0.07	0.09	0.02	1.00		
11	0.00	0.00	0.00	0.00	0.01	0.01	0.00	0.20	0.07	0.02	1.00	
12	0.02	0.02	0.01	0.01	0.00	0.00	0.00	0.01	0.08	0.12	0.24	1.00
13	0.01	0.04	0.00	0.00	0.07	0.00	0.04	0.00	0.02	0.02	0.00	0.26
14	0.00	0.06	0.06	0.01	0.72	0.00	0.47	0.01	0.00	0.02	0.01	0.00
15	0.01	0.00	0.00	0.01	0.01	0.03	0.00	0.01	0.00	0.03	0.03	0.00
16	0.03	0.05	0.04	0.12	0.00	0.02	0.01	0.04	0.01	0.00	0.00	0.01
17	0.00	0.03	0.01	0.03	0.00	0.00	0.00	0.00	0.00	0.00	0.00	0.00
18	0.04	0.09	0.03	0.05	0.07	0.11	0.03	0.00	0.00	0.01	0.02	0.02
19	0.02	0.05	0.05	0.07	0.01	0.05	0.00	0.02	0.00	0.01	0.07	0.00
22	0.07	0.03	0.00	0.03	0.00	0.00	0.00	0.00	0.00	0.00	0.00	0.00

```
                 *** AUTO M.A.C. MODE NO. ***
```

MODE	20	21	25	26	27	28	29	30	31	32	33	36	37
20	1.00												
21	0.39	1.00											
25	0.01	0.02	1.00										
26	0.00	0.00	0.00	1.00									
27	0.09	0.04	0.00	0.01	1.00								
28	0.01	0.09	0.01	0.00	0.10	1.00							
29	0.00	0.00	0.03	0.01	0.00	0.06	1.00						
30	0.00	0.02	0.00	0.02	0.02	0.04	0.06	1.00					
31	0.00	0.00	0.00	0.00	0.00	0.05	0.02	0.00	1.00				
32	0.01	0.05	0.01	0.04	0.07	0.01	0.30	0.35	0.10	1.00			
33	0.00	0.00	0.07	0.00	0.00	0.00	0.57	0.05	0.13	0.27	1.00		
36	0.05	0.04	0.02	0.19	0.09	0.14	0.00	0.01	0.02	0.06	0.01	1.00	
37	0.02	0.01	0.00	0.01	0.00	0.01	0.08	0.00	0.03	0.11	0.11	0.08	1.00

```
                 *** AUTO M.A.C. MODE NO. ***
```

MODE	13	14	15	16	17	18	19	22
1	0.01	0.00	0.01	0.03	0.00	0.04	0.02	0.07
2	0.04	0.06	0.00	0.05	0.03	0.09	0.05	0.03
3	0.00	0.06	0.00	0.04	0.01	0.03	0.05	0.00
4	0.00	0.01	0.01	0.12	0.03	0.05	0.07	0.03
5	0.07	0.72	0.01	0.00	0.00	0.07	0.01	0.00
6	0.00	0.00	0.03	0.02	0.00	0.11	0.05	0.00
7	0.04	0.47	0.00	0.01	0.00	0.03	0.00	0.00
8	0.00	0.01	0.01	0.04	0.00	0.00	0.02	0.00
9	0.02	0.00	0.00	0.01	0.00	0.00	0.00	0.00
10	0.02	0.02	0.03	0.00	0.00	0.01	0.01	0.00
11	0.00	0.01	0.03	0.00	0.00	0.02	0.07	0.00
12	0.26	0.00	0.00	0.01	0.00	0.02	0.00	0.00
13	1.00	0.06	0.03	0.00	0.00	0.02	0.01	0.01
14		1.00	0.02	0.00	0.00	0.00	0.00	0.02
15			1.00	0.02	0.00	0.00	0.00	0.01
16				1.00	0.19	0.00	0.02	0.08
17					1.00	0.15	0.19	0.07
18						1.00	0.52	0.06
19							1.00	0.00
22								1.00

MODE	23	24
23	1.00	0.04
24		1.00

MODE	34	35
34	1.00	0.11
35		1.00

<u>tab.3</u> Cross correlations of mode shapes
(Results of AUTO MAC - calculations)

EFFECTS OF DIFFERENT DAMAGES ON DYNAMIC PROPERTIES

The method is a <u>linear vibration problem</u> due to the small amplitudes used in the tests. Only results for same point of excitation and same force level should be compared directly.

<u>influence from moment of inertia</u>

damage alters mom. of inertia and damping

<u>influence from modulus of elasticity</u>

possible change of tang. modulus due to change of static stresses

	occurence of constraint stresses	secondary effect	decrease of stiffness	increase of damping	influence on stiffness	influence on damping
bending cracks			X	X		
shear cracks (seldom ev. composite constructions)			X	X		
torsional cr. (open cross section)			X	X		
change of pre-stressing force					X	X
foundation settlement	X				X	X
changes of system's centerline	X				X	X
changes of static stresses						
temperature modes with vectorcomponents in longitudinal direction	X				X	X

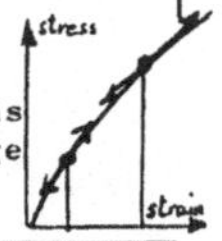

<u>fig. 9</u>

EXAMPLE OF FREQUENCY CHANGES IN PAST PROJECTS

Project	type of investigation	length of element	change of stiffness	resulting Δf [Hz]
Raach	meas.+calc.	3,0m	-37,5% middle field	-0,018
Raach	calc.	3,0m	-87,5% side field (open constr. joint)	-0,034 ÷ -0,242
Gänstor	meas.+calc.	13,0m	+20% (tension members)	+0,045
Gänstor	meas.+calc.	3,0m	+24% crack > 1mm +12% crack < 1mm }	+0,031
Lavant	meas.+SDM	40,0m	-10% ("roving" negative stiffener)	-0,01 ÷ -0,457

<u>tab. 4</u>

System Identification Approaches in Structural Safety Evaluation

H. G. Natke[1] and J. T. P. Yao[2]

1. INTRODUCTION

In system identification studies, the system characteristics as represented by the equation of motion are estimated using recorded input (excitation) and output (response) data. Most available methods to-date are applicable for structures with linear (or slightly nonlinear) behavior, because the test loads are always kept at low amplitudes in order to avoid permanent deformation and/or damage.

In practice, existing structures may be investigated by experienced engineers who can decide to (a) inspect the structure (b) examine its design documents, (c) conduct nondestructive evaluation, (d) perform laboratory tests of material samples taken from the structure, and/or (e) make additional structural analyses. Results of all these studies may be summarized by the engineer using his/her experience and judgement. Frequently, such evaluation processes are conducted without system identification studies.

It is believed that system identification techniques can be extended to (a) systematically process all the available information concerning the

* Supported in part by NATO Research Grant No. 625/84 and NSF Grant No. ECE-8412569.

1 Professor of Vibrations and Measurements and Director Curt-Risch-Institute of Dynamics, Acoustics and Measurements, University of Hannover, Fed. Rep. Germany

2 Professor of Civil Engineering, Purdue University, West Lafayette, IN 47907.

structure (objective or subjective, crisp or fuzzy), (b) estimate system characteristics as well as limit states which are needed for further reliability analysis, and (c) include the development of various expert systems each consisting of a knowledge base and an inference machine. The extended system identification methodology as shown in Figure 1 can be useful for the identification of complex and highly nonlinear structures.

The present paper is a continuation and extension of a related paper by the authors (Natke and Yao, 1986). Therefore, the principles from the earlier paper are repeated at first to be followed by an extension with detailed discussion.

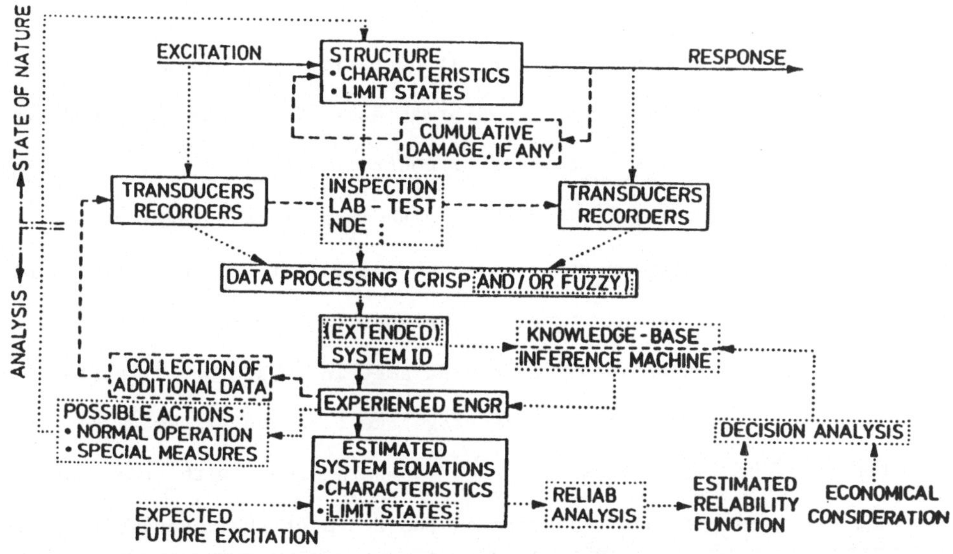

Fig. 1 Extended System Identification Methodology
for Complex and Highly Nonlinear Structures

2. ROLE OF IDENTIFICATION IN SAFETY EVALUATION

2.1 General

The dynamic behavior of the mechanical system in a virgin state must be known with a given level of confidence. It is important to have such information in order to be able to define significant deviations from this behavior as a pattern. Starting with the determination of a mathematical model which may be erroneous and gives an uncertain prediction, it must be improved with a first test and initial identification (Natke 1983; Cottin, Felgenhauer, and Natke 1984; Cottin, and Natke 1986) as shown in Fig. 2. This procedure leads to a pattern of the undamaged system including known estimated variances of the parameters and states. The result indicates that the model in general is defined in a condenced form (variations in matrix elements do not correspond directly to the changes of design variables of system elements) and that the model takes into account only one given environmental condition.

The additional procedure may consist of periodical dynamical testing and the observation of permanent deformations. The states can be determined by (robust) observers (coming from control-theory) or by filtering (e.g. using the Kalman filter). This approach has disadvantages when it is applied to civil engineering structures. Therefore, it is not considered herein. Periodical testing results in either (a) a parametric test model using modal quantities or in (b) a non-parametric test model estimating input and output quantities. At this stage the question arises as to which parameters must be estimated in order to compare them with the initial ones. These parameters must be chosen on the basis of their sensitivity due to a priori and unknown failures. We are interested in an early failure detection in terms of small and local damage, which affects mainly normal modes with higher eigenfrequencies of the system. Consequently, it is preferable to consider the dynamic stiffness matrix which contains the higher modes in a more sensitive way than the frequency response function matrix (using Fourier transform or Laplace transform) or the corresponding physical parameter matrices (see their expansions with regard to the eigenmodes).

Failure detection requires the comparison of the selected quantities of the updated model at the stage of inspection with the corresponding quantities of the initial model within the limitation of their uncertainties. A significant deviation in these parameters indicates the possible occurrence of damage. The determination of the location and measure of the damage are discussed in Section 2.5.

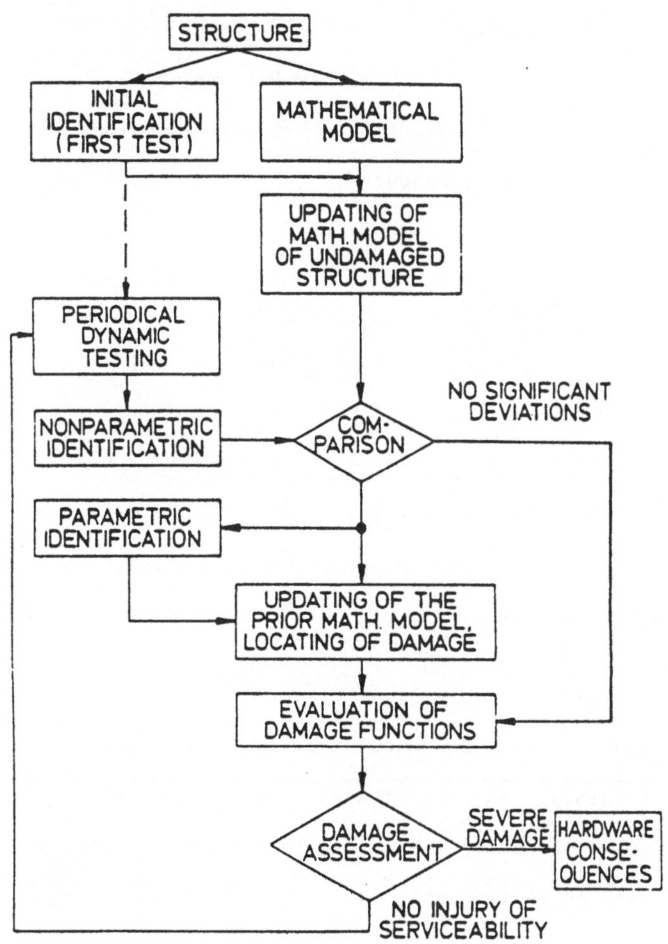

Fig. 2 A Procedure for Structural Damage Evaluation

2.2   Detection by Non-Parametric Identification

Non-parametric identification as defined herein uses the following equation in the frequency domain with zero initial conditions,

$$S(jw) \ U(jw) \ = \ P(jw) \tag{1}$$

where $S(jw)$ denotes the dynamic stiffness matrix, $U(jw)$ denotes the Fourier transformed output, $P(jw)$ denotes input. The updated initial model as described in Section 2.3, denoted with the subscript "0", is used for combining with the measured and transformed signals $U_i(jw)$, $P_i(jw) = P_0(jw)$ at the time τ_i of periodical tests:

$$S_i(jw)U_i(jw) \ = \ P_0(jw), \tag{2}$$

$$S_i(jw) \ = \ : \ S_0(jw) \ + \ \Delta S_i(jw) \tag{3}$$

$$S_0(jw) \ U_i(jw) \ = \ P_0(jw) \ - \ \Delta S_i(jw) \ U_i(jw). \tag{4}$$

Eq. (4) is interpreted as follows: Inserting the measurement $U_i(jw)$ for a given $P_0(jw)$ into the prior model as characterized by $S_0(jw)$ gives in general an additional force

$$\Delta \ S_i(jw) \ U_i(jw) \ = \ : \ \Delta P_i(jw) \tag{5}$$

If the modulus of the components of $\Delta P_i(jw)$ vary within a defined range $[0, \ \Delta P_0]$ (scalar), no significant damage of the structure are indicated. However, the moduli exceeding this limiting range indicates the occurrence of structural damage (see Fig. 2).

This approach is comparable with that published in (ANCO Rep., 1985) with the following differences: (a) instead of estimating the frequency response function matrix $F_i(jw) = S_i^{-1}(jw)$, the dynamic stiffness matrix is used herein; (b) instead of measuring frequency response functions, the measured input and outputs are used (transformed and, if appropriate, their spectral densities).

Using this approach, it is not necessary to perform expensive parametric identification in order to detect significant changes in dynamic behavior. The assumption to be made is a suitable excitation that the dynamic answer contains the influence of the damage as indicated by the dynamic stiffnesses. The moduli and phases of $\Delta P_i(jw)$ must be investigated due to significant changes in damping ratios and stiffnesses. The dynamic stiffness matrix may be expanded as a Taylor series with respect to the damage parameters (ANCO Rep. 1985). Also, an indicator function may be defined using the frequency response functions.

2.3 Parametric Identification Methods

In order to locate the cause of change in the dynamic behavior and to update the initial mathematical model, a structured (parametric) mathematical model is needed. A parametric model is also required when identifying modal quantities from measured inputs and outputs. The latter is well known as experimental modal analysis and not discussed here [e.g., see (Kozin and Natke (1986), Natke (1983)]. The methods for improving structured mathematical models are based on parameter estimation using input/output measurements (Figs. 3 and 4) as well as estimated modal quantities. They are described in detail by Natke (1983), and Cottin et al. (1984,1986) and summarized by Kozin and Natke (1986) and Natke (1982).

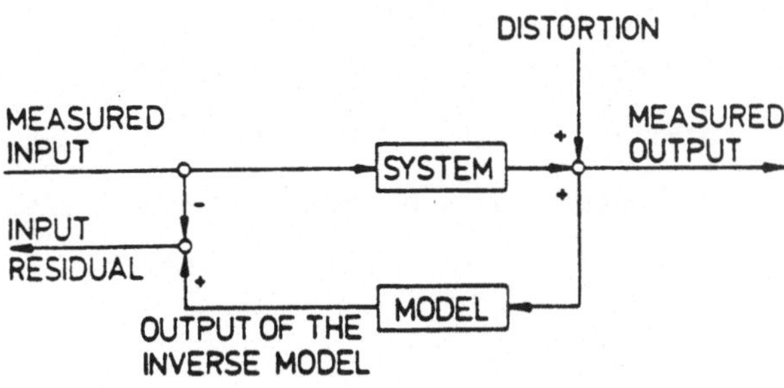

Fig. 3 Input Residual Approach

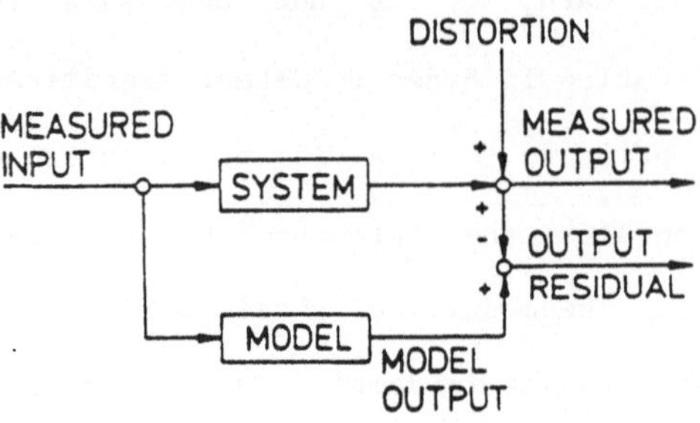

Fig. 4 Output Residual Approach

These updating procedures lead to equivalent linear models though the system may behave nonlinearly. If the damage affect the system to such an extent that an equivalent linear model is no longer applicable for the prediction of actual behavior, it may still be used for the detection of damage. If a structure of the model is assumed, the identification reduces to parameter estimation. If this is not possible, several methods mainly concerned with system elements are discussed by Natke and Yao (1986).

2.4 <u>Introduction of Damage Functions (Estimation of Structural Damage)</u>

If significant damage is indicated as a result of periodical testing, then the improved parameters differs as described for a generalized single degree of freedom system (k-th coordinate at time τ_i), i.e.,

$$m_{ik}\ddot{q}_k(t) + b_{ik}\dot{q}_k(t) + c_{ik}q_k(t) = f_{ik}(t). \tag{6}$$

If nonlinear terms are introduced into Eq. (6), they must also be considered in a similar manner. The effect of cumulative damage is represented by a diagnostic index D which may now be defined by (with m_{ik} not modified)

$$b_{ik} = (1 + D_{b_k}^{(i)})^{\alpha_{b_k}^{(i)}} b_{ok} \tag{7}$$

$$c_{ik} = (1 - D_{c_k}^{(i)})^{\alpha_{c_k}^{(i)}} \, c_{ok} \tag{8}$$

With increasing damage, damping may be increased and stiffness may be decreased in general. Using Equations (6), (7), and (8), parameters D and α may be determined from structural response records. In certain cases, the estimated damage D may be compared with those as obtained by Stephens et al. (1987). As can be seen changing from a linear model to a nonlinear one the damage functions are depending on amplitudes of displacement as well as velocity response. The interrelationship among these parameters D´s and α´s will be studied for the estimation of associated structural damage.

In addition to these above-mentioned definitions, one can produce the relationship between them and the Taylor series expansion, of the coefficients under consideration with respect to some damage parameters (physical or geometrical property of the system). In so doing, the exponents α of Eqs. (7) and (8) follow by the highest powers which are not truncated in the series.

An additional advantage of the parameters introduced is given by their possibly amplified sensitivities compared with the conventional parameter changes. However these properties have to be investigated further.

2.5 Localization of Damage

The additional force $\Delta P_i(j\omega)$ in Eq. (5) is a global quantity with respect to local effects. The sensitivity of this term to early damage detection is examined herein.

The kinetic energy of the free-decay response is a superposition of the kinetic energies of the generalized coordinates. Choosing modal coordinates, the first selection (separation) with respect to damage localization is to find the proper degree of freedom: $\omega_{oi}^2 \, m_{gi}$ from the total kinetic energy with ω_{oi} denoting the eigenfrequency and m_{gi} the generalized mass of the i-th generalized coordinate. The second step in selection can be done using the subsystem-modeling combined with the correction factors within the improvement (identification) approaches (see Section 2.3). Let the cumulative damage be described by a change in the stiffness of subsystem "1" corresponding to the matrix K_1: $K^{ch} = K + a_1 K_1$. No damage implies that $a_1 = 1$. Here a first localization by a prior knowledge of subsystem "1" is used. With $\hat{u}_0$ - eigenvector of the associated undamped system, it follows that

$$\frac{\partial \omega_{0i}^2}{\partial a_1} = \hat{u}_{0i}^T \, K_1 \hat{u}_{0i} = : k_{gi1} \tag{9}$$

For simplicity, the "undamped" eigenvalues are used herein. As can be seen, $\Delta \omega_{0i}^2 \simeq k_{gi1} \Delta a_1$, a small deviation in ω_{oi}^2 can result in a larger deviation of a_1 dependent on the value of the amplification factor k_{gi1}.

The subsystem modeling as used in the improvement approaches (global factors a´s) is one possible way to locate or indicate structural modification including sensitivity investigations (see Natke et al. 1974, Natke 1983). It is used in connection with the known quantities $S_0(j\omega), U_0(j\omega), P_0(j\omega)$. For the excitation P_0 applied after time τ_i, gives the measured response $U_i(j\omega)$, Eq. (4) yields by calculation: $\Delta P_i = P_0 - S_0 U_i$. Then by using Eq. (5), $\Delta S_i U_i = \Delta P_i$ is also calculated. Introducing parameters $\Delta a^{(i)}$ so that these equations may be solved:

$$S_i(j\omega) = -\omega^2 M_i + j\omega B_i + K_i \tag{10}$$

where,

$$M_i = \sum_{\sigma} a_{M\sigma}^{(i)} \, M_\sigma$$

$$B_i = \sum_\rho a_{B\rho}^{(i)} B_\rho$$

$$K_i = \sum_v a_{Kv}^{(i)} K_v$$

$$\{a_\gamma^{(i)}\} = (\{a_{M\sigma}^{(i)}\}^T, \ldots)^T, \text{ if } \equiv 1 \text{ it follows that } S_0(j\omega) = S_i(j\omega)$$

$$a_\gamma^{(i)} = : 1 + \Delta a_\gamma^{(i)}$$

It follows that $S_i(j\omega) = S_0(j\omega) + \Delta S_i(j\omega)$, and

$$\Delta S_i(j\omega) = \Delta S(j\omega, \Delta a_\gamma^{(i)}),$$

which is linear in the parameters of interest:

$$\Delta S(j\omega, \Delta a_\gamma^{(i)}) = -\omega^2 \sum_{\sigma=1}^{S} \Delta a_{M\sigma}^{(i)} M_\sigma$$

$$+ j\omega \sum_{\rho=1}^{R} \Delta a_{B\rho}^{(i)} B_\rho \tag{11}$$

$$+ \sum_{v=1}^{I} \Delta a_{Kv}^{(i)} K_v$$

The system of linear equations reads

$$\Delta S(j\omega, \Delta a_\gamma^{(i)}) \, U_i(j\omega) = \Delta P_i(j\omega) \tag{12}$$

The number $J: = S + R + I$ of unknowns is in general smaller than the number n
of degrees of freedom (assumption!). These equations may be solved with
the use of weighted least squares.

For sensitivity studies must the following investigations be performed.

$$U_i = U_0 + \Delta U_i$$

$$S_0 U_i = P_0 - \Delta P_i$$

$$S_0 U_0 + S_0 \Delta U_i = P_0 - \Delta P_i \tag{13}$$

$$S_0 \Delta U_i = -\Delta P_i \simeq -\Delta S_i U_0 \quad \text{(linearized)}$$

$$\Delta U_i = -S_0^{-1} \Delta S_i U_0$$

For example:

$$\frac{\partial \Delta U_i}{\partial \Delta a_{Kv}^{(i)}} = -S_0^{-1} K_v U_0 \tag{14}$$

Here the amplification is done with

$$S_0^{-1} = F_0(j\omega), \tag{15}$$

the frequency response matrix (transfer matrix) in the virgin state. Depending on a priori knowledge the sensitivity investigations can be performed without using measurements of later tests which can be influenced by damage. If damage occurs in the subsystem of magnitude Δa_γ then we can expect a modified dynamic response as given by Natke (1983). If we use U_i instead of ΔU_i the resulting equation reads as follows:

$$(S_0 + \Delta S_i) U_i = P_0, U_i = (S_0 + \Delta S_i)^{-1} P_0 \tag{16}$$

with an ω - dependent inverse and nonlinear equation in $\Delta a_\gamma^{(i)}$.

Assuming that the degrees of freedom with higher eigenfrequencies will be influenced by the damage, one can measure acceleration $(\omega^2 U(j\omega))$ in order to amplify the effects of modification. However, in consideration of the dependency between velocity and stresses (in simple cases), velocity measurement may be preferable.

lf the nonlinear behavior is measurable (as deviation from the linear one), the damage is detectable. Although the behavior can not be described by the biased linear model, the bias can be the basis for damage detection. What is the problem with localization using a biased model? Detection and localization with the "correction" factors based on subsystem - superposition are done by comparing with the improved model of the virgin system. We are comparing the "detection" factors which are the result of a quality criterion concerning the adaption of dynamic responses of given macro-elements. Without additional restriction, non-physical adjustment is possible. As an example (Gedanken - Experiment), in a small frequency range the measured dynamic response may be lower than before. This meaasurement is somewhat of a frequency response curve. This lowered part can be approximated by a larger mass of the appropriate degree of freedom (with resonance frequency prior to this range) as well as by a greater flexibility of the degree of freedom (within this range).

We can exclude mass modifications, we have to adopt stiffnesses and damping ratios. When we have partitioned our system, we may obtain approximate solutions with different subsystems (increasing in number) and search for the effects to the a´s (preselection by a prior-knowledge). The formalism is that of regression, therefore the result (with respect to localization) depends on how good the ansatz (the partitioning into subsystems) fits the physics (that means that the most sensitive subsystem must be chosen).

3. CONCLUSIONS

A general approach to damage detection is presented and discussed using non-parametric identification during periodical tests with artificial excitation. If damage occurs and it changes the dynamic response to a given excitation significantly by using an input error (i.e., using the dynamic stiffness matrix of the system) one has to establish corresponding parametrical models with known confidence to obtain damage functions. This problem must be investigated further. Problems of localization of damages and their direct calculation are also discussed.

ACKNOWLEDGMENT

The joint research project was supported in part by NATO Research Grant No. 625/84 at University of Hannover and NSF Grant No. ECE-8412569 at Purdue University. A part of this paper was presented (without publication) at the ASCE Structural Engineering Congress in New Orleans, LA, U.S.A., on 15-18 September 1986.

REFERENCES

ANCO Report (1985), Providing Structural Modules with Self-Integrity Monitoring, ANCO Engineers 1311.05

Chen, J.-C., Garba, J.A. (1986), Structural Damage Assessment Using a System Identification Technique, private communication

Cottin, N., Felgenhauer, H.-P., Natke, H.G. (1984), On the Parameter Identification of Elastomechanical Systems Using Input and Output Residuals, Ingenieur-Archiv 54, pp. 378-387

Cottin, N., Natke, H.G. (1986), On the Parameter Identification of Elastomechanical Systems Using Weighted Input and Output Residuals, Ingenieur-Archiv 56, pp. 106-113

Kozin, F., Natke, H.G. (1986), System Identification Techniques, Structural Safety, Vol. 3, Nos. 3 & 4, pp. 269-316.

Natke, H.G., Collmann, D., Zimmermann, H. (1974), Beitrag zur Korrektur des Rechenmodells einers elastomechanischen Systems anhand von Versuchsergebnissen, VDI-Bericht, No. 221, pp. 23-32.

Natke, H.G., Editor, (1982), Identification of Vibrating Structures, CISM Courses and Lectures, No. 272, Springer-Verlag Wien, New York

Natke, H.G. (1983), Einfuhrung in Theorie und Praxis der Zeitreihen- und Modalanalyse, Vieweg Verlay Braunschweig, Wiesbaden.

Natke, H.G. (1983a), Deliberations on the Improvement of the Computational Model with Measured Eigenmagnitudes, Revue Roumaine des Sciences Techniques, Mecanique Appliquee 28, pp. 159-173

Natke, H.G., Yao, J.T.P. (1986), Research Topics in Structural Identification, _Dynamic Response of Structures, Proceedings of the 3rd Conf. on Dynamic Response of Structures ASCE_, March 31-April 2, 1986, Los Angeles, CA, pp. 542-550.

Natke, H.G., Yao, J.T.P., (1986a), System Identification Approach in Structural Daamage Evaluation, Structures Congress '86, Preprint 17-1, ASCE.

Stephen, J.E., and Yao, J.T.P., (1987), "Damage Assessment Using Response Measurements," _Journal of Structural Engineering_, ASCE, Vol. 113, No. 4, pp. 787-801.

Structural Damage Assessment Using a System Identification Technique

Jay-Chung Chen and John A. Garba

Applied Technologies Section
Jet Propulsion Laboratory
California Institute of Technology
Pasadena, California 91109

Abstract

The need for monitoring the dynamic characteristics of large structural systems for purposes of assessing the potential degradation of structural properties has been established. This paper develops a theory for assessing the occurrence, location, and extent of potential damage utilizing on-orbit response measurements. Feasibility of the method is demonstrated using a simple structural system as an example.

INTRODUCTION

Most load carrying structural systems such as aircraft, spacecraft, high rise buildings, and offshore platforms continuously accumulate damage during their service environment. For purposes of assuring safety it is most desirable that this damage be monitored as to its occurrence, its location and as to the extent of the damage.

Large space structures in earth orbit or on interplanetary missions are apt to suffer structural damage over their service time caused by such adverse events as docking, impact by foreign objects and other hostile actions, and from environmental effects due to long time exposure to space such as thermal vacuum, radiation, and ultra violet light effects. Damage which is not detected and not corrected may potentially cause more damage and eventually catastrophic structural failure. Therefore, a methodology for the monitoring of the structural integrity in order to rapidly detect the occurrence and identify the location of the damage will be instrumental in assuring the safety of the structural system. In particular, such detection can permit real time corrective action or a change in operational procedures to minimize the possibility of causing further damage.

As configurations of space structures become larger and more complex, the use of exotic materials will become more common, and the functional requirements of the structural systems will become more complex and critical. Hence a rapid and remote structural damage detection capability will be more essential. In the case of Space Station, one common attribute of all proposed configurations is the complexity of the structural system. A typical configuration consists of many bays connected in a complicated manner, each of which is a 3-dimensional structure. Components include antennas, solar arrays, docking structures, fluid containers and many other complex systems. Should one or more structural components incur damage in such a large system, it may be virtually impossible to detect the presence of the damage, let alone locate it and quantify the extent. For major structural failures such as rupture of major truss members, visual inspection may be sufficient to locate and assess the damage. However, it is very difficult to visually observe any damage due to material degradation, since the surface appearance due to such damage will most likely remain unchanged. It is therefore necessary that, for the damage detection and assessment, the characteristics of the structure, which includes the load carrying capability as its inherent property, be monitored.

It becomes clear that continuous monitoring of structural integrity and the detection/assessment of damage and its subsequent compensation, repair, and control are important considerations for large flexible space structures. The ultimate future requirement will be a remote data acquisition system with rapid on-board analysis for almost real time assessment. The present study will focus on the feasibility of such a methodology which will provide a damage assessment capability for large space structures.

APPROACH

Techniques of using experimentally measured data for determining the parameters in the equations of motion of a system are commonly called system identification. A typical procedure involves the modal test of the structural system during which the responses due to external excitations are measured. From the response data, the dynamic characteristics of the system such as the natural frequencies and mode shapes can be determined directly or through data processing techniques depending on the test method employed. Because the natural frequencies and mode shapes of a structural

stiffness, these system parameters may be "identified" by comparing those determined by test to those dynamic characteristics predicted from the mathematical model. Based on this concept, attempts have recently been made to develop techniques for using vibration measurements to evaluate the structural integrity of offshore oil and gas platforms, (References 1 to 3). Because of the adverse conditions in acquiring these measurements, the high redundancy of the structural system and the lack of sufficient instrumentation, the methodology of detection of structural failures did not become mature. The technology is not being implemented by the platform industry at the present time. On the other hand, in the aerospace industry, modal tests are performed on extensively instrumented spacecraft using precisely controlled excitations for determining natural frequencies, mode shapes and damping. Here, the objective is to verify the mathematical model to be used in loads analysis by comparing the dynamic characteristics obtained by test to those predicted by the analytical model. The differences between the test and analysis results are then used to modify the mathematical model so that the model accurately predicts the test results, (References 4 to 8). Furthermore, it is very likely that on-orbit modal testing is necessary for·the large space structures because of the adaptive control requirement. These modal test results can be made available for the purpose of on-orbit damage assessment.

In the present study, the investigation will be centered around developing methodology to use the test measured data or test determined dynamic characteristics not to verify the mathematical model but rather to identify the damage of the structure in terms of its location and the extent, i.e., the reduction of stiffness or load carrying capability.

THEORETICAL DEVELOPMENT

The fundamental questions for damage assessment are whether it is feasible to identify the occurrence, location and extent of the damage from given measured structural dynamic characteristics. Therefore, the relationship between the physical parameters, such as the mass and stiffness, and the dynamic characteristics, such as the eigenvalues and eigenvectors or natural frequency and mode shape, must be established. The governing equations for the structural dynamic system in finite element representation can be written as

477

$$[M]\{\ddot{x}\} + [C]\{\dot{x}\} + [K]\{x\} = \{f(t)\} \tag{1}$$

where the matrices $[M]$, $[C]$ and $[K]$ represent the discretized mass and inertia, damping and stiffness distribution. $\{x\}$, $\{x\}$, and $\{x\}$ are the acceleration, velocity and displacement vectors of the degrees-of-freedom being modelled and $\{f(t)\}$ is the external forcing function vector. The homogeneous solutions to Equation (1) are the eigenvalues and the eigen-vectors. For simplicity, the damping terms will be ignored at present time, thus

$$[M]\{\ddot{x}\} + [K]\{x\} = 0 \tag{2}$$

let

$$\{x\} = \{\phi\}_i \sin \omega_i t \tag{3}$$

where ω_i is the ith eigenvalue and $\{\phi\}$ is the corresponding eigenvector. Upon substitution, into Equation (2) the relationship between the physical parameters $[M]$ and $[K]$ and the dynamic characteristics ω_i and $\{\phi\}$ can be established as

$$[K]\{\phi\}_i - \omega_i^2[M]\{\phi\}_i = 0 \tag{4}$$

It is clear that values of ω_i and $\{\phi\}_i$ are functions of the mass $[M]$ and $[K]$ of the system. In other words, any changes in $[M]$ and $[K]$ due to the loss of mass or loss of stiffness of certain parts of the structural system will be reflected in its natural frequency and mode shape measurements. A discovery of a deviation of the measured natural frequency and mode shape with respect to those previously measured when the system was in an undamaged condition is an indication of the occurrence of damage. Certain changes in the physical parameters will effect certain modes but not others. The occurrence of damage will not be detected if only those modes which are not affected by the changes are measured. To establish damage assessment a wide range of measurements should be made to ensure the inclusion of the effected modes. It should be noted that only frequency measurements will be sufficient.

Next, the feasibility of locating the damage will be investigated. First, it will be postulated that the mass distribution of the system $[M]$ remains either unchanged or is changed by only a known quantity. This is a reason-able assumption in that most structural damage for the large space struc-tures will result in stiffness losses instead of complete separation or breakage with a loss of mass. Also, for certain large space structures such as the Space Station, the major contribution to the mass matrix comes

from non-load carrying components such as instrument packages, fluid
containers and power generation units. These weights can be accurately

Multiplying Equation (4) by a diagonal matrix $\lceil \phi \rfloor_i$ one obtains the following:

$$\lceil \phi \rfloor_i [K]\{\phi\}_i = \omega_i^2 \lceil \phi \rfloor_i [M]\{\phi\}_i \tag{5}$$

Without loss of generality, the mass matrix [M] will be assumed to be
diagonal, then the right hand side of Equation (5) will be as follows.

$$\omega_i^2 \lceil \phi \rfloor_i \lceil M_{jj} \rfloor \{\phi_{ji}\} = \omega_i^2 \left\{ M_{jj}\phi_{ji}^2 \right\} \tag{6}$$

The expression on the right hand side of Equation (6) clearly represents
kinetic energy. Equations (5) and (6) indicate that the kinetic energy
distribution at each degree-of-freedom for the ith mode is equal to the
potential energy distribution. The potential energy distribution is a
function only of the stiffness matrix and the modal displacements. For a
localized damage, the mode shape should be similar to or only slightly
deviated from that of the undamaged system. Hence the potential energy
distribution will be similar except for those degrees-of-freedom associated
with the damaged component. The location of the damage can be found by
identifying those degrees-of-freedom whose kinetic energies are different
from those of the undamaged system. Since the correct stiffness matrix for
the damaged system is unknown, the kinetic energy distribution can be used
for this purpose instead. It should be noted from Equation (6) that the
kinetic energy distribution is a function of the mass matrix and the modal
displacement. The frequency does not effect the relative distribution
among each degree-of- freedom. Unlike the detection of damage occurrence,
frequency measurement alone is not sufficient and mode shape measurements
are required for locating the damage.

The next task is to quantify the extent of the damage. For this purpose,
Equation (4) will be used in which the stiffness and mass matrices as well
as the eigenvalue and eigenvector are all assumed to be of the damaged system. The stiffness matrix will be decomposed into the following.

$$[K] = [K_0] + [\Delta K] \tag{7}$$

where $[K_0]$ is the stiffness matrix for the undamaged system and $[\Delta K]$ is the
perturbation due to the structural damage.

Substituting Equation (7) into Equation (4), one obtains

$$[\Delta K]\{\phi\}_i = \left(\omega_i^2[M] - [K_o]\right)\{\phi\}_i \tag{8}$$

knowns are the elements in the $[\Delta K]$ matrix which will be denoted as $[\Delta k_{ij}]$. Let

$$[\Delta K]\{\phi\}_i = [\Delta k_{ij}]\{\phi\}_i = [C]_i\{\Delta k_{ij}\} \tag{9}$$

where $[C]_i$ is defined as the connectivity matrix for the ith mode. The elements in the connectivity matrix are a function of the elements from the ith modal displacement $\{\phi\}_i$. Substituting Equation (9) into Equation (8), one obtains

$$[C]_i\{\Delta k_{ij}\} = \{y\}_i \tag{10}$$

where

$$\{y\}_i = \omega_i^2[M] - [K_o] \{\phi\}_i \tag{11}$$

It should be noted that the dimensions of the vectors $\{y\}_i$ and $\{\Delta k_{ij}\}$ are different. The dimension of $\{y\}_i$ is equal to the number of the degrees--of-freedom of the system and the dimension of $\{\Delta k_{ij}\}$ will be equal to the number of independent elements in the stiffness matrix. Although the stiffness matrix is highly banded, the number of independent elements is usually larger than the number of the degrees-of-freedom of the system. In otherwords, Equation (10) is a set of algebraic equations in which there are more unknowns, the k_{ij}'s, than there are equations. In principle, there are an infinite number of $\{\Delta k_{ij}\}$ vectors which will satisfy Equation (10).

Equation (10) is obtained by using measurements from a single mode namely, the ith mode. For using multiple modes, the governing equations can be written as

$$\begin{bmatrix} [C]_1 \\ [C]_2 \\ \cdot \\ \cdot \\ \cdot \\ [C]_n \end{bmatrix} \{\Delta k_{ij}\} = \begin{Bmatrix} \{y\}_1 \\ \{y\}_2 \\ \cdot \\ \cdot \\ \cdot \\ \{y\}_n \end{Bmatrix} \tag{12}$$

The number of unknowns in Equation (12) remain same as that of Equation (10), however, the number of equations have been increased. It is possible that the number of equations becomes larger than the number of unknowns. In principle, a solution to Equation (12) might not exist. The procedure for solving for $\{\Delta k_{ij}\}$ in either Equations (10) or (12) will be discussed next. The equations to be solved can be written as

$$\text{Dimension} = \underset{[C]}{(N{\times}M)} \quad \underset{\{\Delta k_{ij}\}}{(M{\times}1)} = \underset{\{Y\}}{(N{\times}1)} \tag{13}$$

where

N = number of equations

M = number of Δk_{ij}'s, the unknown

The solution procedure will be outlined for three different cases.

CASE I, $M > N$

If the number of unknowns is greater than the number of equations, Equation (13) yields an infinite number of solutions. Due to the nature of the problem, namely to seek the changes in the stiffness structural elements as represented by the quantities Δk_{ij}, it is reasonable to postulate that the "optimal" solution is the one with the smallest Euclidian norm. Thus, the solution procedure becomes a constrained minimization subject to Equation (13). The procedure can be formulated as follows:

$$E = \frac{1}{2}\|\Delta k_{ij}\| + \{\lambda\}^T(\{Y\} - [C]\{\Delta k_{ij}\}) \tag{14}$$

where

$$\left.\begin{array}{l} \|\Delta k_{ij}\| = \{\Delta k_{ij}\}^T\{\Delta k_{ij}\} \quad \text{Euclidian Norm} \\ \{\lambda\} = \text{vector of Lagrange multipliers} \end{array}\right\} \tag{15}$$

and

$$\frac{\partial E}{\partial(\Delta k_{ij})} = \{\Delta k_{ij}\} - [C]^T\{\lambda\} = 0 \tag{16}$$

$$\left\{\frac{\partial E}{\partial \lambda}\right\} = \{Y\} - [C]\{\Delta k_{ij}\} = 0 \tag{17}$$

From Equation (16), one obtains

$$\{\Delta k_{ij}\} = [C]^T\{\lambda\} \tag{18}$$

Substituting Equation (18) into Equation (17), one obtains

$$\{Y\} = [C][C]^T\{\lambda\} \tag{19}$$

and

$$\{\lambda\} = \left([C][C]^T\right)^{-1}\{Y\} \tag{20}$$

From Equations (20) and (18), the optimal solution becomes

$$\{\Delta k_{ij}\} = [C]^T\left([C][C]^T\right)^{-1}\{Y\} \tag{21}$$

CASE II, $N > M$

If the number of equations is greater than the number of unknowns, no exact solution exists. Therefore, one may look for an approximate solution in which the Euclidian norm of the error is minimized. Let

$$\{\varepsilon\} = \{Y\} - [C]\{\Delta k_{ij}\} \tag{22}$$

and

$$\|\varepsilon^2\| = \{\varepsilon\}^T\{\varepsilon\} \tag{23}$$

$$= \left(\{Y\} - [C]\{\Delta k_{ij}\}\right)^T\left(\{Y\} - [C]\{\Delta k_{ij}\}\right)$$

The minimization procedure will provide the following,

Thus

$$\left\{\frac{\partial\|\varepsilon^2\|}{\partial(\Delta k_{ij})}\right\} = -2[C]^T\{Y\} + 2[C]^T[C]\{\Delta k_{ij}\} = 0 \tag{24}$$

$$\{\Delta k_{ij}\} = \left([C]^T[C]\right)^{-1}[C]^T\{Y\} \tag{25}$$

CASE III, N = M

When the number of equations and the number of unknowns are equal, an exact solution can be obtained, thus

$$\{\Delta k_{ij}\} = [C]^{-1}\{Y\} \tag{26}$$

It should be noted that the norm minimization of $\|\Delta k_{ij}\|$ and $\|\epsilon^2\|$ of Equations (14) and (23), respectively, can be performed for "weighted" norms. The purpose of the weighted norm is to choose a relative emphasis of the components of the vector norm being minimized. The weighting matrix [Q] should be compatible and positive definite. Equations (14) and (23) can be written as

$$E = \frac{1}{2}\{\Delta k_{ij}\}^T[Q]\{\Delta k_{ij}\}$$
$$+ \{\lambda\}(\{Y\} - [C]\{\Delta k_{ij}\}) \tag{27}$$

and

$$\|\varepsilon^2\| = \{\varepsilon\}^T[Q]\{\varepsilon\} = \left(\{Y\} - [C]\{\Delta k_{ij}\}\right)^T x$$
$$[Q](\{Y\} - [C]\{\Delta k_{ij}\}) \tag{28}$$

respectively.

The corresponding solutions for CASE I and II will be as follows:

$$\{\Delta k_{ij}\} = [Q]^{-1}[C]^T\left([C][Q]^{-1}[C]^T\right)^{-1}\{Y\} \tag{29}$$

and

$$\{\Delta k_{ij}\} = \left([C]^T[Q][C]\right)^{-1}[C]^T[Q]\{Y\} \qquad (30)$$

Thus consistent and unified approach for estimating the structural damage has been derived. The next task is to demonstrate the procedure outlined above using an example.

ILLUSTRATIVE EXAMPLE

In order to illustrate the proposed methodology, a MAST beam will be used as an example. Fig. 1 shows the schematic of the space beam with the node numbers for the finite element analysis. For the damaged structure, the elastic modulus of two of the longerons are reduced almost to zero. Table 1 shows the dimensions and the material properties of the space beam. Table 2 shows the modal analysis results for the undamaged and damaged systems. For the first ten modes, only the first mode shows substantial change infrequency which is an indication that something has happened. Based on an evaluation of the effective mass, only mode 1 and mode 5 are of global nature and the other modes are local modes because of the small effective mass. The mode shapes of these two global modes are compared in Tables 3 and 4 for mode 1 and mode 5, respectively. The amplitudes of the modal displacements are normalized such that the maximum amplitude is unity in the x or y direction. The normalization is performed separately for the x and y displacements. Comparison of the damaged and undamaged shapes for mode 1 indicates that the displacements in the y-direction are very similar, however, substantial differences exist for the displacement in x-direction. It should be noted that mode 1 is predominantly a y-direction mode such that the modal displacements in y-direction are orders of magnitude greater than those of the x-direction. Yet the differences are concentrated in the small amplitude x-direction displacements. For mode 5, exactly the opposite is found. Fig. 2 shows graphically the mode shape comparison for some of the degrees-of-freedom along the longerons. In spite of the substantial differences in frequency, mode shapes and effective masses, the location of the damage cannot be definitively pinpointed. Next, the kinetic energy distribution will be examined. But first, the kinetic energy change ratio will be defined as

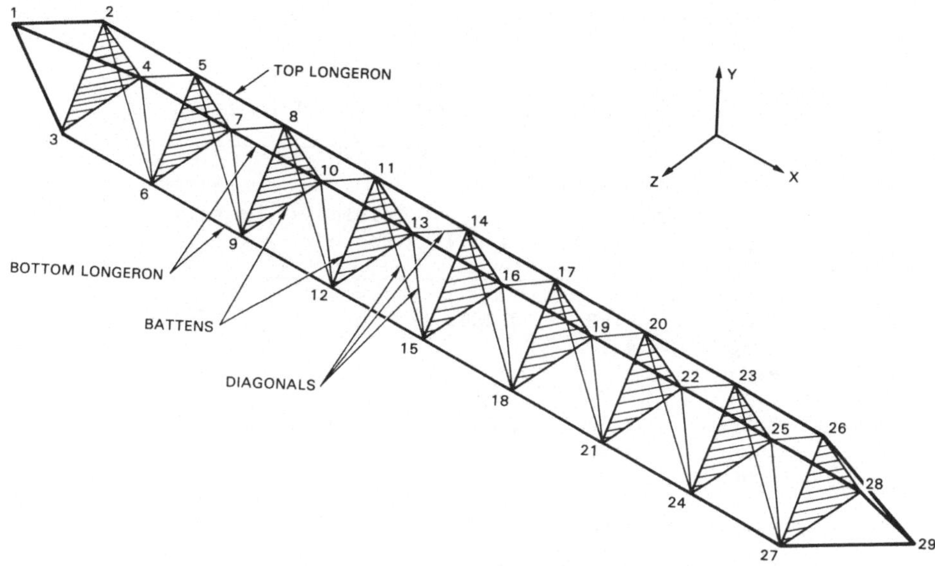

Figure 1. Schematic of MAST Truss

$$\text{Kinetic Energy Change Ratio} = \frac{(\text{K.E.})_{\text{undamaged}} - (\text{K.E.})_{\text{damaged}}}{(\text{K.E.})_{\text{undamaged}}} \tag{31}$$

Table 5 shows the kinetic energy change ratio for mode 1 and mode 5. At node 14, the kinetic energy change ratios are several orders of magnitude greater than at the rest of the nodes. This is an indication that elements connected to node 14 maybe damaged. However, no noticeable changes in kinetic energy are shown at either node 21 or node 24 where the second damaged element is located.

Next, the damage of the structural member will be quantified by the procedure outlined previously. Only the modal data of the first mode will be used which implies that the number of equations is equal to the number of degrees-of-freedom in Eqs. (8) or (10). The number of unknowns, namely the elements in the $[\Delta K]$ matrix, consist of the diagonal elements and the banded off-diagonal terms. Therefore, there are more unknowns than equations and the minimization procedure outlined in Case I will be applied. The resulting stiffness perturbation Δk_{ij} is listed in ratios of the estimated

Table 1. Dimensions and Material Properties

Element	Elastic Modulus E (psi)	Area A (in^2)	Poisson's Ratio ν	Mass*** Density ρ (lb/in^3)
Top Long-erons	9.62 x 10^6	0.2805	0.3	0.07841
Bottom Long-erons	9.62 x 10^6	0.2192	0.3	0.07841
Diag-onals	9.62 x 10^6	0.06487	0.3	0.1604
Battens	9.62 x 10^6	0.0352	0.3	0.05954
Dam-aged* Long-eron	1.00	0.2805	0.3	0.07841
Dam-aged** Long-eron	1.00	0.2192	0.3	0.07841

*Element between nodes 11 and 14

**Element between nodes 21 and 24

***Concentrated mass of 0.787 lbs. is added to each node representing the joint mass.

Table 2. System Modal Data

Undamaged Case

Mode Number	Frequency (Hz)	Effective Mass (%)	
		X DOF	Y DOF
1	0.884861	0.000	85.165
2	2.462949	0.007	0.000
3	4.005897	0.000	7.404
4	5.290891	0.046	0.000
5	5.985745	90.855	0.000
6	6.545278	0.000	2.596
7	7.039045	0.184	0.000
8	8.226030	0.041	0.000
9	8.231417	0.000	0.262
10	9.171545	0.076	0.000

Damaged Case

Mode Number	Frequency (Hz)	Effective Mass (%)	
		X DOF	Y DOF
1	0.459477	0.029	83.105
2	2.300969	0.184	0.266
3	3.870547	0.061	8.790
4	5.181898	0.066	0.026
5	5.656655	85.241	0.076
6	6.096672	0.946	1.176
7	6.962743	0.564	0.218
8	7.780087	0.002	0.962
9	8.136252	0.593	0.087
10	8.657618	0.008	1.650

Table 3. Mode Shape(1) Comparison

| | Undamaged | | Damaged | |
Node	X-DOF	Y-DOF	X-DOF	Y-DOF
2	-1.000	0.223	-1.000	0.230
3	-0.523	0.216	-0.402	0.216
4	-0.587	0.263	-0.403	0.226
5	0.851	0.526	0.999	0.466
6	-0.493	0.482	-0.363	0.457
7	-0.468	0.577	-0.341	0.477
8	0.617	0.777	0.980	0.696
9	-0.400	0.744	-0.310	0.686
10	-0.308	0.817	-0.273	0.704
11	0.325	0.939	0.948	0.903
12	-0.253	0.924	-0.244	0.897
13	-0.122	0.963	-0.203	0.964
14	0.000	0.995	-0.867	0.943
15	-0.072	1.000	-0.167	1.000
16	0.072	1.000	-0.086	0.945
17	-0.325	0.939	-0.903	0.816
18	0.122	0.963	-0.134	0.823
19	0.253	0.924	0.035	0.816
20	-0.618	0.777	-0.930	0.662
21	0.308	0.817	-0.109	0.676
22	0.400	0.744	0.169	0.646
23	-0.851	0.526	-0.953	0.447
24	0.468	0.577	0.402	0.451
25	0.493	0.482	0.262	0.432
26	-1.000	0.223	-0.956	0.220
27	0.587	0.263	0.423	0.219
28	0.523	0.216	0.324	0.196

487

Table 4. Mode Shape(5) Comparison

	Undamaged		Damaged	
Node	X–DOF	Y–DOF	X–DOF	Y–DOF
2	0.578	-0.876	0.352	-0.368
3	0.454	0.828	0.409	1.000
4	0.521	1.000	0.408	0.737
5	0.766	0.056	0.485	-0.017
6	0.651	-0.386	0.561	0.057
7	0.731	0.287	0.584	-0.198
8	0.897	-0.034	0.561	-0.256
9	0.805	0.065	0.697	0.071
10	0.879	0.038	0.717	-0.398
11	0.975	-0.129	0.599	-0.032
12	0.916	-0.107	0.819	0.180
13	0.962	-0.055	0.792	0.048
14	1.000	0.000	0.853	-0.017
15	0.972	-0.086	0.910	0.399
16	0.972	0.086	0.803	-0.156
17	0.975	0.129	0.801	-0.234
18	0.962	0.055	0.970	-0.044
19	0.916	0.107	0.752	-0.274
20	0.897	0.035	0.702	-0.015
21	0.879	-0.038	1.000	0.124
22	0.805	-0.065	0.623	-0.191
23	0.766	-0.056	0.578	-0.192
24	0.731	-0.287	0.322	-0.698
25	0.651	0.386	0.492	0.113
26	0.578	0.875	0.430	0.607
27	0.521	-1.000	0.263	-0.600
28	0.455	-0.829	0.342	-0.575

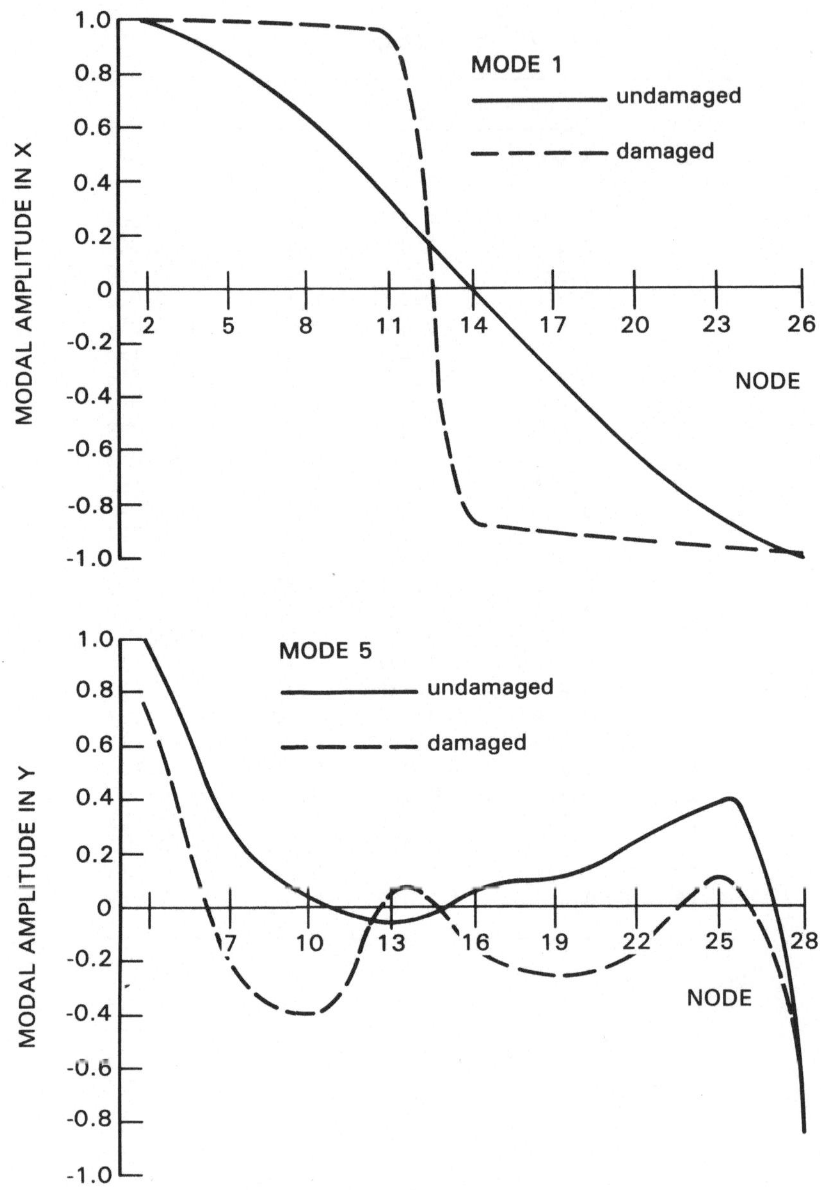

Figure 2. Modal Amplitude Comparison

Table 5. Kinetic Energy Change Ratios

	Mode 1		Mode 5	
Node	X-DOF	Y-DOF	X-DOF	Y-DOF
2	0.793+0	0.270+0	-0.442+0	-0.701+0
3	0.618-1	0.191+0	0.214+0	0.146+1
4	-0.155+0	-0.116+0	-0.783-1	-0.804-1
5	0.147+1	-0.639-1	-0.399+0	-0.851+0
6	-0.271-1	0.704-1	0.112+0	-0.963+0
7	-0.502-1	-0.185+0	-0.418-1	-0.196+0
8	0.351+1	-0.449-1	-0.413+0	0.924+2
9	0.816-1	0.135-1	0.124+0	0.100+1
10	0.406+0	-0.115+0	-0.311-2	0.187+3
11	0.142+2	0.989-1	-0.434+0	-0.896+0
12	0.664+0	0.121+0	0.201+0	0.378+1
13	0.397+1	0.192+0	0.168-1	0.272+0
14	1.377+9	0.680-1	0.921-1	0.140+8
15	0.856+1	0.189+0	0.314+0	0.357+2
16	0.150+1	0.616-1	0.243-1	0.464+1
17	0.128+2	-0.101+0	0.136-1	0.454+1
18	0.116+1	-0.131+0	0.524+0	0.960-1
19	-0.965+0	-0.727-1	0.120-1	0.100+2
20	0.306+1	-0.135+0	-0.808-1	-0.668+0
21	-0.777+0	-0.185+0	0.940+0	0.174+2
22	-0.680+0	-0.101+0	-0.100+0	0.136+2
23	0.124+1	-0.141+0	-0.147+0	0.191+2
24	0.322+0	-0.273+0	-0.708+0	0.900+1
25	-0.493+0	-0.444-1	-0.141+0	-0.855+0
26	0.640+0	0.158+0	-0.170+0	-0.187+0
27	-0.677-1	-0.173+0	-0.617+0	-0.390+0
28	-0.309+0	-0.128-1	-0.151+0	-0.184+0

values and the corresponding undamaged quantities as shown in Table 6 under First Iteration. It is important that the following points be noted when the results are examined. First, it is assumed that damage occurs by breakage, i.e., reduction in stiffness, therefore, an increase in the stiffness is not allowed. Secondly, because of the numerical round-off errors, any estimated stiffness change which is less than 10% of the undamaged value will be ignored. For instance, No. 20, the element $\Delta k_{13,19}$ is 0.36 which according to the model is the stiffness between nodes 8 and 11 in x-direction and increment of 36% is certainly inconsistent with the expected results. Similar results are found in No. 38 $\Delta k_{25,31}$, No. 50 $\Delta k_{33,39}$, No. 68 $\Delta k_{45,51}$ and No. 75 $\Delta k_{51,51}$. These elements will be "constrained" in the next iteration of estimation in such a way that no positive increment is allowed. The exception is since the $\Delta k_{51,51}$ since the 17.4% increase is considered small. Two other element is, No. 29 $\Delta k_{19,25}$ and No. 59 $\Delta k_{39,45}$ show a stiffness reduction of more than 100% which is again physically impossible. However, since the values are not too much greater than 100%, they will not be constrained in the second iteration.

The results of the second iteration are also listed in Table 6. The stiffness reductions for $\Delta k_{19,25}$ and $\Delta k_{39,45}$ are greater than those of the first iteration. Therefore, they will be constrainted to a value of 100% for the next iteration. For $\Delta k_{13,13}$ and $\Delta k_{51,51}$, the stiffness increases become negligible. The results of the third iteration can also be found in Table 6 which lists the exact solution in the last column. The only significant stiffness reductions are limited to $\Delta k_{19,19}$, $\Delta k_{19,25}$, $\Delta k_{25,25}$, $\Delta k_{39,39}$, $\Delta k_{39,45}$ and $\Delta k_{45,45}$. According to the finite element model, degrees-of-freedom (19,25) and (39,45) are associated with nodes 11, 14 and 21, 24 in x-direction, respectively. The reductions of these elements in the stiffness matrix clearly indicates that the structural elements between these nodes have been severely damaged. The exact solution is also listed in the last column of Table 6 and a comparison indicates that the results from the third iteration are very accurate except for those values which are insignificantly small.

The damages to the structural system for the sample problem have been identified and the extent of the damage quantified. In fact, the results are very encouraging.

Table 6. Estimated Stiffness Changes

		$\Delta K(i,j)/K(i,j)$						$\Delta K(i,j)/K(i,j)$			
	D.O.F.	No. Of Iteration			Exact		D.O.F.	No. Of Iteration			Exact
No.	i, j	1st.	2nd.	3rd	Soln.	No.	i, j	1st.	2nd.	3rd	Soln.
1	1, 1	0.257-1	-0.183-5	-0.183-5	-0.387-4	40	27,27	-0.117-1	-0.138-5	-0.138-5	-0.285-4
2	1, 7	0.434-1	0.347-5	0.347-5	0.491-5	41	27,33	-0.893-1	0.242-5	0.242-5	0.416-5
3	2, 2	-0.947-6	-0.947-6	-0.947-6	-0.668-5	42	28,28	0.786-9	0.786-9	0.786-9	0.000+0
4	3, 3	-0.520-3	0.373-6	0.373-6	-0.508-5	43	29,29	0.310-7	0.310-7	0.310-7	-0.380-4
5	3, 9	-0.946-3	-0.491-6	-0.491-6	0.628-5	44	29,35	-0.605-7	-0.605-7	-0.605-7	0.416-5
6	4, 4	0.149-6	0.149-6	0.149-6	-0.578-5	45	30,30	-0.156-6	-0.156-6	-0.156-6	-0.101-5
7	5, 5	0.138-6	0.138-6	0.138-6	0.000+0	46	31,31	0.956-1	-0.493-7	-0.493-7	0.760-5
8	5,11	0.110-5	0.110-5	0.110-5	0.628-5	47	31,37	-0.134+0	0.655-6	0.655-6	0.491-5
9	6, 6	-0.304-5	-0.304-5	-0.304-5	-0.158-4	48	32,32	-0.441-6	-0.441-6	-0.441-6	0.207-4
10	7, 7	-0.403-1	-0.179-6	-0.179-6	0.760-5	49	33,33	0.399-1	0.530-8	0.530-8	-0.285-4
11	7,13	-0.132+0	-0.257-5	-0.257-5	0.491-5	50	33,39	0.245+0	0.000+0	0.000+0	0.628-5
12	8, 8	-0.938-6	-0.938-6	-0.938-6	0.207-4	51	34,34	-0.325-8	-0.325-8	-0.325-8	-0.101-5
13	9, 9	0.704-3	-0.249-7	-0.249-7	-0.285-4	52	35,35	-0.165-7	-0.165-7	-0.165-7	-0.285-4
14	9,15	0.304-2	0.224-5	0.224-5	0.628-5	53	35,41	0.508-7	0.508-7	0.508-7	0.628-5
15	10,10	-0.883-6	-0.883-6	-0.883-6	0.000+0	54	36,36	-0.138-6	-0.138-6	-0.138-6	0.000+0
16	11,11	-0.505-6	-0.505-6	-0.505-6	-0.285-4	55	37,37	-0.369-1	-0.261-6	-0.261-6	0.760-5
17	11,17	-0.127-5	-0.127-5	-0.127-5	0.628-5	56	37,43	0.498-1	0.106-6	0.106-6	0.491-5
18	12,12	-0.686-7	-0.686-7	-0.686-7	-0.101-5	57	38,38	0.372-7	0.372-7	0.372-7	0.207-4
19	13,13	0.993-1	0.134-5	0.134-5	0.000+0	58	39,39	-0.117+0	-0.116+0	-0.454+0	-0.454+0
20	13,19	0.360+0	0.000+0	0.000+0	0.491-5	59	39,45	-0.111+1	-0.120+1	-0.100+1	-0.100+1
21	14,14	-0.171-5	-0.171-5	-0.171-5	-0.156-4	60	40,40	0.158-6	0.158-6	0.158-6	0.000+0
22	15,15	-0.169-2	-0.854-6	-0.854-6	-0.190-4	61	41,41	0.269-6	0.269-6	0.269-6	0.380-4
23	15,21	-0.926-2	0.103-5	0.103-5	0.212-5	62	41,47	-0.274-5	-0.274-5	-0.274-5	-0.629-5
24	16,16	-0.273-6	-0.273-6	-0.273-6	-0.101-5	63	42,42	-0.141-6	-0.141-6	-0.141-6	0.000+0
25	17,17	0.786-6	0.786-6	0.786-6	-0.285-4	64	43,43	0.150-1	0.224-6	0.224-6	0.760-5
26	17,23	0.631-6	0.631-6	0.631-6	0.212-5	65	43,49	-0.162-1	0.171-5	0.171-5	0.491-5
27	18,18	-0.475-7	-0.475-7	-0.475-7	0.000+0	66	44,44	-0.187-7	-0.187-7	-0.187-7	0.207-4
28	19,19	-0.255+0	-0.323+0	-0.465+0	-0.465+0	67	45,45	-0.277+0	-0.429+0	-0.454+0	-0.454+0
29	19,25	-0.108+1	-0.133+1	-0.100+1	-0.100+1	68	45,51	0.338+0	0.000+0	0.000+0	0.628-5
30	20,20	0.130-6	0.130-6	0.130-6	0.207-4	69	46,46	0.295-6	0.295-6	0.295-6	0.000+0
31	21,21	0.435-2	0.160-6	0.160-6	-0.380-4	70	47,47	0.128-5	0.128-5	0.128-5	0.285-4
32	21,27	0.311-1	0.421-5	0.421-5	0.833-5	71	47,53	-0.156-5	-0.156-5	-0.156-5	-0.416-5
33	22,22	-0.192-6	-0.192-6	-0.192-6	-0.314-5	72	48,48	0.868-6	0.868-6	0.868-6	0.314-5
34	23,23	-0.647-6	-0.647-6	-0.647-6	-0.190-4	73	49,49	-0.962-2	-0.131-5	-0.131-5	0.387-4
35	23,29	0.440-6	0.440-6	0.440-6	0.833-5	74	50,50	0.111-5	0.111-5	0.111-5	-0.134-4
36	24,24	0.147-8	0.147-8	0.147-8	0.202-5	75	51,51	0.174+0	-0.736-6	-0.736-6	0.337-5
37	25,25	-0.248+0	-0.296+0	-0.465+0	-0.465+0	76	52,52	-0.151-5	-0.151-5	-0.151-5	-0.158-4
38	25,31	0.358+0	0.000+0	0.000+0	0.491-5	77	53,53	-0.145-5	-0.145-5	-0.145-5	0.258-5
39	26,26	0.412-6	0.412-6	0.412-6	0.517-5	78	54,54	0.155-6	0.155-6	0.155-6	-0.578-5

CONCLUDING REMARKS

Although the feasibility of using the proposed procedure for on-board structural damage assessment based on the measured modal data has been demonstrated by the example problem, many related issues have not yet been addressed. The questions regarding the accuracy requirement of the measurement, the completeness of the measurement with respect to the number of degrees-of-freedom in the analytical model and the selection of which modal data to be used are just some of these issues which are of critical importance for the actual application of the methodology. Furthermore, laboratory experiments should be performed to verify the procedure as part of the process to make the proposed methodology reach maturity.

Acknowledgement

The research was carried out by the Jet Propulsion Laboratory, California Institute of Technology, under a contract with NASA. This task was sponsored by Samuel L. Venneri, NASA Office of Aeronautics and Space Technology, Code RM.

References

1. Vandiver, J. K., "Detection of Structural Failure on Fixed Platforms by Measurement of Dynamic Response," Proceedings of the Offshore Technology Conference, Vol. 2, Paper OTC 2267, May, 1975, pp. 243-252.

2. Wojnarowski, M. E., Stiansen, S. C., and Reddy, N. E., "Structural Integrity Evaluation of a Fixed Platform Using Vibration Criteria," Proceedings of the Offshore Technology Conference, Vol. 3 Paper OTC 2909, May, 1977, pp. 247-256.

3. Coppolino, R. N., and Rubin, S., "Detectability of Structural Failures in Offshore Platforms by Ambient Vibration Monitoring," Proceeding of the Offshore Technology Conference, Vol. 1, Paper OTC 3865, May, 1980, pp. 101-110.

4. Thoren, A. R., "Derivation of Mass and Stiffness Matrices from Dynamic Test Data," AIAA/ASME/SAI 13th Structures, Structural Dynamics and Materials Conference, San Antonio, Texas, April 10-12, 1972, AIAA Paper 72-346.

5. Berman, A. and Flannelly, W. G., "Theory of Incomplete Models of Dynamics Structures," AIAA Journal, Vol. 9, No. 8, August, 1971, pp. 1481-1487.

**Summary of
Workshop Discussion
and
Concluding Remarks and
Recommendations**

Summary of Workshop Discussion

J. L. Beck*
Earthquake Engineering Research Laboratory
California Institute of Technology
Pasadena, California, 91125, U.S.A.

J. Yao moderated the discussion in the final session. He began the session by presenting the state of the art on the topics addressed in the workshop. This report was prepared jointly with G. Natke and was based on a framework provided by the list of topics in Fig. 3 of the Preface. This state of the art is summarized in the final chapter: Concluding Remarks and Recommendations.

In this chapter, discussion by the workshop participants in the closing session is summarized using the same unifying framework.

1. Damage Descriptions and Basic Requirements

F. Kozin raised the question of whether the purview of the discussion was limited to damage detection using only vibrational characteristics, such as changes in stiffness and damping properties, or whether it extended to other nondestructive techniques, such as acoustic emissions, ultrasonic sound, and so on. G. Natke replied that it encompassed any form of data as long as system identification was involved in extracting information, since this was the unifying theme of the workshop. S. Masri mentioned that he was organizing an International Workshop on Non-Destructive Evaluation for Performance of Civil Structures which would encompass all NDE techniques and would be held at the University of Southern California on February 2-3, 1988.

G. Lallement noted that the current emphasis in damage detection is to use modal analysis, but that identified modal parameters are not very sensitive to localized damage. An approach is needed which is sensitive to the pre-determined critical areas of the structure. P. Ibanez gave an example of an analysis relating to damage detection for an off-shore platform which illustrated this lack of sensitivity. A finite-element model of the platform showed that a large fraction of the structural members could be broken without the fundamental period changing by more than 2%, but changes in the fundamental period of this size were found from test to test in the absence of damage (from

---

*Associate Professor of Civil Engineering

changes in liquid mass on the platform, nonlinear effects exhibited because of different wave heights and so on). J. Beck pointed out that another type of sensitivity problem is that a detected reduction in stiffness need not imply a corresponding reduction in strength, particularly in buildings, because of the behavior of nonstructural components. For civil structures, it is reduction in strength, not reduction in stiffness, which is the primary concern in safety evaluation.

As far as test requirements are concerned, G. Natke and R. Ciesielski emphasized the importance of attempting to optimize the test procedure. Further work is required to determine the optimal quantities to measure and the optimal choice of instrument locations based on prior knowledge of the critical areas of a structure.

2. System Identification

M. Torkamani pointed out that there have been many algorithms developed for system identification and that it would be useful to run a benchmark study in which they were applied to the same problem. G. Natke commented that this had not been very successful when different groups had tested the same structure. J. Beck proposed using simulated test data where the "system" is a finite-element model and therefore known. M. Link said that such a benchmark study had, in fact, been done recently under the auspices of GARTEUR (Group for Aeronautical Research & Technology in Europe). Different system identification groups had been presented with a finite-element model which represented an unmodified simple two-dimensional frame structure, and which had several hundred degrees of freedom. They were also given incomplete modal information for the "structure" after modifications. The goal was to identify the structural modifications, which were unknown to the participants in the study. The results had not yet been released. J. Chen mentioned that NASA had sponsored independent modal analysis tests of the Gallileo spacecraft by several different companies. There was general agreement in the identified modal properties but also some significant differences. J. Beck commented that this is the value of such "benchmark" tests. We can learn a lot from exploring the differences in the results. V. Matzen also strongly supported the idea of benchmark studies.

W. Iwan drew attention to the lack of discussion of instrumentation in the workshop. He felt that the "software people" should make it clear to the "hardware people" what type of instrumentation and performance is required, rather than trying to make do with what is available. P. Ibanez commented that cross-axis errors in accelerometers can be as large as 5%, and that such measurement noise can have a significant effect on the parameter estimates from system identification.

3. System Identification and Damage Evaluation

The major share of the discussion in the final session was devoted to this topic.

P. Ibanez felt that to make further progress in damage detection using system identification, we had to make better use of prior knowledge. Damage is usually localized but when performing the inverse problem, current techniques in estimating the stiffness matrix change, ΔK, allow the damage to "spread out." Because of this diffusing of the stiffness change, the severity of the damage may not be detected. J. Beck noted that there are some forms of damage which are distributed, such as possible damage to structural members in the proposed NASA space station due to ultra-violet radiation. This is another example where prior knowledge of the possible types of damage could be utilized. J. Chen said that in making the study he presented in his workshop paper, he had misgivings about using the minimization of a matrix norm to identify stiffness changes. However, it seemed to work well for the truss problem when he imposed the "physical" constraints that the local stiffness should not increase and it should not decrease by more than its initial value.

S. Ibrahim wondered if damage detection by system identification is to some degree premature, since it is currently based on modal analysis and this is a technology that is still developing, particularly as regards the determination of structural modifications from changes in the modal properties.

W. Iwan suggested that we take the point of view of a physician and use system identification and vibrational characteristics as a diagnostic tool to identify the possible existence, and perhaps general location, of damage. If a positive indication should occur, then special tests could be performed which are more focussed to detect the precise location and severity of the damage.

In a similar vein, P. Ibanez commented that modal properties may be insufficient for reliable damage detection, and that we must consider other measurable quantities. C. Cempel agreed. He felt that we need special condition symptoms that can be monitored using "working" (or ambient) signals to detect the presence of damage. Special test procedures could then be applied to gain further information. He also felt that we need special inference procedures since measurements are always performed in a noisy environment. J. Beck noted that Bayesian probability provides such a methodology and, in addition, allows the incorporation of prior knowledge relating to the model being used in the identification.

M. Link observed that initial analytical models are frequently inadequate, so conclusions on structural modifications from damage are conditional on this uncertainty. There is a problem in quantifying the confidence level for initial models. Before these models are used for damage detection, they should ideally be "calibrated" by applying system identification to tests of the system prior to any possibility of damage.

4. Concepts

There was not much discussion on this subject, possibly because the workshop was drawing to a close. J. Chen pointed out that damage assessment is clearly probabilistic in nature. Furthermore, the consequences of a particular type of damage need to be stated by the engineer to help the decision maker. Along these lines, damage detection might benefit from the application of decision theory. F. Kozin suggested that strategical network theory, such as that used for locating failures in a large power distribution system, may have application in damage detection.

Following these comments, J. Yao closed the Workshop by thanking G. Natke for his outstanding leadership and generosity; C. Hoff for handling arrangements at Lambrecht; J. Beck for his efforts in obtaining financial support for the U.S. participants; and the sponsoring agencies (see Preface) for their generous support.

Concluding Remarks and Recommendations

H. G. Natke and J. T. P. Yao

The state of the art is summarized in the order of the items listed in the table of contents.

1. Damage descriptions and basic requirements: Structural modelling is generally done for spatial behaviour using various constitutive laws etc. without the consideration of limit states and micro-behaviour. From a practical point of view, the parameters of such models e. g. finite-element models, in general are not the most sensitive to damage. Several damage indices have been proposed. However, a general definition of damage due to structural modifications is lacking. Instrumentation and testing are performed often by experience and a priori knowledge of the structural behaviour, although optimal experimental design techniques are available.

2. In system identification, many approaches exist. They are mostly developed for linear systems using a pre-determined form so that the identification is reduced to parameter estimation. Experimental modal analysis is very popular and applied successfully at present. However, such results may be incomplete, and error estimations have not been made in practice to-date. The progress from modal to physical parameter estimation including the knowledge of the prior computational model is apparent: up-dating of computational models with known confidence. It is necessary to have an adequate computational model for the problem under consideration. Error localization within the computational model is problematic. The treatment of a large number of degrees of freedom and of a large number of parameters to be estimated is still difficult. The identification of non-linear structures is under-developed to-date. Data processing techniques are

available, but they have to be applied carefully. It is desirable to improve the available hardware, software, and their use in identification to meet the needs of those investigators who conduct such studies.

3. There exists little or no interaction between system identification and safety (or damage) evaluation. The results of system identification, e. g. the estimates of the structure modifications, have to be assessed. It is thought that probabilistic methods, fuzzy sets and pattern recognition can be useful. An extended system identification is proposed with the use of expert systems with continuously updated data bases and improved inference machines in safety (or damage) evaluations. Certain applications of deterministic and probabilistic theory are practical at present.

4. Concepts of safety evaluation in the sense of the workshop exist only for special purposes. In addition, only proposals are mentioned, such as that based on the extended system identification with a knowledge base for use by experienced engineers.

From an engineering viewpoint, the state of the art in system identification and safety evaluation may be summarized as in Table 1. In system identification studies, loads (or excitation) as well as the structural responses are measured and recorded. These data are then processed for use in the estimation of structural characteristics. Inherent in each structure are various limit states (or resistances) which are functions of the configuration, materials, methods of fabrication and construction,damage, and environmental conditions (e. g. temparature). At present, no consideration is given to the estimation (or identification) of these limit states. Other quantities such as damage which are important to engineers are not considered in current studies of system identification.

In present studies of safety evaluation of existing structures, future loading conditions are extrapolated from past records. Structural characteristics may be assumed, calculated, measured or estimated from system identification studies. The structural response may be calculated with the use of simple or sophisticated methods. The actual behaviour of structures may also be observed or estimated. Meanwhile, structural resistance or limit states may be roughly estimated from basic strengths of materials. Several empirical models have been suggested for the evaluation of the damage. On the other hand, highly idealized mathematical models have been used for the calculation of structural reliability. The levels of accuracy in computation and measurements and the degree of sophistication in the use of mathematical models have not been consistent.

Based on the workshop discussions as summarized by Dr.James Beck and given in the previous chapter and Table 1, the following recommendations are made for further studies in structural safety evaluation based on system identification approaches:

- Damage definition and description including the relationship between damage and structural modification have to be considered intensively. Measures (indices) of damage need further consideration.

- Detection of structural modifications caused by damage using the most sensitive parameters needs extensive further work using models which are physically interpretable.

- Test requirements (for inspection as well as for identification) should be established by prior knowledge of structural analysis and failure prediction.

- Updating approaches for the computational models should be extended to treat cases involving large numbers of degrees of freedom as well as parameters to be estimated. Determination of the confidence of the prior (or initial) computational model (more than a guess) should be investigated.

- Structure (of the model) identification of non-linear systems needs extensive consideration. In addition, problems of observability, controllability and identifiability must be investigated.

- Benchmark (or calibration) studies of available methods using simulation as well as test results are recommended. Although difficulties exist in conducting such studies, it is believed that one of the advantages of conducting benchmark studies is to stimulate discussion leading to possible improvement and wider acceptance of available methods.

- Optimum design of experiments for the detection of structural modifications is highly desirable.

- Extended system identification needs to be developed not only considering a computational model with known confidence but also including the estimation of limit states and the assessment of structural damage.

- The assessment of structural safety using deterministic and probabilistic theories may be further extended. Moreover, practical applications of additional tools such as pattern recognition and fuzzy sets should be explored. Some overall (all encompassing) concepts are needed. These should be combined with the development of expert systems. Theories such as subsystem analyses should also be investigated.

	SYSTEM ID	SAFETY EVALUATION
Loads	Measured & Recorded Data Processed	Extrapolation of Past Records
Structural Characteristics	Estimated -Linear Systems (Excellent) -Slightly Nonlinear Systems (Good)	Assumed or Calculated or Estimated from System ID
Structural Response (Behaviour)	Measured & Recorded Data Processed	Response Calculated (Simple - Sophisticated Method) Behaviour Observed
Structural Resistance (Limit States)	Not Considered at present	Roughly Estimated from Strength of Materials
Damage	Not Considered at present	Empirical Models
Reliability	Not Considered at present	Highly Idealized Math. Model

Table 1: An Engineering Viewpoint of System Identification and
Safety Evaluation